CONFIDENTIALITY AND ITS DISCONTENTS

CONFIDENTIALITY AND ITS DISCONTENTS

Dilemmas of Privacy in Psychotherapy

Paul W. Mosher and Jeffrey Berman

Fordham University Press
New York 2015

Fordham University Press has no responsibility for the persistence or accuracy of URLs for external or third-party Internet websites referred to in this publication and does not guarantee that any content on such websites is, or will remain, accurate or appropriate.

Fordham University Press also publishes its books in a variety of electronic formats. Some content that appears in print may not be available in electronic books.

Visit us online at www.fordhampress.com.

Library of Congress Cataloging-in-Publication Data

Mosher, Paul W., author.
    Confidentiality and its discontents : dilemmas of privacy in psychotherapy / Paul W. Mosher and Jeffrey Berman.
        p. ; cm.—(Psychoanalytic interventions)
    Includes bibliographical references and index.
    ISBN 978-0-8232-6509-1 (cloth : alk. paper)—ISBN 978-0-8232-6510-7 (pbk. : alk. paper)
    I. Berman, Jeffrey, 1945–, author.   II. Title.   III. Series: Psychoanalytic interventions.
    [DNLM: 1. Confidentiality.   2. Privacy.   3. Psychotherapy.   4. Professional-Patient Relations. WM 33.1]
    RC480
    616.89'14—dc23

                                                           2014030499

Printed in the United States of America

17 16 15   5 4 3 2 1

First edition

*We dedicate this book to the memory of Jerome Beigler, MD,
psychoanalyst and courageous defender of confidentiality
in psychotherapy and psychoanalysis.*

Contents

Acknowledgments

Paul Mosher and Jeffrey Berman have been friends and neighbors for forty years, but this is their first book together. Paul is a psychoanalyst in private practice in Albany, New York, a Clinical Professor of psychiatry at Albany Medical College, and a widely recognized expert on the subject of privacy in psychotherapy. In the 1990s he was chair of the Committee on Confidentiality of the American Psychoanalytic Association, and he has also served as a member of the Confidentiality Committee of the American Psychiatric Association. He coordinated the preparation of an amici curiae brief on behalf of four psychoanalytic organizations at the time of the Supreme Court's deliberations on the *Jaffee v. Redmond* case. He also maintains the *Jaffee v. Redmond* website. He has written several articles and book chapters on the subject of privacy in psychotherapy. Paul was also the proposer of a digitized archive of psychoanalytic literature and one of the creators of Psychoanalytic Electronic Publishing (PEP), a joint venture of the American Psychoanalytic Association and the Institute of Psychoanalysis in London, which digitized psychoanalytic journals, articles, and books, transforming psychoanalytic scholarship worldwide. Jeffrey Berman is Distinguished Teaching Professor of English at the University at Albany, an honorary member of the American Psychoanalytic Association, and the author of thirteen books on literature and psychoanalysis, self-disclosing writing in the classroom, and death education. His most recent book is *Dying in Character: Memoirs on the End of Life*. His teaching and writing have been featured in the *Chronicle of Higher Education, National Public Radio*, and the *Boston Globe*. The *Princeton Review* selected him as one of America's top professors.

It was Paul's idea to create a book of this sort, and he did much of the psychiatric and legal research during his long professional career. He enlisted Jeff's help as a coauthor to convey the human interest in privacy stories. Paul believes he never could have written a book like this on his own. Jeff has long been an armchair analyst, and it has been a pleasure for him to work with a real psychoanalyst. Jeff has never been in psychoanalysis, preferring instead to muddle through life, but if he could begin his life over again, he would want to be in analysis with Paul. Paul appreciates the compliment but isn't accepting new patients.

Part of the research for our book involved interviewing several of the people who played key roles in the famous psychotherapy privacy stories on which we focus. We are grateful for their willingness to share their impressions of experiences that were at the time painful, even traumatic. Joseph Lifschutz, a San Francisco psychoanalyst who was arrested and briefly jailed in 1969 for refusing to testify about a former patient, gave us details that cast new light on one of our earliest privacy stories. Frank Werner spoke to us about his mother's shock and horror upon discovering that her psychiatrist had betrayed confidentiality by publishing in 1973, without permission, a thousand-page case study, *In Search of a Response*. The resulting lawsuit, *Jane Doe v. Joan Roe and Peter Poe*, is surely one of the oddest and most outrageous privacy stories in the history of psychotherapy. We interviewed two of the people who were centrally involved in the landmark 1996 Supreme Court decision *Jaffee v. Redmond*, which created for the first time a federal psychotherapist-patient privilege: Karen Beyer, a licensed social worker whose refusal to turn over her psychotherapy notes, as ordered by a Chicago judge, initiated the case, and her attorney, Sandra Nye. Alayne Katz offered us her personal impressions of Robert Bierenbaum's brutal murder of her sister, Gail, in 1985 along with her perspective, as a lawyer, of the complex legal and psychiatric issues arising over whether the therapists' testimony could be used in the trial. And former New York Chief Judge Sol Wachtler candidly offered us his impressions about the stigma of mental illness that contributed to his bizarre behavior in 1992, arrest, and imprisonment.

One of the most serendipitous consequences of our conversations with the above people is that some of them implied, after they read the final version of the chapters in which their interviews appeared, that they found the experience "cathartic and positive," to quote one of them. We speculate that they felt this way because we encouraged them to tell their stories, from their own points of view, without interrupting or contradicting them. We then showed them exactly how we intended to use and contextualize their words. Some of their comments were "off the record," private disclosures we have promised will remain confidential. Other comments were "for the record," but information the interviewees had never publicly revealed. Often the interviewees were surprised by their willingness to open up to us. We gave them an opportunity to tell their stories and to see their own words on paper, where they took on heightened significance.

We are also grateful for the help of several people who read one or more chapters of our book and made constructive comments. They include Judge Joseph W. Bellacosa, Julie Berman, Norman Clemens, Lana Fishkin, Ralph Fishkin, Suzanne Hicks, Barry Landau, Paula Mosher, Rebecca Mosher, and Judith Schachter. Thanks also to Seth P. Stein, executive director of the New York State Psychiatric Association, and Rachel A. Fernbach, staff attorney, New York State Psychiatric Association.

Professors Peter L. Rudnytsky and Esther Rashkin expressed early interest in including our book in their Psychoanalytic Interventions series, published by Fordham University Press. We are grateful for their enthusiasm and support. We are also grateful to the three "anonymous" readers of the entire manuscript who allowed their names to be known to us—Paul Brinich, Norman Clemens, and Sander L. Gilman—and to Robert Fellman for his superb copy editing of the manuscript. Their criticisms have strengthened this book; we alone are responsible for lingering problems.

Sections of Chapter 1 have appeared, in different form, in Paul W. Mosher, "We Have Met the Enemy and He (Is) Was Us," in *Confidentiality: Ethical Perspectives and Clinical Dilemmas*, ed. Charles Levin, Allannah Furlong, and Mary Kay O'Neil (Hillsdale, N.J.: Analytic Press, 2003), 229–249. Sections of Chapters 4 and 5 have appeared, in different form, in Jeffrey Berman, *The Talking Cure: Literary Representations of Psychoanalysis* (New York: NYU Press, 1985); Jeffrey Berman, "Review of *Tales from the Couch: Writers on Therapy*, edited by Jason Shinder," *Psychoanalytic Psychology* 18 (2001): 743–755; and "Revisiting Philip Roth's Psychoanalysts," in *The Cambridge Companion to Philip Roth*, ed. Timothy Parrish (Cambridge: Cambridge University Press, 2007), 94–110. Sections of Chapter 8 have appeared, in different form, in Jeffrey Berman, *Surviving Literary Suicide* (Amherst: University of Massachusetts Press, 1999); and Paul W. Mosher and Jeffrey Berman, "Book Review: *An Accident of Hope: The Therapy Tapes of Anne Sexton*, by Dawn M. Skorczewski," *Journal of the American Psychoanalytic Association* 61 (2013): 848–854.

Introduction

Confidentiality matters. We all define ourselves by those aspects of self-hood that we have shown to others and by those other aspects of selfhood that remain private and hidden. In fact, it is the latter part of selfhood, our "private" self, that sets us apart from other people, giving us a sense of uniqueness and difference. No one, not even in the most brutal totalitarian society, can control what we think and feel so long as we choose to keep those thoughts and feelings to ourselves.

Confidentiality matters even more to those of us who, in seeking the help of another person with our inner or interpersonal difficulties, choose to expose our inner world to that "other" in the hope of coming to a better understanding of ourselves. This is the goal of psychoanalysis and similar forms of psychotherapy, where we drop the barrier of personal secrecy and risk making intimate disclosures to a person we trust. The quality of help we derive from such an experience depends on our willingness to place our trust in such a person.

But beyond that, confidentiality matters even more when we engage with another person in a process of self-discovery and through that process discover aspects of our inner lives that even we have never been able to access. Those aspects of ourselves have been walled off from us because they are in one way or another so unacceptable in our role as civilized beings that an active psychological process causes them to be unconscious.

When Sigmund Freud abandoned the use of hypnosis as a method to discover the unconscious inner lives of his patients and adopted instead new methods of psychological treatment, eventually to be called psycho-analysis, the most significant change he made was the introduction into the clinical situation of a device called "free association." In this method of treatment, the patient is instructed to look inward at his or her stream of thought and to report out loud to the psychotherapist every thought that comes to mind. In fact, although Freud was for the most part quite relaxed in his instructions to practitioners taking up his new method, generally inviting them to experiment, he was specific about the instructions to the patient about free association. He even went so far as to provide the psycho-therapist with a script (in quotes!) as to what to say to the patient:

> "One more thing before you start. What you tell me must differ in one respect from an ordinary conversation. Ordinarily you rightly try to keep a connecting thread running through your remarks and you exclude any intrusive ideas that may occur to you and any side-issues, so as not to wander too far from the point. But in this case you must proceed differ-ently. You will notice that as you relate things various thoughts will occur to you which you would like to put aside on the ground of certain criticisms and objections. You will be tempted to say to yourself that this or that is irrelevant here, or is quite unimportant, or nonsensical, so that there is no need to say it. You must never give in to these criticisms, but must say it in spite of them—indeed, you must say it precisely *because* you feel an aversion to doing so. Later on you will find out and learn to understand the reason for this injunction, which is really the only one you have to follow. So say whatever goes through your mind. Act as though, for instance, you were a traveler sitting next to the window of a railway carriage and describing to someone inside the carriage the changing views which you see outside. Finally, never forget that you have promised to be absolutely honest, and never leave anything out because, for some reason or other, it is unpleasant to tell it." (Freud, "On Beginning the Treatment," 134–135)

"Unpleasant to tell it," indeed! Freud recognizes here that what takes place behind the scenes in our inner lives is a conflict between what we might be inclined to say and what might seem improper to reveal. It is easy to understand, then, that patients would be unlikely to risk carrying out Freud's instructions without having a good reason to believe that tolerat-ing this unpleasantness would serve a useful purpose and that the thera-pist, the person to whom they are making themselves vulnerable by baring their inner life, is absolutely ethical and trustworthy. And, of course, one of the most important ways that a patient must trust a therapist is to be convinced that what is spoken in the consulting room will stay in the con-sulting room. Freud insisted on a patient's full disclosure and promised com-plete confidentiality in return, as Joseph Wortis confirms in *Fragments of an Analysis with Freud.* "He then went on to speak of the fundamental condition for an analysis: absolute honesty—I was to tell literally everything that went through my head: whether important, unimportant, painful, irrelevant, absurd, or insulting. He for his part would guarantee absolute privacy, regardless of what I revealed: murder, theft, treachery or the like" (20).

Freud anticipated some of the challenges to maintaining strict confidentiality, and in a famous footnote to "On Beginning the Treatment" he refers to treating a government official who imposed limits on his self-disclosures:

> It is very remarkable how the whole task becomes impossible if a reservation is allowed at any single place. But we have only to reflect what would happen if the right of asylum existed at any one point in a town; how long would it be before all the riff-raff of the town had collected there? I once treated a high official who was bound by his oath of office not to communicate certain things because they were state secrets, and the analysis came to grief as a consequence of this restriction. *Psychoanalytic treatment must have no regard for any consideration, because the neurosis and its resistances are themselves without any such regard.* (note 1, 135–136)

It is now a century since Freud first laid out the "fundamental rule" of full disclosure in psychoanalysis, and much has changed. The rule itself is no longer taken as literally as it once was; the idea of giving patients "rules" to follow at all has been challenged as psychotherapy has come to be viewed as a more egalitarian "two-person" enterprise. Nonetheless, the need for confidentiality implied by the fundamental rule is not diminished because of the irreducible requirement to strive for openness and honesty in psychotherapeutic sessions, at least in "insight-oriented" (learning about one's inner self as a path to emotional well-being) forms of psychotherapy.

The Therapist's Dual Allegiance

In Freud's day the patient's faith in the integrity of the doctor was based on the esteem in which physicians were held and the well-known promise in the physician's Hippocratic Oath to keep inviolate secrets learned from patients, an important element of the ethical canons of the medical profession. However, about halfway through the past century, the physician's promise began to erode as a result of changes in the way the health professions were organized and regulated and the fact that these professions were understood to have a dual allegiance to patient and society. If a psychotherapy patient's conflict is viewed, at least in part, as a tension or struggle between his or her inner life and the necessary strictures of a civilized society, the conflict is intensified by the therapist's dual allegiance. And so, as the twentieth century progressed, the psychotherapist's ethical commitment to strict confidentiality became increasingly problematic.

Confidentiality and Its Discontents focuses mainly on the human stories arising from the psychotherapist's dual allegiance to patient and society, but before we discuss these stories, we provide a brief historical account of the major events and legal cases that have contributed to a loss of confidentiality in the patient-therapist relationship. The title of our opening chapter, "We Have Met the Enemy and He (Is) Was Us," suggests that psychotherapists must accept some responsibility for this loss of confidentiality. Freud pledged to protect his patients' secrecy, but, surprisingly, this promise was at times more honored in the breach than in the observance, creating a precedent that has had an unfortunate effect on later analysts. Changing attitudes toward psychotherapy and the law have further eroded confidentiality. Increasingly, psychotherapists find themselves in a situation where they encourage patients to say whatever comes to mind but then are forced to issue a "Miranda Warning" implying, in effect, that anything (or at least some things) patients say can be used against them. There have been some positive developments, however, notably, a new absolute federal privilege for psychotherapy.

The Buried Bodies Case

We then look at the story of Robert Garrow, the "Buried Bodies case," which has become a watershed in establishing the principle of confidentiality in the client-lawyer relationship. Accused in 1973 of stabbing to death a young man who was camping with friends in upstate New York, Garrow confessed to three additional brutal murders to his two lawyers, who then found themselves confronted with an agonizing decision. Should they conceal privileged information from the prosecution to help their client avoid a long prison term? If so, they risked disbarment and even criminal prosecution for withholding evidence. The Buried Bodies case is fascinating for its insights into the ambiguities of privileged information; the story has become iconic in legal ethics instruction.

Many of the vexing issues in the Buried Bodies case involving secrecy, privacy, confidentiality, and trust also appear in psychotherapy, where therapists sometimes wonder how far they should go in protecting a patient's disclosure. The Buried Bodies case highlights, albeit in a different realm, what might be called "Confidentiality and Its Discontents," where therapists often find themselves under intense pressure to reveal the details of a patient's therapy. Confidentiality has both ethical and clinical implications. Breaking confidentiality often means losing confidence in the therapeutic process.

The Buried Bodies case is engaging for its human drama—a story about a man who risked everything he held sacred to defend a legal and ethical principle. Therapists are understandably reluctant to describe how they feel when confronted with the decision to break confidentiality. The lead lawyer in the Buried Bodies case, Frank Armani, offers us, by analogy, an insight into the therapist's feelings. How does one resolve a conflict when personal and professional obligations are in conflict? What is the price one pays—to one's family, career, community, and personal well-being—for preserving a dark secret?

The Buried Bodies case has another intriguing and paradoxical dimension. The story reveals the difficulty of writing about privileged information without betraying confidentiality, the same difficulty every psychotherapist confronts when writing about clinical material. Unlocking one secret may require the creation of another secret, suggesting that the major details of a private story may not emerge until decades later.

The Case of Joseph Lifschutz

The next chapter highlights the story of a San Francisco psychoanalyst who was jailed briefly in 1969 because of his decision to defy a court order requiring him to testify about a former patient. Joseph Lifschutz, now an octogenarian, revealed to us in a rare telephone interview the details of his incarceration, which has become a defining moment in the struggle to maintain the confidentiality of the therapist-patient relationship. We also discuss briefly the story of Anne Hayman, a British psychoanalyst who similarly refused to testify about a patient when she was subpoenaed in the middle 1960s.

"The Angry Act": The Psychoanalyst's Breach of Confidentiality in Philip Roth's Life and Art

Nowhere is the breach of confidentiality in the patient-analyst relationship more evident than in the next chapter. Philip Roth was in analysis in the mid-1960s when, without his knowledge or permission, his psychoanalyst,

Dr. Hans J. Kleinschmidt, published a thirty-page article about the relationship between creativity and psychopathology. Dr. Kleinschmidt disguised his patient's identity and assumed that Roth would never discover that his analyst had written about him. Roth was horrified when he came across the article, for it contained a highly embarrassing experience the novelist had confided in therapy that he also used in an early chapter of his celebrated novel *Portnoy's Complaint*. Roth rightly feared that anyone who read his novel and the psychoanalyst's article would realize he was Dr. Kleinschmidt's patient. The title of the psychoanalyst's article could not have been more ironic: "The Angry Act: The Role of Aggression in Creativity." Roth's anger turned to fury when, astonishingly, Dr. Kleinschmidt denied any wrongdoing and insisted that his patient was overreacting to the accidental breach of confidentiality. If Roth could not let go of his rage, the psychoanalyst finally told him, he should find another analyst.

Roth did neither. Instead, he wrote a long, convoluted novel, *My Life as a Man*, about a fictional novelist who, like Roth himself, is in analysis for several years and finds himself victimized by an analyst who breaches confidentiality by writing about him and then denies responsibility when the patient discovers the professional transgression. Otto Spielvogel, the psychoanalyst who has only a brief comic punch line at the end of *Portnoy's Complaint*—"So [*said the doctor*]. Now vee may perhaps to begin? Yes?"—is now developed into a major character. Dr. Spielvogel's denial of responsibility becomes, if anything, a greater betrayal than the initial breach of confidentiality. How can a patient trust an analyst who is so impervious to criticism, who is so indifferent to a patient's alarm about the loss of privacy, and who interprets a patient's justifiable anger as a narcissistic melodrama? Roth's fictional novelist cannot understand his psychoanalyst's insensitivity to confidentiality. Nor could the real novelist. Roth goes out of his way in *My Life as a Man* to create parallels between himself and his thinly veiled fictional novelist. A funhouse filled with mirrors, *My Life as a Man* both invites and defies readers to make connections between Roth's life and art.

In the early 1980s Jeff Berman was writing *The Talking Cure: Literary Representations of Psychoanalysis*, and as he read *My Life as a Man*, he began to wonder whether Roth was writing about an experience of his own and a breach of confidentiality by his psychoanalyst. Roth supplies the clues in the novel that allowed Jeff to find the real psychoanalytic article. To authenticate his discovery, Jeff decided to send a copy of his chapter to Dr. Kleinschmidt. The story of Jeff's stormy interview with Dr. Kleinschmidt appears in our chapter "'The Angry Act': The Psychoanalyst's Breach of Confidentiality in Philip Roth's Life and Art," where we discuss the hazards of publishing a case report without the patient's permission. What are a psychoanalyst's motives in writing about a patient? What are a literary critic's motives in writing about both analyst and patient? We also discuss Jeff's ethical dilemma. Would writing about Dr. Kleinschmidt's breach of confidentiality be another violation of Roth's privacy?

Roth has never gone public with his anger toward Dr. Kleinschmidt, preferring to find revenge not through the court system or the court of public opinion but through his fiction, where writers who have a grudge against a psychotherapist always have the last word—and where comic novelists have the last laugh. Nor did Dr. Kleinschmidt comment publicly on the case, though he did threaten Jeff with a lawsuit if he attempted to publish *The Talking Cure*, which appeared in 1985. "The Angry Act" raises important privacy questions about psychiatric case studies of living writers. Dr. Kleinschmidt's breach of confidentiality in the patient-therapist relationship never made its way into the court system, but psychoanalysts and literary critics continue to discuss "The Angry Act" as a cautionary tale.

Angry Acts and Counteracts in Philip Roth's Life and Art

To what extent did Dr. Kleinschmidt's breach of confidentiality in "The Angry Act" constitute a traumatic injury that reverberates throughout Roth's writings? We raise this and related privacy questions in the next chapter, "Angry Acts and Counteracts in Philip Roth's Life and Art." Roth repeatedly uses highly personal material from his own life and the lives of others in his fictional and nonfictional writings. Unlike therapists, novelists do not need permission to write about real people, but Roth's stories raise troubling ethical questions about autobiographically transparent fictions or memoirs about men or women, living or dead, who are the objects of a novelist's or memoirist's attention. When does writing about a real person become an invasion of privacy? What are a novelist's responsibilities when writing about privacy matters? In her memoir *Leaving a Doll's House*, Claire Bloom writes about her consternation and shame upon discovering that her husband, Philip Roth, had characterized her scathingly in his thinly disguised novel *Deception*, where he refers to her as "Claire." She felt betrayed by her husband's breach of trust and confidentiality, suggesting that Roth victimized her in his version of "The Angry Act" in the same way the novelist himself had been victimized by his psychoanalyst years earlier. Roth's long and distinguished career as a novelist reveals that "The Angry Act" plays a central role in his stories, offering us insights into his characters' lives and the blurred line between reality and fantasy.

The Case of Jane Doe v. Joan Roe and Peter Poe

Our next chapter explores another psychiatric case study published without the patients' approval, one that *did* arouse intense legal attention in state and federal courts. The psychiatrist Leida Berg and her psychologist husband, Harold Steinberg, treated "Helena" and "Henry" for several years. Dr. Berg began writing her case study, *In Search of a Response*, but the two patients were dismayed when she showed them the manuscript. Helena felt that Dr. Berg was partial to Henry's point of view because the therapist devoted twice as much space to describing his feelings. Helena was frightened when she read the case study, realizing perhaps for the first time the depth of her husband's psychological disorganization. Reading the manuscript magnified Helena's feelings of jealousy, abandonment, and rejection. Reading the manuscript was no less disturbing to Henry: he was outraged by the loss of his privacy. He also felt the therapist was financially exploiting them. As an attorney, he believed that psychiatry and the law were incompatible. "Everyone ought to be allowed to plead the Fifth Amendment about their feelings," he bitterly told Dr. Berg. He viewed her as a "mad scientist" who was engaged in a sinister experiment. Curiously, Dr. Berg dismissed her patients' criticisms, and few readers of the case study will come away believing that the therapist understood their fears about the loss of privacy.

Dr. Berg claimed to have received permission from her patients to write about them, a claim denied by Helena, who was married to Henry while they were both in therapy. Henry divorced Helena, remarried, and died before the case study was published, but his widow stated that he had given permission for the case study to appear in print. Dismayed when she discovered the case study was about to be published, Helena immediately sought an injunction to block its distribution, fearing the authors did not sufficiently disguise her identity. Everything is odd about the case of *Jane Doe v. Joan Roe and Peter Poe*, including its singsong name. The case focused on competing claims, a patient's right to privacy versus a therapist's right to publish a scientific work.

Our interest in *In Search of a Response* lies mainly in its human drama: a maverick psychiatrist who believed she found a way to cure schizophrenia; a male patient who remained in psychotherapy for years despite his contempt for the psychiatrist who treated him; a female patient who, becoming a psychiatric social worker, felt betrayed by the woman in whom she had invested her confidence and trust; and the patients' son, who enigmatically commented on the book on the Amazon website and then allowed us to interview him about the case study. A psychiatrist's publication of a case study over a patient's strenuous objections constitutes a form of "treachery," to use Henry's word, not only for the patient but for psychotherapy itself, which cannot exist without the guarantee of privacy and confidentiality.

The Anne Sexton Controversy

Our next chapter examines a different type of breach in the patient-therapist relationship. The Pulitzer Prize–winning poet Anne Sexton committed suicide in 1974 at the age of forty-five. A "confessional" poet, she was also a highly controversial one, in life and in death. After her death, her daughter and literary executor, Linda Gray Sexton, requested that Dr. Martin Orne, the psychiatrist who had treated Anne Sexton from 1956 through 1964, turn over to the poet's biographer, Diane Wood Middlebrook, hundreds of hours of audiotapes he had made of his patient's therapy, along with his therapy notes and her unpublished poems. Believing that Anne Sexton would have wanted him to do so—many of her poems describe, with startling candor, the details of her therapy—Dr. Orne complied with the request, to the dismay of the mental health community, who believed it would further erode the privacy and confidentiality necessary for psychotherapy.

The Anne Sexton controversy raises a host of legal, psychiatric, and literary questions. Should therapists be compelled to hand over therapy tapes and notes to a deceased patient's family, as the law now requires in some states? Does a patient's right to confidentiality end with his or her death? What are the privacy rights of relatives and friends who find their reputations besmirched by a biographer's use of therapy tapes? Will Orne's willingness to turn over Sexton's tapes to Middlebrook, along with the controversy surrounding her biography, discourage future patients from entering psychotherapy? Do the ethics of psychotherapy rest upon absolute or relative guidelines? Might there be a positive use of therapy tapes and notes, allowing us to see how therapy has helped a patient and how the treatment might have been more effective in light of current research?

The Anne Sexton controversy also reveals the clash between the rights of a patient in therapy and the demands of science and art for public discussion. The problem long bedeviled Freud, as Peter Gay suggests:

> No doubt the twin aims of psychoanalysis—to provide therapy and to generate theory—are usually compatible and interdependent. But at times they clash: the rights of the patient to privacy may conflict with the demands of science for public discussion. It was a difficulty Freud would confront again, and not with his patients alone; as his own most revealing analysand, he found self-disclosure at once painful and necessary. The compromises he engineered were never wholly satisfactory, either to him or to his readers. (74)

Apart from these questions, the story of Anne Sexton is captivating because of the many parallels between the talking cure and the writing cure. The sins of the fathers—and the mothers—are strikingly evident in Anne Sexton's and Linda Gray Sexton's lives. Reading their stories allows us to see how they tried to come to terms with intergenerational conflicts. Writing about shame may result in the loss of privacy, to oneself and others, along with public criticism, but it may also lead to heightened self-knowledge, which is the goal of both the literary writer and the therapist.

The Tarasoff Case

In the next chapter we discuss another way confidentiality can be breached: when a patient's threat of committing violence compels a psychotherapist to contact the police. The Tarasoff case explores one of the most famous events in psychotherapy, resulting in two landmark rulings that have profoundly changed the patient-therapist relationship. Tatiana (Tanya) Tarasoff was a nineteen-year-old undergraduate at the University of California, Berkeley, when a graduate student from India, Prosenjit Poddar, became obsessed with her. In June 1969 he began reluctantly seeing a Berkeley psychologist, Dr. Lawrence Moore, to help him break the obsession. Alarmed by his patient's rapid mental deterioration and his threats of violence against Tanya, Dr. Moore contacted the Berkeley campus police who concluded, after briefly speaking with Prosenjit, that he was not a threat. Feeling betrayed by his therapist, the patient terminated treatment. Four months later he stabbed the young woman to death. Tanya's parents then initiated a lawsuit that resulted in two historic rulings by the California Supreme Court, which decreed that a psychotherapist incurs an obligation to use reasonable care to protect an intended victim against a patient's threat of violence.

Our interest in the Tarasoff story lies in its theme of fatal attraction and in the unpredictable consequences of limiting the protective privilege of the patient-psychotherapist relationship. We're especially interested in Dr. Moore's dilemma. Hoping that an act of violence might be averted if he contacted the police, he knew that almost certainly his therapeutic relationship with Prosenjit would be destroyed by the breach of confidentiality. Ironically, the psychologist's decision to notify the police not only failed to avert the tragedy but also resulted in his colleagues' censure of him for violating the ethics of his profession.

To breach or not to breach? That is the question not only for patient and therapist but for society as a whole. Dr. Moore's Hamletian dilemma strikes at the heart of psychotherapy. Many if not most patients struggle with destructive or self-destructive impulses at one time or another, and no one can accurately predict whether a patient will carry out a verbal threat. Some therapists fear that without the guarantee of absolute confidentiality, the truly violent patient may avoid psychotherapy altogether. Without the guarantee of absolute confidentiality, other patients may hesitate to talk about destructive or self-destructive feelings, thus preventing violent feelings or impulses from being understood and worked through. Most clinicians believe that violent patients are less destructive to others and themselves when they remain in treatment. Even when a therapist issues a "Tarasoff" warning, the patient may later act on violent feelings or impulses, as Poddar did. Research has shown, as Allannah Furlong and others have pointed out, that third-party warnings generally do not avert violence. Would Dr. Moore have been able to help Prosenjit defuse his violence by remaining in therapy? We will never know. The Tarasoff story remains fascinating because of the many "what ifs" it raises about a therapist's duty to a patient and to society.

Jaffee v. Redmond

Our next chapter focuses on *Jaffee v. Redmond*, which remains perhaps the most significant ruling in the history of psychotherapist-patient privacy in the United States. The 1996 decision of the U.S. Supreme Court established the patient-psychotherapist privilege in the federal courts. *Jaffee v. Redmond* was the culmination of a near half-century effort to establish the principle that communication between patients and their psychotherapists is in need of a high level of protection, protection similar to that given to the communication between clients and their lawyers. Few people outside of mental health professionals have heard of *Jaffee*, however, and even therapists are familiar only with the outcome of the case, not the details. The human story is fascinating. We begin with the details of the shooting in a Chicago suburb, and then we focus on how the case proceeded from a federal district court through an appellate court to the Supreme Court. Then we discuss three of the individuals who were centrally involved with the case: Justice Antonin Scalia, whose mocking dissent, along with his other writings, reveals a deep mistrust of psychoanalysis; Karen Beyer, the social worker who refused a court order to testify and turn over her psychotherapy notes; and her lawyer, Sandra Nye.

The People v. Robert Bierenbaum

In the next chapter we discuss a famous psychotherapy case where patient-therapist confidentiality was *almost* breached. Gail Katz Bierenbaum disappeared in New York City on July 7, 1985, without leaving a trace. The only suspect was her physician-husband, Robert Bierenbaum. Three psychotherapists had evaluated one or both of the Bierenbaums. Recognizing the husband's explosive temper and history of violence, the therapists warned the wife she was in grave danger. After her mysterious disappearance, Michael Stone, the psychiatrist who had earlier given Robert Bierenbaum several unprecedented preconditions to be accepted into treatment, opined to the media that the suspect was a "psychopath" who had brutally murdered his wife. The press sensationalized the story. The case was closed after a nine-month investigation because of the absence of any solid clues. Gail Katz Bierenbaum was declared a "missing person," not a homicide, and media attention slowly faded.

Years passed. Robert Bierenbaum moved to another state, remarried, and earned a reputation as an exemplary surgeon. Gail Katz Bierenbaum's family, particularly her lawyer-sister, Alayne Katz, continued to insist that she had been murdered by her husband, but there was still no evidence of a crime. After more than a decade the case was reopened, and the disappearance was now treated as a homicide. The prosecution sought the therapists' help in establishing Bierenbaum's guilt. The therapists were torn between maintaining the confidentiality of the patient-therapist relationship and serving justice.

Like the Buried Bodies case, the Disappearing Wife case is notable because of the legal implications of confidentiality and privileged information. The case was so significant to the future of psychotherapy that leading professional mental health organizations filed an amici brief urging the judge to respect the patient-therapist privilege. The story is fascinating for several reasons. First, the case rested entirely on circumstantial evidence. Bierenbaum maintained a discreet silence before, during, and after the trial, never admitting any wrongdoing. Second, the rumors swirling around the story proved irresistible not only to the tabloids and to the *New York Times* but also to such magazines as the *New Yorker* and *Vanity Fair*, who viewed the case as a cautionary tale about the seductive dangers of innuendos and half-truths. Many people reached conclusions in print about Bierenbaum's role in his wife's disappearance that were contradicted by the jury's verdict. Third, the case was a landmark decision in New York, affirming patient-therapist confidentiality even when a Tarasoff warning is offered. Finally, the story is intriguing for its impact on two people. Michael Stone became a "murderologist," writing about the combination of evil and psychopathy, and Alayne Katz, whom we interviewed, has devoted her law practice to defending victims of domestic violence.

United States v. Sol Wachtler

Of all the stories related to the need for privacy and confidentiality in psychotherapy, none is more bizarre than the one we offer in the next chapter. It is a story of a man whose fear of the *stigma* of mental illness prevented him from seeking treatment with a qualified mental health professional, who he was afraid might betray confidentiality through an indiscretion. Instead, he sought alternative treatment, with disastrous results. In 1972 Sol Wachtler was elected to the Court of Appeals, the youngest judge to sit on New York's highest court. In 1985 he was appointed chief judge by Governor Mario Cuomo. Wachtler wielded enormous power, and he was widely believed to be the likely Republican candidate for governor in 1994 and perhaps even a future candidate for president of the United States. But then came a sudden fall from grace that was, politically and personally, almost unimaginable. While driving home from work, Wachtler was forced off the road by three vehicles containing FBI agents. He was arrested for a myriad of crimes related to extortion and threatening to kidnap and harm his former mistress, a wealthy socialite who was his wife's cousin. It was apparent to many people that Wachtler was not in his right mind when he made the threats, but the U.S. attorney contended otherwise and zealously prosecuted the case.

Wachtler accepted responsibility for his irrational behavior, and several of the psychiatrists who examined him in a hospital and in prison concluded that he was suffering from a severe mood disorder exacerbated by the staggering amount of medication he was taking, which induced a Jekyll-Hyde double existence. Wachtler's relatives, friends, and colleagues had recognized there was something wrong with him, but he had disregarded their concerns and continued to live a double existence until disaster struck.

Would prolonged psychotherapy, along with appropriate medication, have helped Wachtler avoid his personal and political nightmare? No one can say. He implies in *After the Madness: A Judge's Own Prison Memoir* that he would have gone to a psychotherapist if he were not worried about the public learning about his state of mind. His fear of the loss of privacy in the patient-therapist relationship contributed to his undoing. The prosecution's refusal to accept the consideration of "diminished capacity" led to an unusually harsh sentence, and despite the sentencing judge's recommendation that he serve his time at a low-security federal camp, Wachtler was inexplicably sent to a prison designed for hardened criminals, where he was often placed in isolation for his own "protection." He began to experience at first hand many of the criminal justice system's indignities, of which he had warned in his many judicial decisions, including the routine use of strip searches as a form of humiliation and the grave psychological harm of solitary confinement.

Sol Wachtler's time in prison proved to be a transformative learning experience. In our interview with him, he discussed his feelings about the harrowing case and his ongoing efforts to destigmatize mental illness and psychotherapy.

Focusing on Inner and Outer Privacy Stories

The word "person," Ralph Slovenko reminds us in *Psychotherapy and Confidentiality*, derives from the Latin word *persona*, a person's public mask. Slovenko, who has written authoritatively about the relationship between psychiatry and the law, reveals a fundamental difference between the two professions.

> The mask is what we daily see; it conceals the inner nature and conflicts of the individual. The mask is like the outer layer of the Russian wooden doll with successively smaller ones fitted into it. In psychotherapy, for purposes of therapy, the mask or outer doll is removed, but in law the

outer layer, by and large, is what is relevant and determinative of the outcome of litigation. (38)

In *Confidentiality and Its Discontents* we focus on both the inner and outer stories of the characters, their private and public lives, along with the ways in which psychiatry and the law complement and sometimes clash with each other.

Psychotherapists are a self-selected group of professionals who by nature tend to be much more interested in people than they are in dry legal confrontations. In this book we hope to raise the awareness, with respect to these events, of psychotherapists and others with an interest in this subject by telling the human stories behind these events and cases. Many of these stories are human dramas involving heroes and villains, victims and perpetrators, the stuff of compelling fiction. But these stories are real. We hope that the intrinsic interest of the stories will sensitize readers, therapists and nontherapists alike, to privacy issues and the way in which breaches of confidentiality have harmed people's lives. We provide "hooks" that will make it easier for psychotherapists and others to gain greater awareness of the principles involved in the new legal structures that have come into being to help protect the confidentiality of psychotherapeutic work.

Confidentiality and Its Discontents explores the human stories where confidentiality has been carelessly breached to the extent that a patient's identity has become publicly disclosed, or where changing laws mandate psychotherapists to break confidentiality to report those suspected of certain crimes. Breaching confidentiality often has the unintended consequence of ending therapy, which may also end the possibility of therapeutic recovery. Freud coins the expression *compromise formation* to describe how an unconscious conflict forms a symptom that represents a partial solution to the conflict. The stories in our book reveal *failed* compromise formations, where the confidentiality of the patient-therapist relationship has been irrevocably shattered. In an age of growing state and computer surveillance and loss of personal and professional privacy, where health insurance companies routinely require detailed information about a patient's therapy, our concern is that compromising confidentiality in the patient-therapist relationship may compromise the quality—even the existence—of psychotherapy.

The Challenge of Writing About Privacy Stories

"He who fights with monsters should be careful lest he thereby become a monster," Nietzsche warns in *Beyond Good and Evil*. "And if thou gaze long into an abyss, the abyss will also gaze into thee" (466). To apply Nietzsche's admonition to our project, gazing into privacy stories may be beguiling, seducing the gazers into uncovering forbidden secrets, in this case, confidentiality secrets. We have repeatedly confronted this paradox, which is probably true of all privacy stories. The more information one discovers, the more fascinating the story becomes and the greater the temptation to reveal what should remain concealed. Journalists know that the most fascinating information they receive from those they interview is often "off the record."

We are aware of the irony that although we affirm throughout our book the importance of privacy in psychotherapy, we inevitably run the risk, in a few limited instances, of further embarrassing some of those who, still alive, have been victimized by particular breaches. We have tried to present their privacy stories as accurately and sensitively as possible, telling their psychotherapy stories without further breaching their privacy. We were able to locate some of the individuals involved, directly or indirectly, with these privacy stories. We then sent copies of our chapters to them, solicited their responses, and interviewed them. These people included the psychoanalyst Joseph Lifschutz; Harriet and Steven Werner's son, Frank, called "Jan" in Leida Berg's case study; Gail Katz Bierenbaum's sister, Alayne Katz; the social worker Karen Beyer and her lawyer, Sandra Nye; and former New York Chief Judge Sol Wachtler. They described to us, in their own words, how they felt about these privacy stories. To avoid potential problems, we showed these individuals exactly how we were using their comments to make sure that we accurately contextualized and conveyed their points of view. We are grateful for their willingness to share their experiences with us.

To return to Freud's metaphor quoted in the beginning of this introduction, we aim to be travelers, sitting next to a window, describing to someone inside the train, our readers, the changing views of the human stories behind noted psychotherapy privacy cases and breaches. Or to paraphrase a remark sometimes attributed to Albert Einstein, we have tried "to make everything as simple as possible but no simpler."

1.   We Have Met the Enemy, and He (Is) Was Us

To understand the human stories behind noted confidentiality breaches and dilemmas in psychotherapy, we need to understand first the historical reasons for the ongoing loss of confidentiality in the patient-therapist relationship. Neither Freud nor anyone else could have foreseen the historical factors responsible for the therapist's increasingly complicated dual allegiance to patient and society. And so we present briefly here this changing history, mindful of Julian Barnes's observation in *The Sense of an Ending* that "History is that certainty produced at the point where the imperfections of memory meet the inadequacies of documentation" (65).

"The Duty of Medical Discretion"

Freud himself did not always follow the excellent advice he offered to other psychoanalysts about maintaining confidentiality. The words "breach of confidence [or confidentiality]" rarely appeared in the psychoanalytic literature during the first half of the twentieth century. The first use appears in the preface to the first edition of *Studies on Hysteria* in 1895, the book generally viewed as the beginning of psychoanalysis. Acknowledging the difficulty of writing about their patients' intimate lives, Breuer and Freud declare: "It would be a grave breach of confidence to publish material of this kind, with the risk of the patients being recognized and their acquaintances becoming informed of facts which were confided only to the physician. It has therefore been impossible for us to make use of some of the most instructive and convincing of our observations" (xxix). These observations, the authors add, refer "especially to all those cases in which sexual and marital relations play an important aetiological part."

If, as Breuer and Freud argue, hysteria is caused by sexual repression, they must have felt something akin to intellectual repression as a result of their inability to include this material because of confidentiality concerns. Yet Freud succeeded remarkably well in hinting at his patients' erotic conflicts, as when he suggests in the "Fräulein Elisabeth von R." chapter that her illness was caused by thwarted love for her brother-in-law. Freud's language rises to the occasion in evoking these conflicts. He admits, with a combination of modesty and disingenuousness, that he still finds it "strange that the case histories I write should read like short stories and that, as one might say, they lack the serious stamp of science. I must console myself with the reflection that the nature of the subject is evidently responsible for this, rather than any preference of my own" (*Studies on Hysteria*, 160). Notwithstanding his literary modesty, thirty-five years later Freud received the coveted Goethe Prize, honoring his contributions to literature.

James Strachey reveals in a footnote near the end of "Fräulein Anna O." that Freud told him something of crucial importance that Breuer left out of the case study of the first patient in psychoanalysis, an "untoward event" in which, in Strachey's words, "when the treatment had apparently reached a successful end, the patient suddenly made manifest to Breuer the presence of a strong unanalysed positive transference of an unmistakably sexual nature. It was this occurrence, Freud believed, that caused Breuer to hold back the publication of the case history for so many years and that led ultimately to his abandonment of all further collaboration in Freud's researches" (*Studies on Hysteria*, note 1, 40–41). Freud later expanded on the details of this untoward event in *On the History of the Psycho-Analytic Movement*, where he discloses that Breuer "must have discovered from further indications the sexual motivation of this transference" (12).

Freud's first major biographer, Ernest Jones, points out that the sharp temperamental differences between Freud and Breuer complicated and probably doomed their personal and professional relationship. "Breuer was in his work reserved, cautious, averse to any generalization, realistic, and above all vacillating in his ambivalence" (1:297). Without the reticent Breuer as his coauthor, Freud later allowed himself to write more candidly about the intimate aspects of his patients' lives. To be sure, Freud exercised "discretion" when writing about his patients, which is what he claimed he had the right to do in the name of "science." In exercising this discretion, he was selective in choosing the patients about whom he wrote. He was also selective in writing about himself and his own inner experience.

Freud never used the words "breach of confidence" again. Instead, he used the more euphemistic term, "medical discretion," in *Fragment of an*

Analysis of a Case of Hysteria—better known as the story of Dora—in 1905. In the "Prefatory Remarks" Freud, who came to view himself as the conquistador of the unconscious, or, changing metaphors, the disturber of the world's sleep, now associates medical discretion with the forces antithetical to science. "If it is true that the causes of hysterical disorders are to be found in the intimacies of the patients' psychosexual life, and that hysterical symptoms are the expression of their most secret and repressed wishes, then the complete elucidation of a case of hysteria is bound to involve the revelation of those intimacies and the betrayal of those secrets" (7–8).

Freud offers here an ingenious conceptualization of the major dilemmas of psychoanalysis, a dilemma that has never been entirely solved. Patients would never speak freely and openly if they suspected that their "admissions" might be put to "scientific uses." Nor would patients, in Freud's view, give him permission to write about them. Paradoxically, Freud then characterizes analysts who would try to preserve the very kind of confidentiality he has just described as essential as being "persons of delicacy" or "merely timid." As the psychoanalyst Barry Landau observes,

> the fact that the identities of all of Freud's published cases have been found out and disclosed illustrates the problem of trying to balance confidentiality with other strongly held values. Confidentiality can be very inconvenient and often does conflict with other important values. However, the attempt to balance confidentiality with these other values puts confidentiality at risk. To the extent that confidentiality is a necessary condition that makes psychoanalysis possible, such balancing can put the entire psychoanalytic enterprise at risk to a degree that could ultimately be untenable. (personal communication, August 17, 2013)

Freud's blindness thus set the stage for later analysts not only to identify with and protect the creator of psychoanalysis but also to disclose confidential information about their patients while at the same time espousing the necessity for confidentiality.

Freud leaves little doubt where his allegiance lies. "Thus it becomes the physician's duty to publish what he believes he knows of the causes and structure of hysteria, and it becomes a disgraceful piece of cowardice on his part to neglect doing so, as long as he can avoid causing direct personal injury to the single patient concerned" (*Fragment of an Analysis of a Case of Hysteria*, 8). After claiming that he has taken every precaution to prevent his patient from suffering any injury, Freud vows to protect his patients' privacy while writing truthfully about their psychopathology:

I am aware that—in this city, at least—there are many physicians who (revolting though it may seem) choose to read a case history of this kind not as a contribution to the psychopathology of the neuroses, but as a roman à clef designed for their private delectation. I can assure readers of this species that every case history which I may have occasion to publish in the future will be secured against their perspicacity by similar guarantees of secrecy, even though this resolution is bound to put quite extraordinary restrictions upon my choice of material. (9)

"No Mortal Can Keep a Secret"

Despite these assurances, Freud did not entirely fulfill his pledge of secrecy, partly because of his defensive thinking, which made it possible to rationalize the disclosure of confidential information for the purpose of scientific research, and partly because he felt that betraying secrets was human nature and therefore inevitable, as he vividly writes in *Fragment of an Analysis of a Case of Hysteria*. "He that has eyes to see and ears to hear may convince himself that no mortal can keep a secret. If his lips are silent, he chatters with his finger-tips; betrayal oozes out of him at every pore" (77–78). If no mortal can keep a secret, then Freud cannot be criticized for failing to live up to his own ethical imperatives to maintain confidentiality. It is true that he waited four years before he published the case study, but we know the names and backgrounds of most of his patients, including the identity of "Dora," Ida Bauer, who later discovered that she was the subject of Freud's case history and, like his other patients, appeared to be proud of it. Throughout the "Prefatory Remarks" in the case study, Freud anxiously anticipated so many of the considerations and rationalizations that would be used by analysts over the course of the next century. He was never able to resolve the conflict between preserving patient confidentiality and pursuing scientific research. He concealed various facts about his patients' lives to show his efforts to preserve confidentiality, but his views tipped more in the direction of disclosure than of concealment.

According to Ernest Jones, Freud was highly secretive about the details of his own life, apart from what he revealed in his writings, but, "oddly enough, Freud was not a man who found it easy to keep someone else's secrets. He had indeed the reputation of being distinctly indiscreet" (2:409–410). Nor did Freud always practice what he preached. The psychiatrists David J. Lynn and George E. Vaillant have shown, in a review of forty-three of Freud's case studies from 1907 to 1939, the striking disparity between his recommendations on psychoanalytic technique and his actual methods. "Freud communicated with others about analysands in 23 (53%) of these cases. These communications were to people known to the analysand and included Freud's identification of the analysand." An "interesting additional finding," Lynn and Vaillant observe, "is that no fewer than 20 (47%) of the 43 analysands in our series received information from Freud about other analysands" (165). As Louis Breger points out, "Freud almost never followed his recommendation of confidentiality, certainly not when he had some personal or political interest at stake" (370). Or as John Forrester remarks, "Freud is as active in the transgression of his own implicit rules, for his own good reasons, as he is in their observance" (22). Analysts have had many positive reasons to identify with the creator of psychoanalysis, but identification with Freud's disclosure of patients' secrets is not something that is in the best interest of analysts' patients, reputations, or profession.

"The Gossiping Psychoanalyst"

Freud's belief that no mortal can keep a secret anticipates "The Gossiping Psychoanalyst," the title of Stanley L. Olinick's article in a 1980 issue of the *International Review of Psycho-Analysis*. He begins his article by raising a fundamental question. "How does gossiping about patients become an issue among a group of people whose professional lives are otherwise dedicated to fostering the autonomy and inviolability of the individuals who are their patients?" (439). Noting that at that time the psychoanalytic bibliography was "scant and scattered," Olinick calls attention to one of the thorny difficulties confronting anyone writing about this subject, including himself: "adequate clinical discussion of gossip would require breaches of confidentiality that, by the very nature of the topic, would be easily traceable" (439). Without citing clinical or statistical evidence, Olinick nevertheless believes it is no exaggeration to say that "almost every analyst has gossiped about his patients at some time, and that some engage in it with high frequency" (439). In his view, psychoanalysts gossip about those patients whom they envy, those patients with a "narcissistic personality, with grandiosity well-disguised," with a flair for flaunting their "overtly successful social, economic, and marital (or, at any rate, sexual) adjustments" (439). Narcissistic patients thus become objects of envy to the gossiping psychoanalyst. Olinick contends that contrary to what people may believe, isolation and stress are not the primary reasons for psychoanalysts' gossip. Rather, the operative factor for the gossiper is the need to "bask in the gratified curiosity of his listener"; the giver and receiver of gossip "are united in a common action, and both are reassured and confirmed through the reciprocal kinship. The need for union and kinship is satisfied through gratified curiosity, assuaged loneliness, and shared derogating of the envied person" (440).

Wealthy or prominent patients with a sense of entitlement can stir up powerful countertransference feelings in a psychoanalyst. Additional complications arise if, as Ira Brenner suggests, the psychoanalyst feels proud about treating a celebrity patient and begins to gossip as a way to enhance his or her status (10). This is an enduring problem. In 1972 Markowitz had noted "the not infrequent occurrence of casual allusion and gossip on the part of analysts about their patients, asserting vigorously that psychoanalysts have the duty to deny themselves the privilege of enhancing their status by associating themselves publicly with their patients" (Watson, 157). Such gossip always, of course, violates the patient's right to privacy and the psychoanalyst's obligation of confidentiality.

Olinick, Brenner, and Markowitz don't explore the devastating psychological consequences that arise when a psychoanalyst breaches confidentiality by talking about one patient to another patient, but this is the focus of Jane Burka's article "Psychic Fallout from Breach of Confidentiality," appearing in a 2008 issue of *Contemporary Psychoanalysis*. She is an exception to the rule that victims of breach of confidentiality do not publish in professional literature. A psychoanalyst, she writes about how her analyst breached the confidentiality of her training analysis by revealing privileged information about her to a patient with whom he was having a sexual relationship. The experience was traumatic for Burka, undermining her sense of trust, security, and self-esteem. The breach of confidentiality also called into question the value of her training analysis. To make matters worse, her analyst never admitted wrongdoing. Reluctant at first to file a lawsuit because it would mean further public exposure and humiliation, Burka made the wrenching decision to take her analyst to court. Three years after she filed the lawsuit and four and a half years after she ended her analysis, the case came to trial. The court found her analyst liable for negligence and breach of fiduciary duty. Burka concludes by saying that for the psychoanalytic community, "breach of confidentiality represents an apostasy that undermines the foundation on which our profession is based" (197).

Marilyn Monroe is the most dramatic example of a prominent patient who was the victim of breach of confidentiality and invasion of privacy. Ralph Greenson was considered the "dean" of psychoanalysis in southern

California, the "analyst to the stars," counting among his patients Frank Sinatra, Tony Curtis, and Vivien Leigh. He treated Marilyn Monroe during the last two years of her life, when her marriage to Arthur Miller was falling apart. Believing that traditional psychoanalysis had failed her, Greenson treated her at his posh home in Santa Monica, invited her for dinner after her daily therapy session, encouraged his children to help her, and made her feel a part of his family. As Christopher Turner recounts in an article in the London *Telegraph*, Greenson diagnosed her as a "borderline paranoid addictive personality," took on the role of "good father" to her, and sought to create what a colleague called a "foster-home fantasy of a haven where all hurts are mended."

Marilyn Monroe's depression continued to deepen, however, and she died of a drug overdose, accidental or deliberate, in 1962, a death that conspiracy theorists regard as a murder planned by Jack Kennedy, Robert Kennedy, or Greenson himself. Turner quotes Monroe's biographer Donald Spoto accusing Greenson of having "betrayed every ethic and responsibility to his family, his profession, and to Marilyn Monroe" in his "egregious mishandling of his most famous patient." Olinick does not cite Greenson as an example of a gossiping psychoanalyst, but there is no question that the iconic actress was a trophy patient who allowed the analyst to bask in her legendary fame. It's ironic that Greenson, the author of a classic textbook, *The Technique and Practice of Psychoanalysis*, deviated so radically from analytic technique. Analysts who make unusual interventions must "retain humility," in Salman Akhtar's words, "and have the moral courage to seek consultation when faced with difficult and puzzling clinical situations" (28). Greenson is reported to have consulted with his office partner about Monroe, but this might have been a biased consultation, a kind of *folie à deux* resulting from their existing ties to each other. Greenson's treatment of Marilyn Monroe at his home is an example of what Mark Moore calls the "use of unorthodox space gone awry" (43). It's hard to disagree with Moore's conclusion that the "use of unorthodox space in this treatment left open too many unexamined questions and stands as an example of a failure to ensure authenticity and reliability" (44).

It remains unclear whether Greenson's invasion of Marilyn Monroe's privacy and breach of confidentiality contributed to her death; his rationalization was that she would have died sooner if she were hospitalized. Greenson wrote a paper in 1978 entitled "Special Problems in Psychotherapy with the Rich and Famous" that remains unpublished and unavailable to the public, but in light of his problematic countertransference relationship to

Marilyn Monroe, perhaps the title should have been "A Psychotherapist's Special Problems with the Rich and Famous."

"Half-Tamed Demons"

Freud grappled with the problem of what to reveal and conceal throughout his life, and he was prepared to sacrifice the privacy of the patient he knew best—himself. "Breaching" his own confidentiality is, of course, vastly different from breaching his patients' confidentiality. Freud's willingness to sacrifice his privacy in the quest for truth affirms his own commitment to the fundamental rule of full disclosure. To be sure, sometimes he writes about himself in disguise, as in his 1899 essay "Screen Memories," where he refers to curing the "slight phobia" of a thirty-eight-year-old man of university education who, as Strachey points out in his "Editor's Note," was Freud himself (302).

Other times, as in *The Interpretation of Dreams*, Freud does not use disguise, though his intense ambivalence toward self-disclosure is evident, as when he declares in the preface to the first edition that

> if I was to report my own dreams, it inevitably followed that I should
> have to reveal to the public gaze more of the intimacies of my mental life
> than I liked, or than is normally necessary for any writer who is a man of
> science and not a poet. Such was the painful but unavoidable necessity;
> and I have submitted to it rather than totally abandon the possibility of
> giving the evidence for my psychological findings. Naturally, however,
> I have been unable to resist the temptation of taking the edge off some
> of my indiscretions by omissions and substitutions. (xxiii–xxiv)

Sometimes his edginess is apparent, as when he observes, after analyzing one of his most self-revealing dreams, "If anyone should feel tempted to express a hasty condemnation of my reticence, I would advise him to make the experiment of being franker than I am" (121).

Ambivalent or not, Freud was willing to write more autobiographically than any psychologist before or after him, and he was also willing to pay the price for the loss of privacy. Describing near the end of *Fragment of an Analysis of a Case of Hysteria* Dora's "act of revenge" in breaking off treatment with him, Freud observes prophetically, "No one who, like me, conjures up the most evil of those half-tamed demons that inhabit the human breast, and seeks to wrestle with them, can expect to come through the struggle unscathed" (109).

Ironically, Dora's act of revenge evoked Freud's own countertransfer-ential revengeful behavior toward her. "The resemblance between the two injured parties is striking," Gary Gillard suggests in his 1994 Ph.D. disser-tation: "each has been deceived and deserted; and each takes revenge. Freud's revenge, I believe, is in writing the case study account—with its double betrayal of Dora: firstly, in having told her secrets to anyone who came across the account (though of course she is disguised), and secondly, in attributing to her the writer's own desires and fantasies."

The Erosion of Confidentiality

One need not invoke demons, however, to explain psychotherapy breaches. The erosion of confidentiality protection for psychoanalysis in the United States is common knowledge. As Christopher Bollas and David Sundelson have noted, analysts themselves to some degree have been responsible for this deterioration. The reasons for the erosion of confidentiality protec-tion include analysts' anachronistic reliance on the ethical guidance of other professional groups, a failure to distinguish between the privacy needs of psychoanalytic and medical data, and timidity in taking an assertive stand within the legal system and third-party reimbursement systems. There have been a few positive developments, however, such as the emergence of re-visions of ethics codes, a stringent federal evidentiary privilege for psycho-therapeutic data since 1996, and in 2000 federal privacy regulations that extend special protection for psychotherapy information well beyond that provided for other health care information.

In the United States psychoanalysis and the derivative psychotherapies gained preeminence in the middle fifty years of the twentieth century, largely embedded within the health care professions. Until the end of that period in history, U.S. psychoanalysis appeared to be a subspecialty of psy-chiatry and thus more a branch of medicine than a separate profession. Most psychoanalysts were first trained as physicians and psychiatrists and hence were socialized into the American medical profession. Understandably, the development of the ethics, legal framework, and organization of the health care system in the United States has had an enormous influence on the de-velopment of psychoanalysis (Hale).

Psychoanalysts, as members of their "core" professional organizations, such as the American Medical Association and the American Psychiatric Association, relied on the concomitant ethical precepts of these groups to provide a concept of confidentiality. The one-to-one nature of the practice of psychoanalysis generally mirrored the organization of the entire health care system, because medicine, sometimes described then as a "cottage industry," was practiced mostly by individual general practitioners who had long and trusting relationships with patients. As late as the 1960s, a congenial definition of confidentiality meant the ethical understanding that the physician would not disclose information obtained from the patient to any other person.

Subsequent transformations in the health care system have altered the common usage and meaning of the term "confidentiality." These transformations include the specialization of medical care with shared information among multiple physicians per patient, the increased use of third-party payment, a decrease in the paternalistic role and authority of physicians, an increasingly mobile population, the evolution of health care systems using treatment teams, the development of managed care systems in which confidential information is shared with nonprofessional case managers, and networked computerized record-keeping systems (Siegler).

The health care professions adapted their confidentiality practices to these changes, but "confidentiality" could be maintained only by altering the meaning of that term to encompass the sharing of information among a widening circle of individuals involved with the patient's care. In general, there is little evidence that this changing concept of confidentiality has resulted in an erosion in the overall quality of health care. However, psychoanalytic treatment continues strictly to be based on a trusting *one-to-one* relationship, and, consequently, it is no longer sensible for analysts to rely solely on ethical codes of other health care professions to articulate standards of confidentiality for psychoanalysis. While psychoanalysts have had their own ethical principles and guidelines for many years, they have not aggressively modernized or updated their confidentiality guidelines to take into account the profound changes that have occurred throughout the rest of the health care system.

External pressures have brought increased demands for access to confidential psychotherapeutic information. Information disclosed to psychotherapists forms a rich repository of highly sensitive personal data that, if available, could be used for many purposes unrelated to psychotherapy. For example, psychotherapists have been legally obliged to report past, present, and possible future child abuse to state authorities, to testify in divorce proceedings and child custody hearings, to participate in disability determinations, to assume legal liability if they fail to take measures to protect intended victims of potentially violent patients, and to disclose sensitive information to representatives of third-party payers. As earlier requirements that physicians and others report certain information such as communicable diseases and gunshot wounds to the state have gradually been expanded, psychotherapists have been carried along in a creeping process of expanded disclosures.

The Tarasoff Decisions

One of the earliest and most influential state court cases limiting the protective privilege of the patient-psychotherapist relationship was the Tarasoff decision—or rather, decisions. The two decisions have had a curious history and legacy. *Tarasoff I*, which ruled in 1974 that a therapist has a *duty to warn* those whom a patient threatens to harm, was vacated by the California Supreme Court two years later and replaced by *Tarasoff II*, which held that a therapist has an *obligation to use reasonable care to protect* those whom a patient threatens to harm. The Tarasoff case was finally settled out of court and never went to a full trial. Nor has the Tarasoff doctrine spread to every state. Nevertheless, Tarasoff has become the guiding principle for all American mental health professionals. The Tarasoff rulings are widely misunderstood to "require" psychotherapists to intervene in such instances. In fact, Tarasoff rulings establish the principle that in certain situations psychotherapists can be held liable (that is, sued) by a third-party victim to whom they owed, but did not fulfill, a "duty to protect" (Buckner and Firestone; Slovenko, *Psychotherapy and Confidentiality*, 270–341).

Distinguishing Psychotherapy Information from Medical Information

The way in which confidentiality was conceptualized by the mental health professions began to diverge from the views of other health care professionals in the early stages of the Cold War. The divergence first surfaced in 1950 in response to a request from J. Edgar Hoover, the director of the Federal Bureau of Investigation, published in the *Journal of the American Medical Association* (*JAMA*), that physicians should report their patients' subversive inclinations to the FBI. Psychiatrists and psychoanalysts protested Hoover's request (Slovenko, *Psychotherapy and Confidentiality*, 270). Prior to 1957, American physicians were bound to avoid nonconsensual disclosures of confidential information except to protect the *health* of another individual or group or if *required* by law. In 1957, the ethics code of the American Medical Association was changed to allow disclosures "necessary to protect the *welfare* of the individual or the community." This was interpreted as allowing physicians' disclosure of confidential information for political and "national security" purposes. That code remained in effect until 1984, when the wording was changed to require physicians to "safeguard patient confidences within the constraints of the law" (Graber, Beasley, and Eaddy).

Ten years after Hoover's *JAMA* editorial, a second high-profile Cold War event played a significant role in fostering the divergence of thinking on the subject of professional allegiance. In 1960, Bernon F. Mitchell, a young cryptographer at the top-secret National Security Agency in Washington, defected to the Soviet Union along with a male friend, William Hamilton Martin. Both turned up at a high-profile Moscow press conference a few months later. This incident has been described as the "worst scandal in the history of the NSA." When it was discovered that shortly before his defection Mitchell had consulted Dr. Clarence Schilt, the Silver Springs, Maryland, psychiatrist was interviewed by investigators from the controversial House Un-American Activities Committee. Schilt had gathered no material related to the matter of a planned defection during his three sessions with Mitchell. However, Schilt voluntarily agreed to testify in closed session before the committee, where he disclosed details of his interviews with Mitchell. The next day, many details of the interviews appeared on the front page of the *Washington Post*. Thirty-seven psychiatrist members of the Maryland Medical and Chirurgical Faculty signed a petition raising an ethics complaint about Dr. Schilt. The faculty's Professional Conduct Committee exonerated Dr. Schilt, however, concluding that the interests of the nation were more important than those of the individual.

At the time of their defection, Mitchell and Martin were reputed to be "sexual deviants," code for "homosexuals," often used to characterize traitors. But according to an article written by Rick Anderson appearing in *Seattle Weekly News* on July 17, 2007, declassified NSA information indicated that neither Mitchell nor Martin was gay. David K. Johnson, an authority on the Cold War history of gays in the government, noted that to explain the two defectors' actions, the government "turned to the alleged sexual perversion. That was already associated in the popular imagination with subversion and communism."

Disclosure to Third-Party Payers

After World War II, the growth of health insurance with generous, virtually no-questions-asked coverage of psychoanalytic treatment provided a powerful inducement for psychoanalysts to associate themselves with the health care system. There was little incentive in such a system to establish a separate professional identity for psychoanalysis, one with its own ethical standards and record-keeping standards. Such separate standards would have implied a recognition of the differences between the confidentiality requirements of psychoanalysis and those of medical and psychiatric care.

Disclosure to third-party payers has become a serious factor in the breakdown of the protection of confidentiality in the United States. For psychoanalysts, this disclosure problem can be traced to their attempts to maintain a position within the health care reimbursement scheme. Indeed, published evidence (Goodman, 444–445) supports a view that analytic organizations, toward whom others look for a model of confidentiality protection, actually fostered the development of intensive case management in which sensitive information was being disclosed by telephone to "case managers" on a regular basis throughout the 1990s and early 2000s.

In the early 1970s during the Nixon administration, with the first hint of a publicly supported health care system, the American Psychoanalytic Association made an explicit decision to play down differences between psychoanalysis and other health care modalities to gain third-party coverage for analysis (Goodman, 445). As a result of this decision, the American Psychoanalytic Association developed a very detailed reporting form for third-party utilization management based on an existing protocol intended for a different purpose. That protocol was originally designed as a guide to candidates (trainees in psychoanalysis) who were writing *deidentified* case reports for progression in training or certification. These reports were intended to be seen only by professional colleagues (Gray, personal communication to Paul Mosher via e-mail, 2001). The original hope in adopting this detailed protocol for third-party reporting was to provide a standard mechanism by which a *professional peer, a psychiatrist employed by the insurer,* could carry out confidential case reviews. Even this plan was controversial, because prior to that time any disclosures by analysts of identifiable case material, even to professional colleagues, would have been unthinkable except perhaps for the purpose of gaining "certification" as an analyst. However, the forces within the profession favoring cooperation with the imagined new third-party payment system prevailed.

The American Psychoanalytic Association reporting form was reproduced in a widely circulated peer review manual (American Psychiatric Association, 1976), thereby conveying the message that filing detailed case reports by therapists as part of the third-party payment system had the profession's approval. Before long, what had been intended to be a confidential "peer" review mechanism mutated into a system in which *identifiable* therapists' reports were reviewed within insurance companies by subdoctoral reviewers from professions such as nursing and social work. The reporting forms were retained in files, and eventually information taken from the forms was entered into computer databases. As insurers requested more and more confidential treatment information, a backlash developed among significant numbers of psychotherapists.

Around the time that these events were occurring, an attempt was made to prohibit such disclosures and to create a mechanism for claims review more respectful of confidentiality. In 1978, the Mental Health Information Act became effective in Washington, D.C. (District of Columbia, 1978), prohibiting third-party payers from requesting confidential mental health information. Instead, the new law spelled out a mechanism for confidential reviews of disputed claims by independent professional peers. A similar statute (New Jersey, 1985), applicable only to patients of psychologists, was passed in New Jersey in 1984. Although these statutes continue to be in effect, they are not widely known and have not been duplicated in other jurisdictions.

In the early 1990s, the American Psychoanalytic Association issued a new practice guideline because experience with the prior protocol supported the need to limit the amount of information that could be disclosed (American Psychoanalytic Association, 1992). The American Psychoanalytic Association failed to rescind the old reporting form, however. Finally, as the 1990s reached a close, the American Psychoanalytic Association rescinded the former protocols and reporting form and adopted in their place a utilization review protocol truly protective of patient privacy (American Psychoanalytic Association, 1999). In 2013, the District of Columbia adopted a revised model for review of disputed claims in which either the patient or the therapist could request a confidential independent review, by a professional peer of the therapist, if either patient or therapist believes that the information being requested by the third party is an infringement on the privacy of the treatment (Barry Landau, personal communication).

Reporting Child Abuse

For several decades, health professionals have been required by state statutes to report patients who reveal evidence of child abuse to authorities and have generally been willing to make such reports. These state statutes for the most part resulted from the enactment of a federal law, the Child Abuse Prevention and Treatment Act, first adopted in 1974 and amended several times since then, which requires states to maintain databases and reporting systems for child abuse. Bollas and Sundelson point out that this requirement, which now exists in some form in all fifty states, is one of the most difficult dilemmas to confront psychotherapists. Bollas and Sundelson trace the evolution of these statutes in which the physician first was *permitted* to report, to its contemporary form in which physicians are now *required* to report such evidence. The fundamental rule of psychoanalysis, which asks the patient to say everything that comes to mind and withhold nothing, places the patient in an untenable position. A patient cannot say "everything" if certain self-disclosures require the filing of a report with state authorities. Such reporting requirements imposed on the psychoanalytic relationship cannot be anything but detrimental.

The American Psychoanalytic Association responded by including in its ethics code provisions that *permit* but do *not require* psychoanalysts to make such reports, deferring to the discretion of the analyst, often after confidential collegial consultation (American Psychoanalytic Association, 2001, sec. VIII.10). The reasoning behind this provision is similar to that underlying a similar provision in the code of ethics of the North Carolina Bar Association. That code addresses the question of whether an ethical attorney must report child abuse information obtained in those situations where the North Carolina reporting law overrides the attorney-client privilege:

> [The ethics code] prohibits a lawyer from revealing the confidential information of his or her client except as permitted under [a part of the rules that says a lawyer] "may reveal" the confidential information of his or her client . . . "when . . . required by law or court order." The rule clearly places the decision regarding the disclosure of a client's confidential information within the lawyer's discretion. While that discretion should not be exercised lightly, particularly in the face of a statute compelling disclosure, a lawyer may in good faith conclude that he or she should not reveal confidential information where to do so would substantially undermine the purpose of the representation or substantially damage the interests of his or her client. It is recognized that the ethical rules may not protect a lawyer from criminal prosecution for failure to comply with the reporting statute.

The disagreement between ethics codes and statutes is by no means resolved. Professionals who disobey the law risk, among other things, losing their licenses to practice. A courageous analyst and a dedicated professional organization will be required to confront this disagreement in the courts.

Privileged Information and Privilege Rules

Although professional ethical codes offer the strongest protection to confidentiality, a protective legal framework also exists. In addition to statutes that spell out the obligations of professionals to protect patient confidences, there are privilege rules within the justice system that are meant to shield certain confidential information from disclosure without the patient's consent, particularly as evidence in court proceedings.

The court is seen within the legal system to be crucially reliant on the availability of all information necessary to decide the outcome of a case, and therefore every citizen has a legal obligation to make available information the court requires. The few exceptions to this rule, called privileges, are expressions of a societal consensus that the value of protecting the privacy of certain relationships outweighs the imperative need for truth-seeking in the courts.

Mental health professionals often misunderstand the significance of privilege rules and are frequently warned, for example, that privilege rules do not apply to disclosures to third-party payers, where the relationship is governed by private contractual arrangements. In issuing cautions emphasizing the limits of privilege rules, legal advisors have often erroneously created the impression that privileges have no effect outside the courtroom. This is not the case. Privilege rules tend to be controversial *precisely for the reason* that these rules have ramifications that affect interactions outside the courtroom situation. For instance, participants in privileged conversations tend routinely to take precautions to avoid losing the privileged status. Such precautions might take the form of scrupulously avoiding behavior that could be interpreted as indicating that the conversation did not take place "with an expectation of privacy" or of strengthening professional ethics codes to encourage behavior consistent with the privilege (Imwinkelreid, 514).

Privilege rules can, however, be eroded by an accumulation of loopholes and exceptions. In the United States, about forty-two states currently have a physician-patient privilege statute, but these statutes have so many exceptions that legal scholars tend to view them as relatively unimportant (Slovenko, *Psychotherapy and Confidentiality*, 25–34). The federal courts do not recognize a physician-patient privilege at all.

In 1952, two years after Hoover's editorial appeared in *JAMA*, a local Chicago court ruled in *Binder v. Ruvell* that the information conveyed by a patient to a psychiatrist was *privileged*; that is, unlike almost all other information including ordinary medical information, psychotherapy information should not be subject to compelled disclosure. The court's sophisticated articulation of the special privacy requirements of psychotherapy based on psychoanalytic principles in this landmark case began a trend within the American legal system toward increased protection of psychotherapy information, during the same era in which confidentiality within general health care was eroding.

In the twenty-five years following *Binder*, proposals were made to create a statutory (by legislation) psychiatrist-patient (or psychotherapist-patient) privilege applicable in state courts. These proposals resulted eventually in the passage of privilege statutes in all fifty states. These statutes vary from state to state, some being modeled on the highly respected and centuries-old common-law attorney-client privilege and others on the less certain physician-patient privilege. For example, in the state of New York, patients of psychologists are protected by a privilege explicitly based on the attorney-client model, but patients of psychiatrists are protected only under the general physician-patient privilege.

In 1965 the chief justice of the Supreme Court of the United States appointed an Advisory Committee to initiate a process of rewriting the rules of evidence (including privilege rules) for the federal courts, which have their own set of evidence rules distinct from those in the various state courts. After years of study, the Advisory Committee proposed a set of federal evidence rules including, among others, a set of thirteen privilege rules that contain a specific privilege for confidential communications between psychotherapists and their patients. An analogous wider privilege for patients of physicians in general was explicitly *not* included in the committee's proposal. The proposed rules were sent to Congress in the early 1970s with the expectation of legislative approval.

Instead, the proposed privilege rules became the subject of enormous controversy. Hearings were held, and many interest groups submitted testimony. The proposed privilege rules became so invested with symbolic meaning, pitting professional and other groups against one another, that the controversy threatened to interfere with the passage of all the new evidence rules, of which the proposed privilege rules were only a small part. On the one hand, some argued that, because privilege rules have ramifications that reach far beyond the courtroom, such rules were rightly established through a legislative and political process rather than in the courts. On the other hand, because privilege rules were so controversial, a national political resolution of all the disputes about privileges seemed out of reach.

Consequently, Congress deleted all of the Advisory Committee's proposed privilege rules from the new federal evidence code and instead created a single general privilege rule that in effect empowered the federal courts to evolve privilege rules through a common-law process. Importantly, in adopting this approach, Congress declared that the deletion of the specific privileges should not be perceived as disfavoring a psychotherapist-patient privilege or any of the other specific proposed privileges. The revised rules were signed into law by President Ford in 1973 and became effective in 1975 (Svetanics).

The passage into law of the new rules of evidence set in motion a decades-long process in which the lower federal courts considered the question of a privilege for psychotherapist-patient communications. A few appellate courts, accepting the reasoning in *Binder v. Ruvell*, the relatively uniform approval of *Binder* by legal scholars, and the existence of the privilege in the states, held that such a federal privilege was justified. A greater number of appellate courts evidently thought that the congressional mandate was to recognize only long-standing common-law privileges such as spousal and attorney-client privilege. Over the course of these decades, the Supreme Court demonstrated a conservative attitude toward the creation of new privileges and disapproved virtually all proposed privileges.

Jaffee v. Redmond

Twenty years later, the specific issue of the psychotherapist-patient privilege finally reached the U.S. Supreme Court for the first time, in the case of *Jaffee v. Redmond*. Recognizing that the lower federal courts were clearly in disagreement on the issue of a privilege for psychotherapy patients, the Supreme Court issued its decision in June 1996. By a majority of seven to two, the court ruled that a psychotherapist-patient privilege exists in the federal courts. Had the Supreme Court done nothing more than endorse the holding of the appellate court, *Jaffee* would have been considered a landmark case. The Supreme Court, however, went beyond the appellate court by establishing the foundation of a much stronger privilege. Using highly explicit language, the Supreme Court insisted that the privilege must be "absolute" (that is, not subject to case-by-case balancing). The Supreme

Court placed the new privilege in the same group as the highly respected absolute common-law privileges. We tell the personal stories of some of the people involved in the Jaffee case in Chapter 9 of this book.

In the years following *Jaffee*, the psychotherapy professions in the United States have been slow to understand its importance. *Jaffee* was decided at a time when psychiatry as a medical specialty was undergoing rapid change from a profession dominated in the 1960s by psychotherapy practitioners to a "medicalized" specialty more concerned in the 1990s and 2000s with the presumption of brain pathology as a primary cause of mental disorders and therefore amenable to treatment by physical methods such as psychopharmacology. Because *Jaffee* differentiates the treatment of "physical ailments" from psychotherapy, it has thus intersected sensitive "turf issues."

These turf issues betoken the "quiet revolution" that has been occurring in psychotherapy. Social workers have now displaced psychiatrists and clinical psychologists as the largest single group of psychotherapists, as the headlines of an April 30, 1985, article in the *New York Times* indicated: "Social Workers Vault Into a Leading Role in Psychotherapy." That revolution has continued. According to the Substance Abuse and Mental Health Services Administration (2006) and reported by the National Association of Social Workers in its 2012 Toolkit, "Social workers are the largest group of clinically trained mental health providers in the United States."

Some legal commentators, examining *Jaffee* with a literal eye, asserted that the existence of the privilege in all fifty states prior to *Jaffee* meant that the decision itself would make little difference in most courts. Others, like Ralph Slovenko, believed that the privilege was of little consequence because it would likely be eroded by accumulating exceptions, just as the physician-patient privilege had been weakened in the states.

Ralph Slovenko has been a key figure in the confidentiality debate, and his shifting attitude toward the psychotherapy privilege merits attention. In 1966 Slovenko, a senior assistant district attorney in New Orleans, and Gene L. Usdin, a Tulane psychiatrist, published *Psychotherapy, Confidentiality, and Privileged Communication*. In his preface, U.S. Circuit Judge J. Skelly Wright calls the book the "first serious attempt that has been made in any depth to expose the whole problem of privilege created by the psychotherapist-patient relationship" (viii). Slovenko and Usdin argue that "psychiatric" patients require special protection because of the unique confidentiality needs of psychotherapy, which was the primary treatment modality of psychiatrists at that time. This echoed a similar assertion Slovenko made in his influential 1960 law review article "Psychiatry and a Second Look at the Medical Privilege."

After the publication of his 1966 book, Slovenko became one of the country's leading authorities in the area of forensic psychiatry, being one of the few law professors to have done a psychiatric residency. Although his 1998 book *Psychotherapy and Confidentiality: Testimonial Privileged Communication, Breach of Confidentiality, and Reporting Duties* is advertised as being "the first and only comprehensive work to be published in this field of law," it is, in fact, a greatly expanded and revised version of the 1966 volume. In the years since the earlier book appeared, there have been major developments in this area of law. There have also been significant changes in society's concept of what psychotherapy is and who psychotherapists are. Along the way, Slovenko apparently lost confidence in the value of privilege. In "Psychotherapist-Patient Privilege: A Picture of Misguided Hope," published in 1974, he asserted that the privilege, which had by then been established in several states, was beset by so many exceptions that it was virtually worthless. "The concept of privilege, while it may offer a sense of security, should be abandoned as a means of determining whether disclosure of communication made in psychotherapy should be required" (672).

Whether the psychotherapy privilege will evolve along the lines of the porous state physician-patient privilege rules or the more protective attorney-client privilege is an open question. However, the wording of the *Jaffee* opinion appears to signal the Supreme Court's intent that the new privilege develop in a way that would render it both robust and reliable.

Since *Jaffee*, the lower federal courts have begun to render opinions not only demonstrating how the privilege might evolve but also showing that the courts' view of the privilege is anything but uniform. Generally, lower courts have been rather expansive in applying the privilege. No court has dealt with the issue of whether the privilege extends to all forms of psychotherapy or only to those forms that rely in some way on a psychoanalytically informed approach. The courts appear inclined to view the privilege as applying to all verbal treatments of mental disorders.

The originally proposed federal privilege applied only to psychotherapeutic relationships between licensed physicians (treating a mental disorder) or psychologists and their patients. *Jaffee* extended the privilege to the patients of licensed social workers as well. Lower courts since *Jaffee* appear inclined to extend the privilege even further to include, for example, Employee Assistance Program (EAP) counselors. Considerable difference of opinion exists as to which psychotherapists are included in the federal privilege. In *Jaffee*, the Supreme Court combined the "license-based" approach with a "functional" approach: "confidential communications between a licensed psychotherapist and her patients in the course of diagnosis or treatment." Whether the privilege will be found eventually to extend to psychoanalysts who are not licensed in a mental health discipline is unclear (Dubbleday; Aronowitz; Nelken).

The lower courts have differed on two important issues affecting the *Jaffee* privilege. The first is whether there is a "dangerous patient" exception to the privilege—that is, whether a therapist who has set aside the confidentiality of the relationship to warn or protect an intended victim thereby also makes the contents of the therapy available as evidence in a subsequent court proceeding against the patient (*United States v. Auster*; *United States v. Chase*; *United States v. Ghane*; *United States v. Glass*; Harris; *United States v. Hayes*). The second issue is whether patients who have entered their mental state into a claim or defense thereby automatically waive the privilege protecting all psychotherapeutic conversations in their past (the "patient-litigant exception").

In general, privilege considerations cease to apply when a patient voluntarily "waives" the privilege. For example, if a patient *requests* that a therapist make a disclosure, the therapist may not assert the privilege on his or her own behalf. However, even where the patient requests a disclosure, other considerations, such as ethical constraints, might come into play where the therapist believes that the disclosure could be harmful (Mosher and Swire).

Federal Policy Regarding Confidentiality of Psychotherapy

Despite the initial lack of awareness within the psychotherapeutic professions, the *Jaffee* ruling has had a major influence in shaping the U.S. government's policy toward the confidentiality of psychotherapy information. Until 2001, the United States had no overall federal legal protection for the privacy of most medical information. Congress attempted to enact legislation providing some degree of protection on a federal level, but a decade's worth of such attempts was stalemated by conflicting interest groups that disagreed sharply as to the extent of privacy protection that is warranted and practical. The gathering momentum for computerized health-record systems, however, added a note of urgency to the debate about health care privacy.

On the one hand, some privacy advocates urged the creation of a set of legal standards based on a seemingly idealistic wish to return to a simpler time when true medical record confidentiality could be reasonably expected by patients. More than one professional organization enacted a position statement urging that patients should not be required to have their medical data entered into a computerized record system. On the other hand, management-oriented groups viewed such standards as naïve and impractical. These groups advocated instead the use of so-called fair information practices as a standard for the protection of health data. Such standards are used for the protection of commercial information and are viewed by many privacy advocates as little better than no protection. It seemed that the only way to break the logjam of legislative paralysis created by the polarized positions was to impose a compromise.

A compromise on this issue was finally reached when Congress passed the Health Insurance Portability and Accountability Act of 1996 (HIPAA), one part of which is entitled "Administrative Simplification." The intent of that section is to mandate the standardization, and hence promote the computerization, of the vast amount of largely administrative data exchanged throughout the health care system, in the hope of increasing efficiency and lowering costs. The "Administrative Simplification" section acknowledged the requirement for federal privacy protection for the developing new systems and therefore empowered the executive branch to issue binding federal health privacy rules with the force of law if Congress failed to agree on legislation for the same purpose by 1999. The 1999 deadline passed without congressional agreement. The U.S. Department of Health and Human Services (HHS) then began a herculean two-year effort to create new rules for the federal protection of the privacy of health information. HHS intended that, once in place, these regulations would have the force of federal law and apply to virtually all medical information in the United States (Gostin).

In attempting to create such rules, HHS was placed in the position of achieving the compromise that had eluded the congressional political process for many years. A particularly difficult issue was the status of especially sensitive information (such as AIDS-HIV status, genetic information, and mental health data), each category of which had patient and professional advocacy groups urging that it be treated as an exception from the general rules by being given a higher order of protection. Those responsible for implementing and operating the data systems, however, believed that it would be impractical to designate subsets of information within medical records for special treatment.

Ultimately, HHS decided to treat all health-related information (with a single exception) uniformly. The general rule, which became law in April 2001, provides that most information in a medical record may be disclosed, with no special notification to the patient or specific authorization, for purposes related to what are called "treatment, payment, and health-care operations." Other kinds of disclosures are subject to a complicated set of rules, but in general unauthorized use of a patient's information for nonmedical commercial purposes is prohibited. The entire set of new privacy regulations is referred to as the "HIPAA Privacy Rule."

Psychotherapy Notes

The single exception in the Privacy Rule was the creation of a new category of information called "psychotherapy notes," which remained unchanged throughout the above revisions. Psychotherapy notes must be "separate from the rest of the individual's medical record." Once so segregated, information in these notes is accorded stringent protection unlike that provided for other data. With some narrow exceptions, no disclosure of "psychotherapy notes" for treatment, payment, or "health-care operations" can take place without the patient's specific written authorization. In addition, third-party payers (insurance companies) are prohibited from requiring a patient to disclose such information either when selling an insurance policy or processing an insurance claim. Thus, in regard to insurance coverage, the patient has no incentive to authorize the disclosure of such information to a third-party payer. Finally, an important provision of the rule that requires physicians to grant patients access to their entire medical record does not apply to "psychotherapy notes." In the documents accompanying the publication of the Privacy Rule, HHS gave the following reason for the special protection of psychotherapy information:

> Generally, we have not treated sensitive information differently from other protected health information; however, we have provided additional protections for psychotherapy notes because of *Jaffee v. Redmond* and the unique role of this type of information. There are few reasons why other health-care entities should need access to psychotherapy notes, and in those cases, the individual is in the best position to determine if the notes should be disclosed. (United States Department of Health and Human Services, 2000, 82652)

In December 1999, the surgeon general of the United States issued a major report on the subject of mental health (United States Department of Health and Human Services, 1999a). Chapter 7 of that report is devoted entirely to confidentiality and opens with a passage from the Supreme Court majority opinion in *Jaffee*. The section that deals with disclosure to third-party payers points out that the state laws that permit disclosure of mental health information to insurers were, for the most part, written in an era before managed care and did not anticipate the current demands for disclosure of detailed information. The report goes on to suggest that such permissive statutes need to be revised and offers the New Jersey statute and the similar District of Columbia statutes as a possible model.

Over the decade or more that the federal psychotherapist-patient privilege and the federal rules protecting psychotherapy information have been in place, there have been a number of developments in the way confidential psychotherapy information is protected—and not protected. Unfortunately, the major professional mental health organizations, including the American Psychiatric Association, the American Psychoanalytic Association, and the National Association of Social Workers, have been slow in informing their members about these important developments. Many mental health professionals are still unaware of these changes. Nonetheless, the legal requirements of the Privacy Rule's protection for psychotherapy information have significantly affected the behavior of health insurance companies, which have greatly reduced their demands for sensitive information from the treatment of patients they insure.

Mandatory Reporting Requirements

In contrast to the progress described above at the federal level, state reporting statutes have expanded. Psychotherapists have tended to become more like "state agents" as a result of such requirements. Psychotherapists in various states may be required to report to authorities child abuse (emotional, sexual, or physical) as well as elder abuse, threats to self, and threats to harm others. In addition, recent events of mass violence attributed to individuals who have been diagnosed with mental illnesses have now led to new laws requiring reporting by mental health professionals of patients "likely" to be dangerous to self or others. For example, New York State psychiatrists and medical psychoanalysts face a growing list of reasons for mandatory reporting, including possession of firearms by a patient "likely" to engage in conduct that results in harm to self or others, anyone with a suspected or confirmed case of a communicable disease, child abuse or neglect, abuse of narcotic drugs, positive diagnosis of HIV/AIDS, immunizations given to patients under age eighteen, a gunshot wound, a bullet wound, a powder burn from the discharge of a gun, a knife wound, abuse or neglect of a patient being treated in state facilities, certain burn injuries, etc. Five other states require reporting of "domestic violence," and several other states require reporting of "elder abuse."

Despite the ubiquity of such statutes, there is considerable variation among the states in the definitions of who is a "mandatory reporter" (a person required by state law to make such reports) and under what circumstances such a report must be made. As Erica Wise points out, this situation leaves many psychotherapists uncertain as to whether to make a report in a particular instance and whether such a report is mandatory or discretionary. Additionally, the relationship between reporting requirements, which are state mandates, and the *Jaffee* privilege, which applies in federal courts, is equally confusing and subject to enormous uncertainty (Goldner).

Finally, the mass transition to electronic health records and the abandonment in many practices of paper-based records have posed special problems for the mental health segment of the health-care industry. In some respects, electronic records are more secure than paper records, but this benefit seems at present to be more than outweighed by the potential for the breach of hundreds or thousands of records in a single act, something not possible with old-fashioned paper records. Large-scale security breaches are reported weekly, and many mental health practitioners are unwilling at this point to entrust their records to electronic systems. To cite but one example of breached confidentiality on a massive scale, the Open Security Foundation reports that four million patient names, addresses, dates of birth, and Social Security numbers were contained in four computers stolen from an administrative building ("Dataloss db," August 23, 2013).

Psychotherapy notes, in particular, present a daunting problem because such notes must in some way be segregated from the rest of the electronic record to retain the special protection most recently granted to them in federal law. At the same time, particularly in the psychiatric profession, there has been a recent trend toward wanting to "medicalize" the practice of psychiatry and to integrate it more closely with the rest of medical practice. What this means for the fate of special protection for the clinical data maintained by psychiatrists, especially psychotherapy notes, is only a matter of speculation. "Given the wide variety of training backgrounds, practice styles, and practice settings that now prevail in psychiatry," Norman Clemens points out (personal communication, September 12, 2013), the fate of special protection for clinical data is likely to be "quite variable rather than adhering to a well-defined set of boundaries," a situation that will be "most unfortunate."

Clemens is not opposed to the seismic changes occurring in the creation of an electronic health records system or the attempt to integrate mental health with general health systems. In an article published in the January 2012 issue of the *Journal of Psychiatric Practice*, he warns against a proposed amendment that would weaken the exemplary *Mental Health Information Act of the District of Columbia*, which requires informed consent for the disclosure of mental health information. Clemens insists that patients have a right to know, first, that "there is a significant risk to privacy and confidentiality in maintaining mental health information in electronic networks" and, second, that "consent is theirs to give or refuse" (50). How can informed consent be preserved, he asks, if patients don't have the opportunity to "opt out" of allowing their private mental health records to be shared with other health care providers? Particularly worrisome is whether a psychotherapist or a health care facility will treat a patient who "opts out."

At the same time, the American Psychiatric Association has been backing away from psychotherapy in part as a result of the pharmaceutical industry's growing influence. Once a strong proponent of the confidentiality of psychotherapy information, the association most recently appears to be more concerned with integrating mental health care with medical care in general. Whereas traditional psychoanalytic psychotherapists view themselves as treating the "mind," the terminology, reflecting the so-called psychodynamic-behavior wars of the 1960s, has left a legacy of a "mindless" psychiatry. Psychiatrists now speak of "behavioral health" and "behavioral disorders" instead of "mental health" and "mental illness" (although "mental disorder" is still in use). Psychiatry is trying to redefine itself as "clinical neuroscience" (Insel & Quiron, 2005).

Locating the Enemy

Most psychotherapists probably believe that there is nothing they could have done to prevent the changes that have resulted in the erosion of confidentiality in psychotherapy in the past half-century: the growth of managed care; the proliferation of child and domestic abuse laws; and the growing reliance on electronic medical record systems in which data are exchanged between practitioners, other health care providers seeing the same patient, and insurance companies. Those psychotherapists who acknowledge a problem—and many, we must admit, see no problem at all—believe that therapists' dual allegiance to patients and society evolved naturally and ineluctably. We are the victims of our times, so their argument goes, and there is nothing that a single psychotherapist or a group of professional organizations could have done to resist the loss of confidentiality.

We believe otherwise. As the title of our chapter suggests, we maintain that psychotherapists must accept much if not most of the responsibility for the ongoing erosion of their patients' privacy. Accepting responsibility for the loss of privacy in psychotherapy affirms the importance of agency, the power not only to resist but also reverse the loss of privacy in psychotherapy.

Difficult Decisions

There are many ambiguities and uncertainties in these new federal initiatives, but it appears that after a delay of several decades, during which the confidentiality of psychotherapy has been severely eroded, the tide may be turning. Policy makers, the judicial system, and ordinary citizens are becoming aware of the enormous stakes in the protection of confidentiality in the psychoanalytic and psychotherapeutic process. Nevertheless, many problems remain. Of all the dilemmas arising from psychotherapists' dual allegiance to patient and society, none is more agonizing than how far they should go in protecting a patient's disclosure, especially when the disclosure contains a threat of violence against another person. Psychotherapists rarely write about the problem of determining whether to honor or breach a patient's confidential disclosure. It is for this reason that we turn to the Buried Bodies case, which dramatizes the personal and professional cost, in a different profession, of maintaining privileged information in an emotionally charged setting.

2. The Buried Bodies Case: Lawyers Risk Their Careers to Defend Their Ethical Commitment to Client Privacy

What to look for: With the legal profession's centuries-old tradition of protecting client confidentiality, how far are lawyers willing to go, and what personal and professional risks are they willing to take, to protect the transcendent principle of confidentiality?

Most people understand the need for confidentiality in the lawyer–client relationship. Lawyers cannot provide proper representation unless clients can be completely frank in legal discussions. Lawyers have a responsibility to protect their clients' disclosures, the violation of which is nothing less than a betrayal of trust and can be the cause of disciplinary action. The same confidentiality is essential for psychotherapy. Therapists have a responsibility to protect their patients' disclosures, the violation of which is nothing less than a betrayal of trust. If patients do not trust their therapists, they are not likely to observe Freud's fundamental rule of complete disclosure.

Unlike lawyers, however, psychotherapists do not receive the same extensive professional training about the crucial importance of protecting privileged information, the basis of privilege laws such as "lawyer–client privilege." Privilege laws in psychotherapy are a relatively recent phenomenon compared to laws governing privileging of the lawyer–client relationship. Nor are lawyers subjected to the growing erosion of confidentiality that is occurring in psychotherapy. Consequently, therapists lack lawyers' fierce commitment to confidentiality. Moreover, few therapists are willing to speak and write openly and candidly about the increasing pressure to break confidentiality.

A Landmark Case

Psychotherapists can learn a great deal from lawyers about the need to pre-serve privileged testimony, and they can learn much from the Buried Bod-ies case, which has become a landmark in establishing the principle of confidentiality. The story of Robert Garrow, one of the most notorious mass murderers in New York State's history, appears in *Privileged Informa-tion*, written by Tom Alibrandi with the help of Frank H. Armani, one of the two court lawyers appointed to defend Garrow. Armani, the major pro-tagonist in the book and the character with whom the reader identifies, notes in the prologue that he waited ten years before he was able to tell his story. "The pain of reliving that part of my past seemed too immense" (ix). The reader soon understands the anguish of a man who risked everything—his health, family, and career—to protect his client's horrifying secrets.

The story begins on July 29, 1973, when Garrow, thirty-six, married and the father of two children, stabbed to death the eighteen-year-old Philip Domblewski, who was camping with his three friends in the town of Spec-ulator, deep in the Adirondack Mountains in northern New York State. Domblewski's friends witnessed the murder. There was never any doubt about the factual details. Garrow already had a police record; in 1961 he had been convicted of raping a teenage girl after knocking her boyfriend unconscious with the butt of a pellet gun. Garrow had served eight years in prison for that crime, and he was considered a model parolee after his release. The New York State Crime Commission had regarded him as an example of someone "who broke the pattern of recidivism" (21). At the time of his murder of Domblewski, however, Garrow was also a suspect in the stabbing death of Daniel Porter and the disappearance of Porter's camp-ing companion, Susan Petz, a Boston College student. Garrow was also a prime suspect in the recent disappearance of Alicia Hauck, a high school student.

Armani was chosen to represent the defendant because a year earlier Gar-row had visited the lawyer in his office after a minor car accident. Later Armani was asked to represent Garrow for allegedly molesting two young girls, a crime he denied committing. Garrow failed to show up for his hear-ing, at which time a bench warrant was issued. When Garrow was arrested for the Domblewski murder, following a massive eleven-day manhunt in the Adirondacks during which he was critically wounded, Armani was once again hired to defend him, even though the lawyer had little experience with murder cases. Armani asked his lawyer-friend Francis Belge, who had extensive experience in murder cases, to join the case. The flamboyant and unpredictable Belge agreed to be Armani's co-counsel.

As the case progressed, the two lawyers decided that the best defense for Garrow was to plead not guilty by reason of insanity. On the surface the defendant appeared to be a good husband and father, yet Armani recognized Garrow's dark, sinister side, capable of pitiless violence. The more Armani learned about his client's sordid past, the more he was horrified. Garrow was repeatedly beaten as a child by both his parents; his mother once split his head open with a crowbar. Armani concluded that Garrow was truly insane, but to convince a jury, the lawyer had to produce psychiatrists who were willing to testify to his defendant's insanity. Two psychiatrists declared that Garrow was traumatized when, at age six, he was nearly beaten to death by his father for knocking down a few cornstalks while playing in a field.

A Horrifying Admission

Garrow insisted during his confidential interviews with Armani that he could not remember the details of the Domblewski murder, and he agreed to the lawyer's request to hypnotize him to bring to the surface the repressed information. Under Belge's questioning, Garrow confessed to raping and murdering Alicia Hauck and then stashing her body in a deserted corner of Oakwood Cemetery near Syracuse University. He also confessed to killing Daniel Porter, raping and murdering his companion, Susan Petz, and then hiding her body in the airshaft of a deserted mine in the town of Mineville, forty miles south of Plattsburgh, New York. Afterward, Belge promised Garrow that his confessions were privileged information. "I'm bound by the law not to tell anyone—that is, unless you release me from this confidentiality. Until then, it's like you've told a priest in a confessional. My lips, as are Frank Armani's, are sealed about this" (114).

Garrow's confessions under hypnosis, however, were difficult to believe, and the two lawyers couldn't tell whether he was telling the truth, lying, or hallucinating. Curiously, *Privileged Information* does not disclose the most important reason why the two lawyers set out in search of the bodies. Armani stated, in a panel discussion of the Buried Bodies case (officially called "In re Belge") that took place during the thirty-third ABA National Conference on Professional Responsibility, held in Chicago in 2007, that the main reason he and Belge undertook the search was that "the Petz girl might still have been alive. Because he stashed her up . . . he would leave her tied up and with food. He had chains on her. We learned this within a frame of survival time."

As Armani and Belge pressed Garrow for more details to locate the bodies of the two women, the defendant reluctantly complied, and the two lawyers began their search. They soon discovered Susan Petz's body. "'The rotten creep actually did it,' Armani muttered angrily. 'He killed that poor girl'" (125). They had more difficulty discovering Hauck's body, but when Belge returned to Oakland Cemetery a few days later, without Armani's knowledge, he eventually found Hauck's remains, decomposed and ravaged by animals, her skull lying a few feet away. Belge picked up the victim's skull with a handkerchief, placed it above her shoulders, and photographed the remains.

Privileged Information dramatizes the conflict arising from Armani's dual allegiance to his client and society. Psychotherapists experience a similar conflict when patients threaten to harm themselves or others. Armani swore to maintain the confidence and preserve inviolate the secrets of his client, but he could not help empathizing with the parents of the two slain women who remained unaware of their daughters' fate. He also recalled the tragic loss of his only brother, who had disappeared in 1962 while flying an Air Force reconnaissance mission over the North Sea—a loss from which his mother had never recovered.

"Dean Grant"

Armani soon realized that he had become, inadvertently, another of Garrow's victims. Sickened by the crimes he could not disclose to the authorities or to the victims' families, Armani had never been in a situation like this before. *Privileged Information* gives us an elaborate account of how Armani flew to Logan Airport in Boston to meet with a famous law school official, "Dean Grant," who told him that a lawyer is prohibited from divulging any evidence about his client, with two exceptions: "An attorney was compelled to violate a client's confidence if divulging that information would stop a continuing crime or would prevent a future crime" (140). Since Armani's client was already in custody, the two exceptions do not apply. Therefore, Armani must protect his client's confidentiality.

Dean Grant's legal advice is correct, and Armani's encounter with him appears convincing in *Privileged Information*, the two men engaging in a thoughtful, even lofty discussion of a complex case. The real story, however, as Armani later revealed sheepishly in the Buried Bodies case, is far different. Armani actually had gone for advice to an appellate division judge in Syracuse, Frank Del Vecchio, whom he had known for many years. To protect the judge's identity, Armani disguised Del Vecchio in *Privileged Information* as a law school dean. As we learn from the Buried Bodies case, Armani told Judge Del Vecchio that he had a "little bet going," and he then offered hypothetical details about a lawyer required to give privileged material to a district attorney. What was to Armani an agonizingly difficult legal dilemma was to Judge Del Vecchio a no-brainer. "He says, 'What are you talking about? You know, you're smarter than that. That's Fifth and Sixth Amendment rights.' And his reaction was actually: 'Get the Hell out of here,' you know. He thought I was wasting his time."

The Pressure to Disclose

What is simple to the judge, however, is agonizingly complicated to Armani, who is subjected to almost unimaginable public and police pressure. He can express only sorrow, not the details of Garrow's confession, when Susan Petz's father pleads for information about his daughter's disappearance. Armani is later told that a "contract" has been put out for his death, and a Molotov cocktail is found near his home. He received middle-of-the-night hate calls at the various motels in which he stayed. To protect himself, he began carrying a loaded handgun.

One of the crucial moments in the story occurs when Armani and Belge try to convince the Hamilton County district attorney, William Intemann, to agree to a plea bargain. They will offer information on several unsolved homicides in exchange for a reduction of charges against Garrow for killing Philip Domblewski to manslaughter first degree, based on an insanity plea, plus an insanity judgment on future charges against him for crimes already committed. When Intemann asks Belge how he will solve the homicides, the latter, sensing a trap, answers evasively, "Let's just say that we can direct you to where some bodies are hidden." The district attorney responds, "Does that mean that Alicia Hauck and Susan Petz are dead and not just missing?" "No comment," Belge replies, "until you agree to our terms" (161–162). Intemann rejects the deal and then threatens to seek criminal charges against both lawyers for withholding evidence and obstruction of justice. Armani and Belge realize that, like Garrow, they too are on trial.

Around this time Armani's personal life began to suffer as a result of his involvement with the case. Once a heavy drinker, he had given up alcohol years earlier, but he started drinking again, to his wife's disapproval. When the Garrow trial began, in the summer of 1974, he received daily verbal abuse from the local townspeople as well as the many strangers who were attracted to the high-profile case. His law practice was nearly ruined.

During the trial the prosecution presented three of Domblewski's friends who all testified that they saw Garrow stab him to death. After the prosecution rested its case, Belge stunned everyone in the courtroom, including Armani, by calling as the defense's first witness Garrow himself. On the stand, the defendant recalled the long history of savage beatings he had received as a child and publicly admitted for the first time that he had killed not only Philip Domblewski but also Daniel Porter, Susan Petz, and Alicia Hauck. With these public admissions by Garrow, Armani and Belge were now released from their ethical duty to protect these secrets. Outside the courtroom, Belge told a group of reporters that he and Armani had wanted to disclose this information when they first received it but were duty bound to remain silent. "You guys are going to have a helluva battle on your hands over what you did," (217), a reporter shouts prophetically to them.

Toward the end of the trial, Armani and Belge faced a battle first with each other—their friendship ends in a fistfight—and then a second battle with the Onandaga County district attorney, Jon Holcombe, who during an afternoon recess walked into the courtroom to inform the two lawyers that he intended to bring criminal charges against them for withholding information about the Petz and Hauck murders. The judge, incensed that Holcombe's accusations would prejudice the jury and result in a mistrial, ordered Holcombe out of the county immediately, lest he be arrested.

After less than two hours of deliberation, the jury rejected the lawyer's insanity defense and found Garrow guilty of first-degree murder. The following day, Judge Marthen sentenced him to the maximum sentence allowed by New York State, twenty-five years to life.

Furious Controversy and the Threat of Indictments

By this time, the Garrow case shifted to a more momentous question. Did Armani and Belge act correctly in protecting their client's confidentiality? The case drew national and international attention, with regular articles appearing in *Time*, *Newsweek*, the *New York Times*, *Los Angeles Times*, and the *Christian Science Monitor*. The upstate New York press generally condemned the two lawyers, but the national newspapers and magazines offered more balanced, thoughtful commentary.

The drama continued when Jon Holcombe sought grand jury indictments against both lawyers for illegally withholding knowledge about the Hauck and Petz murders and for violating the New York State law that guarantees the deceased a decent and humane burial. Elliot Taikeff, a highly respected lawyer from New York City, was so convinced of Armani's innocence that he agreed to defend him without a fee. Facing the grand jury, Armani offered a heartfelt defense. The grand jury believed him and decided not to indict him. "The effectiveness of counsel," stated Ormand Gale, the Onondaga County Court judge, "is only as great as the confidentiality of its client-attorney relationship. If the lawyer cannot get all the facts about the case because his client fears that those facts can be divulged later to the press or the authorities, he can only give his client half of a defense" (281). The grand jury indicted Belge, however, perhaps because of his decision to return to Oakwood Cemetery to reassemble and photograph Hauck's body. The indictments were later dropped.

The final pages of *Privileged Information* reveal how the Robert Garrow case played out. Garrow had told Armani after the trial that he was going to "try and make it out of here," a thinly veiled threat to escape from the Clinton Correctional Facility, New York State's largest maximum security prison, in Dannemora. Warning Garrow that there's no confidentiality with that kind of information, Armani concluded that because the threat was "idle, grandiose talk," he had no obligation to report it to the prison authorities (266–267). Garrow was, indeed, thinking about escape, and he agreed to withdraw his $10 million lawsuit against New York State alleging police brutality and receiving improper medical treatment in exchange for transfer to the Fishkill Correctional Facility, a medium security medical-psychiatric facility located near Newburgh. With the aid of his son, who smuggled a gun into the prison, Garrow escaped and once again was the target of a massive manhunt. Armani gave the police a crucial clue that helped them locate Garrow, and after shooting a guard, the convicted murderer was himself shot and killed, on September 11, 1978. Garrow's body is buried in Oakwood Cemetery, only a few hundred feet from where he had hidden Alicia Hauck's body.

The Buried Bodies case is a fascinating story about the human cost of confidential information. Preserving Garrow's privileged information proved nearly deadly to Armani's health. "Keeping the horrible secret was enough to drive you insane" (307), he remarked wryly after Garrow's death. A list was discovered in Garrow's prison cell of the people he planned to murder if he escaped. Armani's name is on the list.

Sworn to Silence

The Buried Bodies case received national attention long after Garrow's death. *Privileged Information* became the basis for the film *Sworn to Silence*, directed by Peter Levin and starring Peter Coyote, Dabney Coleman, and Liam Neeson, which premiered on ABC in 1987. To avoid a libel suit, the story was set in a small, rural Pennsylvania town, and the names of the characters were changed. Armani, played by the good-looking Coyote, is called Sam Fischetti; Belge, played by Dabney Coleman, is Martin Costigan; and the psychotic Garrow, played by Liam Neeson, who rarely performs a sinister role, is Vincent Cauley. Cauley is found guilty of first-degree murder and sent to prison, but there's no mention that he is shot to death after his escape. Nor does the film disclose the prosecutor's efforts to indict the two lawyers for obstruction of justice.

Dabney Coleman received an Emmy for his performance, but the film's most riveting performance comes from Coyote, who is confronted by an agonizing moral dilemma, that of a man who risks martyrdom to defend an unpopular cause. After Cauley's conviction, the community is so outraged by Fischetti that there is a town meeting to remove him from the school board. He makes a dramatic appearance during the stormy meeting and forces the female district attorney to acknowledge publicly that she, too, remained silent about the privileged information the two defense lawyers had revealed to her. In what turns out to be the most poignant line in the film, one that does not appear in the book, Fischetti admits that "the law does not protect us from heartbreak." His logic and eloquence succeed in changing the townspeople's angry jeers to cheers.

Peter Coyote became interested in *Sworn to Silence* because it was about a real human crisis, as he observes on his website, about a man "caught between his instincts for decency . . . and his sworn oath to his client and the Constitution to provide the best defense possible." Coyote acknowledged that his feelings about the story changed as he learned more about its legal implications. "My initial reaction was that humanity demands disclosure to the parents of the victims, but thinking about it made me realize that our whole system of justice is based on protecting the individual from the awesome power of the state. If a venal and corrupt person is not protected under the law, an innocent person won't be protected either."

Privileged Information and *Sworn to Secrecy* depict Armani-Fischetti as defending the principle of confidentiality at all cost. The actual truth of the story, however, is more ambiguous. Armani discloses in the Buried Bodies case that he knew he was violating his client's confidentiality when he told police where to look for Garrow after his prison escape. "I was on a death list," Armani admits. "He had sued me. And my family was threatened, my wife and children. Nothing goes before that" (Lerman et al.).

A Central Case in Legal Ethics Instruction

The Buried Bodies case affirms the principle of confidentiality in the lawyer-client relationship while showing, at the same time, the difficulty of defending reprehensible people and unpopular causes. In reporting Garrow's death, Roger Mudd observed on national television news that the "public and the legal profession are still strongly divided on the issue of how far an attorney may go in protecting his client's right of confidentiality" (*Privileged Information*, 308). Confidentiality remains an emotionally charged issue, one that exposes both the strengths and shortcomings of our legal system. The Buried Bodies case dramatizes one of the most vexing ethical dilemmas of the last half century, and the case is now routinely taught in law schools across the country. As the law professor Lisa Lerman observes, the "case very quickly became a central piece of the professional responsibility teaching canon. Many other cases that were taught at that time have faded from memory, but this case continues to fascinate law students and faculty not only in the United States, but all over the world" (Lerman et al.). Lerman adds that, like Atticus Finch, the beloved lawyer in Harper Lee's Pulitzer Prize–winning novel, *To Kill a Mockingbird*, Armani takes many professional risks in representing an unpopular client. "But unlike Atticus Finch, Frank Armani is a real person."

Should the Buried Bodies case be an indispensable part of a psychotherapist's professional training, as it is with a lawyer's? There are good reasons for believing so. Confidentiality is essential to both psychotherapy and law. Lawyers may not be familiar with the "fundamental rule" of psychoanalysis, but the instructions they give to their clients are the same as those Freud gave to his patients: "never forget that you have promised to be absolutely honest, and never leave anything out because, for some reason or other, it is unpleasant to tell it" ("On Beginning the Treatment," 135). Indeed, lawyers seem to take Freud's instructions more seriously than do psychotherapists! The fundamental rule of full disclosure requires the fundamental guarantee of complete confidentiality. To paraphrase Judge Gale's statement, the effectiveness of a psychotherapist is only as great as the confidentiality of the therapist-patient relationship. Both the lawyer and psychotherapist are sworn to secrecy. Recall Joseph Wortis's statement about Freud's promise to him. "He for his part would guarantee absolute privacy, regardless of what I revealed: murder, theft, treachery or the like" (20). This is literally true of the promise Armani makes to Garrow. Privacy and confidentiality exist in a reciprocal relationship, and a threat to one is a threat to the other. Privacy matters in law and psychotherapy. Those who protect the principle of confidentiality, particularly when it involves a client or patient who stirs up negative feelings in the lawyer or therapist, respectively, or outrage in the community, perform the greatest service to their professions.

Take-home lesson: Members of the legal profession are strongly committed to the principle of attorney-client confidentiality, and a lawyer who acts to protect that principle, even in the face of conflicting legal or ethical requirements, can be seen as a hero within the profession.

3. The Case of Joseph Lifschutz: A Psychoanalyst in Jail

What to look for: Can a psychotherapist refuse an order from a court to testify about a patient's psychotherapy when the patient offers no objection to the psychotherapist's testifying?

It is with disbelief that one hears the story of Dr. Joseph Lifschutz, a San Francisco psychoanalyst who was thrown into the San Mateo County Jail because he took a principled stand against revealing any information about a person who claimed to have once been his patient. Dr. Lifschutz's experience shows the power of the justice system to step in with a heavy hand to enforce a legal principle that conflicts with the ethical judgment of a highly respected psychotherapist. The story, which took place at the end of the 1960s, assumed national prominence because of the clash between the principle of absolute confidentiality in the patient-psychoanalyst relationship and the demand of a legal system that was willing to hold an individual in contempt for refusing to turn over subpoenaed records in a court trial.

Joseph Lifschutz's dilemma was, admittedly, different from Frank Armani's. Dr. Lifschutz was not the recipient of a horrifying confession that awakened moral revulsion in him, nor was he dealing with a psychopathic killer whose actions evoked fear and loathing within the community. The pressure to disclose never reached the feverish pitch experienced by Armani, in part, because the patient did not object to the psychoanalyst's potential disclosure. Nor did Dr. Lifschutz find his personal safety—or his family's—imperiled by the man whose secrets he was guarding. Rather, Dr. Lifschutz found himself confronting an unprecedented legal dilemma whose outcome might affect the future of psychotherapy. In addition, he must have feared that going to prison to defend a central tenet of his profession might result, ironically, in professional ostracism, even in the loss of his license to practice.

During our telephone interviews in late 2012 and early 2013, Dr. Lifschutz recalled the circumstances leading up to his legal difficulties in 1968, including the time when he had begun practice in the San Francisco suburb of Orinda ten years before that, when he was thirty-five. His very first patient, whom he saw for a period of about six months at the beginning of his career, was Joseph Housek, a physics and electronics teacher at Burlingame High School in San Mateo County.

Imagine Dr. Lifschutz's surprise when ten years later he received a subpoena from an attorney demanding that he appear at a deposition to testify about, and to produce his records related to, his treatment of Mr. Housek. As Dr. Lifschutz later found out, in June 1968 Housek had become involved in an altercation at school with a student, John Arabian. The student assaulted his teacher, causing Housek to suffer, as he later claimed, "physical injuries, pain, suffering, and severe mental and emotional distress."

Housek then filed a $175,000 damage suit (about $1,200,000 in present-day funds) against the student. Both sides "lawyered up," and depositions took place. Questioned by Arabian's lawyers whether he ever had any mental health treatment, Housek disclosed that Dr. Lifschutz had treated him ten years earlier. This disclosure then led Arabian's lawyers to request Dr. Lifschutz to appear at a deposition in October 1968. Housek's attorney objected but withdrew his objection when Arabian's attorney pointed out that the raising of the claim of "mental and emotional distress" constituted Housek's waiver of the right to keep his former therapist from testifying. Housek could, of course, have decided instead to withdraw the part of his claim referring to "mental and emotional distress." Dr. Lifschutz appeared for the deposition, but he declined to say whether he had seen Mr. Housek. The psychoanalyst similarly refused to produce any records of such a case.

An Epic Legal Confrontation

Dr. Lifschutz's refusal to cooperate set up an epic legal confrontation between the demand of the court system that all available evidence ("every man's testimony") must be made available to a court so that a fair legal decision based on all available facts can be reached and the conflicting principle that confidential psychotherapy cannot take place in an environment in which the psychotherapist might later be called on to reveal in public details of his treatment of his patients. Without getting into the legal minutia of this case, we nevertheless note that the patient had, in fact, made his mental condition a part of the legal case through his claim of mental distress, which, under existing California law, constituted a "waiver" of the *patient's* privilege. Moreover, Housek had no objection to the questioning of Dr. Lifschutz. On the other hand, Housek had not explicitly agreed to Arabian's lawyers' questioning of the psychoanalyst.

Arabian's lawyers then requested a court order demanding that Dr. Lifschutz testify. Despite the existence of a psychotherapist-patient privilege in California, the court granted the order on the grounds that the privilege had been waived. As noted above, the legal rule says that a psychotherapist may not be compelled to testify about the treatment of a patient against a patient's wishes, but the rule contains an exception for situations in which the patient has brought his or her "mental or emotional" condition into the case, as Housek had done. Nonetheless, Dr. Lifschutz continued to refuse to testify, now defying a court order, and based his refusal on a novel set of arguments we will summarize below. In an interview with Bram Fridhandler, Dr. Lifschutz said that his lawyer could find no legal precedents of a psychotherapist who asserted confidentiality when the patient had not done so. Asked by the lawyer if he was interested in establishing a precedent, Dr. Lifschutz fatefully replied, "sure" (Fridhandler).

Persistent in seeking a legal means to avoid his client's need to testify, Dr. Lifschutz's lawyer appealed the decision to the California Court of Appeals, which refused to reverse the lower court or grant the analyst a hearing on the matter. Finally, Dr. Lifschutz's lawyer petitioned the U.S. Supreme Court to intervene, but that court declined to intervene as well.

Having allowed Dr. Lifschutz every possible avenue of appeal, Arabian's lawyers then went back to the original court to demand enforcement of the court's order. Once again Dr. Lifschutz refused to cooperate. This was now one and a half years after the original incident and after a full year of legal wrangling. On December 5, 1969, Dr. Lifschutz appeared before a California Superior Court judge, James T. O'Keefe Jr., who ordered the psychoanalyst turned over to the custody of the bailiff until such time as he agreed to testify. Being turned over to the bailiff means, of course, being sent to jail, and that is exactly what happened to Dr. Lifschutz. To use a contemporary term, the story then "went viral." The spectacle of a respected and ethical psychoanalyst arrested and jailed out of a motive to protect his patient's confidences shocked the psychiatric profession and the public. The story also stunned Dr. Lifschutz, as he reported to the *New York Times* on December 14, 1969, shortly after his release. "'It was certainly a shocking experience,' Dr. Lifschutz said of his stay in the San Mateo County jail. While there, he said, he read *Phineas Redux* by Anthony

Trollope, which 'kept me occupied and it was a great consolation to me.'
He was in a single cell as a civil prisoner."

During our telephone interviews, Dr. Lifschutz, then eighty-eight and
retired for about ten years, had difficulty recalling some of the details of
his famous court case, but he had no trouble remembering his jail stay:

> I was in a cell all by myself—thank God! But I was in a cell block. In the
> cell you had a toilet seat, you know. There were four cells in the block.
> One was just an ordinary mild criminal. Cell three had a second-degree
> murderer in there serving his sentence for murder. And the fourth cell
> had a psychotic guy. What he did is he stuffed up his toilet bowl in his
> cell—each cell had a toilet bowl—he stuffed it up and flooded. . . . I woke
> up lying on my cot, and I looked down, and my shoes were floating.
> He had stuffed up the toilet, you know, he was psychotic. Se we had an
> adventure—my two and a half days in jail—that was the adventure—my
> shoes were floating away!

Dr. Lifschutz revealed to us many details of his incarceration that he had
never before discussed in public, including the fact that his wife was understandably frightened when he told her he was prepared to go to jail.
"Don't worry," he reassured her, "I'm alright." She was supportive, as were
his friends and colleagues, including the members of the San Francisco Psychoanalytic Society, of which he was president at the time. He never doubted,
then or now, that he made the right decision to maintain the confidentiality of the patient-therapist relationship. "I knew it had to be done." Throughout the interview he was affable and good natured, with a lively sense of
humor. He asked his wife to help him answer some of Paul's questions. He
expressed pride over his defiance in favor of confidentiality. One had the
sense that he had few if any regrets about his life or his principles.

The news of Dr. Lifschutz's arrest created an uproar and made the
front page of the *San Francisco Chronicle*. His patients were proud of him
for defending his beliefs. So too were lawyers impressed by his silence.
He referred to himself wryly as the "doc who didn't talk." Interestingly,
he believed the experience did not hurt his career and may have helped it.

Dr. Lifschutz's "adventure," which has become a defining legend in the
history of the protection of the confidentiality of psychotherapy in the
United States, lasted a mere two and a half days. As soon as he was jailed,
his lawyer filed an appeal to the California Court of Appeals for his release. The appeal was denied, but he was released from jail by the lower
court pending the Court of Appeals' decision on the substance of his claims

justifying his refusal to testify. Deciding the case in April 1970, the court refused to accept any of Dr. Lifschutz's arguments. He had to testify. The court's unanimous decision nevertheless placed great constraints on what he could be required to reveal. He had to answer "at least that question" as to whether Housek had been his patient, particularly since Housek had already told the opposing lawyers that was in fact the case.

So on May 4, 1970, now almost two years after the original incident, Dr. Lifschutz appeared in court and turned over his records of his encounters with Housek to the judge to be examined *in camera* (meaning that the records were reviewed by the judge in her chambers out of public view). Dr. Lifschutz testified in court that Housek had indeed been his patient. The analyst also revealed the dates of the treatment. Dr. Lifschutz disclosed no other information in court. The judge found that the records contained nothing relevant to Housek's current complaint and returned the records to Dr. Lifschutz. As reported in the *New York Times* on June 2, 1970, "The judge held that the psychiatrist, Dr. Joseph Lifschutz, met court requirements by testifying on May 4 that he treated the patient. He gave the dates but no other details."

Did Dr. Lifschutz "win" or "lose" the case for privacy? What precedent was actually set by this landmark case? Ralph Slovenko (*Psychotherapy and Confidentiality*, note 25, 71) and David W. Louisell and Kent Sinclair Jr. offer succinct summaries of the legal issues in the case. The novel legal arguments crafted by Dr. Lifschutz and his lawyer, based largely on claims of a constitutional right to privacy—and all of which the California Court of Appeals rejected—came down to this:

1. The psychoanalyst, independent of the patient, has the right to claim that the information is privileged and should be kept out of court to protect the psychoanalyst's privacy.
2. To deny the privilege to the psychoanalyst is discriminatory because the laws in California grant a privilege to both priests and those confessing to priests, so that each party has the power to refuse to disclose the contents of the confessional, independent of the other. To deny the same privilege to psychoanalysts as granted to priests is to discriminate against the contemporary version of the confessor, the psychoanalyst.
3. To deny the privilege to the psychoanalyst would unconstitutionally cause him economic harm in that it would make it less likely that patients would feel free to consult with psychoanalysts out of fear that their confidences could later be revealed.

Implications of the Landmark Case

The 1970 landmark case affected confidentiality in psychotherapy in the following ways:

1. The psychotherapist-patient privilege belongs only to the patient. The psychotherapist can assert the privilege on the patient's behalf in the patient's absence, but the psychotherapist has no privilege of his own.
2. A psychotherapist has no recognized constitutional privacy right that would support a psychotherapist's refusal to testify.
3. Although a patient's placing his mental condition into a case may constitute a waiver of the patient's privilege protecting previous psychotherapy communications, this waiver is to be construed very narrowly. Only information directly and specifically relevant to the case and the patient's claims may be forced into evidence.

Joseph Lifschutz remains, unapologetically, an absolutist with respect to privacy. "Confidentiality is the backbone of psychiatry and psychoanalysis." His views on confidentiality have not changed in the last forty years. He still believes that an absolute privilege of confidentiality is essential to psychotherapy. He still believes that psychotherapists have a constitutional privilege to remain silent even if their patients waive the privilege. As he wrote in his essay on the Anne Sexton controversy, which we discuss in Chapter 7, "it is inadequate in psychoanalysis to leave the question of disclosure to the holder of the privilege alone. It is entirely legitimate, even ethically required, to oppose a patient's wishes for disclosure" (8). He still believes that psychotherapy is a form of secular confession and that patients have the same rights as penitents. He still believes that psychoanalysts have an ethical responsibility to their patients, to their profession, and to society. He still believes that patient-therapist confidentiality does not end at death: "Some things people shouldn't know about," he bluntly exclaimed. He preceded many of his statements with the expression, "in my opinion," and it was clear that he was uncompromising about those opinions even when they were not shared by other therapists, lawyers, legislators, or judges.

Anne Hayman

Joseph Lifschutz's refusal to testify about any of the details of his treatment of a patient recalls the case of Anne Hayman, a British psychoanalyst who similarly refused to testify about a patient when she was subpoenaed in the mid-1960s. She published an article anonymously in the highly regarded medical journal *The Lancet* in 1965 in which she explained her objections to giving any information about the patient. The article, only two pages long, describes her dilemma over two conflicting moral obligations: the need to obey the law or to follow the rules of professional conduct. Betraying confidentiality, she told the judge presiding over the case, would be not only unethical but also antithetical to the spirit of psychotherapy. "To the Judge's query whether I would still object if 'the patient' gave permission, I answered with an example: suppose a patient had been in treatment for some time and was going through a temporary phase of admiring and depending on me; he might therefore feel it necessary to sacrifice himself and give permission, but it might not be proper for me to act on this" (786).

In Hayman's view, patients who allow a therapist to testify on their behalf are likely to be influenced by unconscious motives, namely, their positive transference to the therapist, which may lead them to actions opposed to their own best interest. A patient's consent may not be the "objective reality" required by the court. Hayman concludes her article with the statement that "Justice, as well as our ethic, is likely to be served by silence" (786). Impressed that the psychoanalyst was acting out of principle, the judge accepted her explanation and decided not to imprison her for contempt of court.

In "A Psychoanalyst Looks at the Witness Stand," which appears as the epilogue to the 2003 volume *Confidentiality*, edited by Levin, Furlong, and O'Neill, Hayman offered a fuller account of her experiences defending the principle of confidentiality. She doesn't explain why she decided not to reveal her identity in *The Lancet* article, but she notes that when she was first subpoenaed, she consulted with five senior analysts, none of whom was able to offer her useful advice about what to do. Instead, she was forced to rely on her own instincts, which proved correct, despite the fact that there was no psychotherapy privilege in England. She hoped that the court's decision would be hailed by others in the legal and psychotherapy communities, but "there were no such grand hurrahs," she observes ruefully (294). Her defense of confidentiality appeared to sink into obscurity.

Few colleagues later sought out Hayman's advice about the protection of confidentiality, though Christopher Bollas and David Sundelson applauded her stance in their 1995 book *The New Informants*. She believes her defense of privacy was unwelcome, "perhaps because it provoked a fantasy of a frightening or degrading example to be avoided, or because it may have invited unwarranted guilt feelings in someone who, for whatever reason, could not follow a similar course" (295).

Like Lifschutz, Hayman remains an "absolutist" when it comes to strict confidentiality. She admits, however, that difficult situations may arise when it might be necessary to consider breaking confidentiality. "In such a situation, the analyst could be faced with a choice between two dreadfully 'wrong' decisions—either of doing nothing to prevent an anticipated harm somewhere or of mistreating and damaging an analysand by some relaxation of the rule of confidentiality" (304). She ends her essay by admitting that she "cannot begin to work out the philosophical dilemma of these conflicting values" (307).

The Hayman case does not establish the privilege of confidentiality in the English legal system, as some therapists incorrectly believe. Rather, it is an instance of *judicial discretion*, not the creation of a formal legal precedent, as Peter Jenkins points out. The exercise of judicial discretion is seldom, if ever, reviewed by an appellate court where a precedent might be set. The only precedent it created, Jenkins admits sardonically, "is for therapists to risk contempt of court, and a possible fine or jail sentence, in defiance of a witness summons to give evidence regarding a client" (7). Nevertheless, like Joseph Lifschutz, Anne Hayman remains a legendary figure in the struggle to preserve confidentiality in psychotherapy. She retired in 1998 after more than forty years' practice as a psychoanalyst of adults and children.

"My Case Amounted to Something"

The passing of time has not changed the two psychoanalysts' views of confidentiality. Dr. Lifschutz looked back at his long and distinguished career with a sense of deep satisfaction and accomplishment. "My case amounted to something," he observed to us, "and I'm very proud of that." He was "very pleased" with the U.S. Supreme Court's decision in *Jaffee v. Redmond*, for which he had submitted an amici brief, as we discuss in Chapter 9. He remains optimistic about the future of psychotherapy despite the increasing challenges to confidentiality. Notwithstanding his occasional memory lapses, he had no difficulty summarizing the principle with which he has become identified: "It's very important to keep your mouth shut!"

Is Dr. Lifschutz a "medical martyr"? This was the question raised during a 1971 meeting of the American Psychoanalytic Association in Washington,

D.C. Many of the analysts attending the meeting disagreed with his arguments for claiming absolute confidentiality, but they admired the courage of his convictions. Dr. Lifschutz expressed gratitude for the extensive support of colleagues and professional organizations, many of which submitted amicus briefs in his defense. Near the end of the meeting, one analyst-attorney, Jay Katz, declaring that the "designation of martyr is not ordinarily bestowed on contemporaries," jokingly concluded, "who knows, perhaps we may one day refer to Dr. Lifschutz as 'Saint Joseph!'" (Watson, 175).

Katz's observation recalls a statement made by another psychoanalyst, the Jungian analyst Johanna von Haller in Robertson Davies's 1972 novel *The Manticore*, that "any theologian understands martyrdom, but only the martyr experiences the fire" (101). Davies's narrator, a criminal lawyer, knows all about confidentiality and the ease with which it is broken. "The priest, the physician, the lawyer—we all know that their lips are sealed by an oath no torture could compel them to break. Strange, then, how many people's secrets become quite well known. Tell nobody anything, and be closeminded even about that, had been my watchword for more than twenty of my forty years" (172). Telling nobody anything about the patient-therapist relationship has been Joseph Lifschutz's watchword for the entirety of his career, and regardless of whether he is sanctified in the future, he remains a stellar example of the principle of silence.

Take-home lesson: In the American legal system, the psychotherapist-patient privilege belongs only to the patient. A psychotherapist has no legal grounds to refuse to testify about a patient's treatment when the patient has had the opportunity to object to the testimony but does not object.

4. "The Angry Act": The Psychoanalyst's Breach of Confidentiality in Philip Roth's Life and Art

What to look for: What are the risks when a psychotherapist publishes highly personal and embarrassing information about a patient's life without first gaining the patient's permission?

Imagine reading a story, by one of the country's most acclaimed writers, about a young fictional novelist who discloses a humiliating childhood experience to his psychoanalyst. Without the fictional novelist's knowledge or permission, the analyst uses the same material, word for word, in an article published on "creativity" in a journal devoted to psychoanalysis and the arts. To make matters worse, the fictional novelist has used the same material in a bestselling confessional novel that has generated extraordinary literary controversy. The psychoanalytic article and the novel are published simultaneously, suggesting that neither the analyst nor the novelist was aware of the other's publication. Discovering that his analyst has appropriated his material, the fictional novelist becomes understandably enraged, fearing that anyone who reads his highly autobiographical novel and the psychoanalytic case study will realize he is the analyst's patient. The fictional novelist accuses the analyst not only of a breach of confidence but also of exposing him to public ridicule. Astonishingly, the analyst denies any wrongdoing, interpreting his patient's anger as narcissistic melodrama.

Now imagine that there are many autobiographical parallels between the real novelist who is writing the story and the fictional novelist who is the central character in the confessional novel described in the preceding paragraph. Both are young Jewish-American writers who have the same ethnic working-class background, the same personality, the same ironic vision, the same complaints, the same transgressions, the same fierce literary ambitions, the same driving obsessions, the same conflicts, the same love-hate relationship with women, the same ambivalence toward their religion and culture. Nor is this all. Both the real and fictional novelists enter into a lengthy psychoanalysis with the same kind of elderly orthodox Freudian psychoanalyst with a German accent. Moreover, the real and fictional novelists are married to women who have tricked them into marriage by claiming to be pregnant, refuse to divorce them, make their lives miserable, and then die, to their husbands' overwhelming relief, in car accidents. Imagine there are, indeed, so many striking parallels between the real and fictional novelists that the reader begins to suspect the former is writing about his actual psychoanalysis—one that has involved a major breach of privacy.

Imagine, further, that the reader is a newly tenured literature professor, studying at a psychoanalytic institute, and writing a book about the intersections between literature and psychoanalysis. Imagine that the real novelist provides all the clues necessary for a psychoanalytically oriented reader, such as the literature professor, to find the real psychoanalytic case study containing the embarrassing childhood material used by the real novelist in his earlier story. Imagine that the literature professor, to his amazement, finds the article that establishes, definitively, the connection between the real novelist and real analyst. The literature professor realizes he has made an unprecedented discovery, one that not only casts much light on the novelist's writings but also demonstrates the way in which the writer transforms the analyst's breach of confidentiality into a fascinating new story.

The literature professor is excited by the literary and biographical implications of his discovery, but he also realizes, to his dismay, that he is in an ethical dilemma. If he includes this new information in his own book, he will further call attention to the analyst's breach of confidentiality, which might be viewed as another violation of the novelist's privacy.

Imagine this situation, and you can begin to understand Jeff Berman's dilemma when he was writing *The Talking Cure: Literary Representations of Psychoanalysis* in the early 1980s. At the time, Philip Roth was his favorite novelist, and he had eagerly read and reread all of his stories. Roth's 1969 tour de force, *Portnoy's Complaint*, about Alex Portnoy's exuberant psychoanalysis, was one of the most famous—and infamous—novels of the age. Alex's outrageous sexual misadventures made him arguably the most analyzed and psychoanalyzed fictional character of the second half of the twentieth century. Nearly every major literary critic—and many minor ones as well—had something to say about Alex's sexual hangups, his conflicted relationship with a domineering mother and passive father, and his tangled feelings toward the woman to whom he refers, with a mixture of affection and contempt, as the "Monkey."

Portnoy's Complaint is a long psychoanalytic monologue in which the protagonist plays both the roles of analysand and analyst. Alex has grown up reading Freud's writings, from which he quotes with Mosaic authority, and he interrogates every aspect of his neurotic life as he lies on his analyst's couch. Few characters in literature are as guilt ridden as Alex—he calls himself the "Raskolnikov of jerking off"—and few are as colorful and verbally dazzling. Alex dominates the novel, analyzing himself with the rigor of a psychoanalytic acolyte. Only at the end of the nearly three hundred page story does his amused and bemused analyst, Dr. Otto Spielvogel, deliver his mocking punch line: "So [*said the doctor*]. Now vee may perhaps to begin. Yes?" (274).

Alex, with Roth's apparent approval, usurps Dr. Spielvogel's authority throughout *Portnoy's Complaint*. Only at the end does the novelist seem to acknowledge, in the punch line, that the psychoanalyst's interpretation of Alex may be different from Alex's self-interpretation. Does Alex accurately understand himself, or does he use psychoanalytic theory to deceive himself? Does knowledge lead to power—or to solipsism? One of the novel's central questions is Roth's relationship to Alex: the problem of narrative distance. There are several possibilities here. Are Roth and Alex essentially the same character? Is Alex a younger version of Roth, as Stephen Dedalus is a younger version of James Joyce in *A Portrait of the Artist as a Young Man*? Or do Roth and Alex have little in common apart from a love for verbal virtuosity? The problem of narrative distance in *Portnoy's Complaint* has long bedeviled readers, who suspect but cannot prove that the novel is deeply autobiographical.

Warden Spielvogel

Roth's attitude toward psychoanalysis radically shifts in his mordant 1974 novel *My Life as a Man*. It is a novel-within-a-novel, filled with so many "useful fictions" that it is hard to know where Roth's life ends and his fictional characters' lives begin. The benumbed hero, Peter Tarnopol, is a troubled Jewish-American writer in his late twenties, the author of a novel, *A Jewish Father*, that parallels Roth's first volume of short stories, *Goodbye, Columbus*. Tarnopol finds himself staring into the abyss, ready to end his life largely because of a disastrous marriage to a woman who, he believes, goes out of her way to torture him. The fictional Maureen Johnson

Tarnopol dies in a car accident in Boston in 1966, paralleling the death of Roth's first wife, Margaret Martinson, in a car accident in Central Park in 1968. Tarnopol seeks psychoanalytic help from the same Dr. Otto Spielvogel who has only that one punch line in *Portnoy's Complaint*. But Roth develops Spielvogel into a substantial character in *My Life as a Man*. Much of the novel focuses on the patient–analyst relationship, which is conveyed with psychoanalytic authenticity. "The doctor he reminded me of most," Tarnopol confides to us, "was Dr. Roger Chillingworth in Hawthorne's *Scarlet Letter*. Appropriate enough, because I sat facing him as full of shameful secrets as the Reverend Arthur Dimmesdale" (203). Spielvogel's specialty is treating "creative" people, and the analyst's judgment of his patient is summarized in one wry sentence: "Mr. Tarnopol is considered by Dr. Spielvogel to be among the nation's top young narcissists in the arts" (100). Tarnopol's judgment is equally ironic, referring to him as "Warden Spielvogel."

Spielvogel argues that Tarnopol's dependency upon his hateful wife Maureen derives from his lack of love as a child. The analyst sees Tarnopol's mother as "phallic threatening" in her emotional coldness toward him. To protect himself from feelings of rejection, Tarnopol adopts narcissistic defenses, idealizing both his mother and childhood. In Spielvogel's view, Tarnopol's need to reduce women to "masturbatory sexual objects" is both a derivative of the narcissism and a symptom of repressed rage toward his mother. Tarnopol remains angry at his weak, ineffectual father, according to Spielvogel, who failed to protect him from the mother's intrusiveness.

Tarnopol adamantly opposes this interpretation, declaring that he loves both parents and that he had a happy childhood. His mother and wife could not be more different, he insists. His mother *adored* him, he tells Spielvogel, an adoration that is reciprocated. "Indeed, it was her enormous belief in my perfection that had very likely helped to spawn and nourish whatever gifts I had" (213). Rejecting Spielvogel's belief that idealization of his mother was a "narcissistic defense" arising from the pain she inflicted upon him in childhood, Tarnopol maintains that his sense of "superiority" represented his "altogether willing acceptance of her estimation of me" (217). He concedes he had a conflicted childhood relationship with his "supernumerary father" (214), but not for the reason the analyst claims. "It wasn't his wife's hostility he had to struggle against, but the world's! And he did it, with splitting headaches to be sure, *but without giving in*" (241). Tarnopol and Spielvogel reach an impasse in analysis, neither agreeing with the other's interpretation.

Tarnopol is in analysis with Spielvogel from 1962 through 1967, but a major crisis develops during the beginning of the third year that undermines their collaborative work and dooms the remaining two years of therapy. Tarnopol sees a journal on Spielvogel's desk, asks to read it, and becomes incensed when he discovers the analyst's biographically transparent discussion of Tarnopol's actual therapy. What most infuriates Tarnopol, apart from the rhetorical and interpretive reductiveness of Spielvogel's case study, is that the analyst relates a traumatic childhood incident in Tarnopol's life that Tarnopol has already described in an autobiographical story published in the *New Yorker*. Tarnopol charges that anyone who reads Spielvogel's article and his own story will realize he is Spielvogel's patient. Spielvogel denies any wrongdoing and insists that he had no way of knowing about the unfortunate coincidence because both were writing simultaneously. Tarnopol's rage is illogical and contradictory, according to Spielvogel, in that Tarnopol accuses Spielvogel of both excessive and at the same time insufficient biographical disguises. Spielvogel also denies he was ethically required to receive Tarnopol's permission to publish the essay. "None of us could write such papers," Spielvogel adds, "none of us could share our findings with one another, if we had to rely upon the permission or the approval of our patients in order to publish" (250–251).

Spielvogel's denial of responsibility only enrages Tarnopol further, and when his girlfriend tells her own analyst, Dr. Golding, about Spielvogel's behavior, Dr. Golding is "appalled." Tarnopol and Spielvogel battle back and forth, neither yielding to the other, and finally Spielvogel gives Tarnopol an ultimatum: either get over his anger or end therapy. Tarnopol reluctantly remains in analysis because he believes he would not have survived his marital crisis without Spielvogel's help. Nevertheless, his faith and trust in Spielvogel are permanently shattered.

"The Angry Act"

Roth tells us in *My Life as a Man* that Spielvogel's article, entitled "Creativity: The Narcissism of the Artist," is thirty pages long, two pages of which are devoted to Tarnopol, who is disguised as a "successful Italian-American poet in his late forties." Roth also tells us that the article is published in a journal called the *American Forum for Psychoanalytic Studies*, appearing in the mid-1960s, in a special number focusing on "The Riddle of Creativity." In actual fact, there were not that many journals devoted to psychoanalysis and the arts in the 1960s, and when Jeff read the spring/

summer 1967 issue of *American Imago*, focusing on "Genius, Psychopathology, and Creativity," he came across a thirty-page essay by Hans J. Kleinschmidt called "The Angry Act: The Role of Aggression in Creativity." The title could not be more ironic in light of Roth's furious retaliation against his psychoanalyst.

Once Jeff read Dr. Kleinschmidt's article, two pages of which describe a patient unmistakably similar to Tarnopol, he knew he had found the evidence for which he was searching, what one might call in another context a smoking gun. A comparison of Kleinschmidt's article and *My Life as a Man* demonstrates that Roth based the clinical interpretation of Tarnopol's illness on "The Angry Act," which was itself based on Roth's actual psychoanalysis. First comes Kleinschmidt's discussion in "The Angry Act." Kleinschmidt: "A successful Southern playwright in his early forties illustrates the interplay of narcissism and aggression while his points of fixation are later [than those of another patient] and his conflicts oedipal rather than pre-oedipal. He came into therapy because of anxiety states experienced as a result of his tremendous ambivalence about leaving his wife, three years his senior" (123). Roth's fictional Spielvogel puts it this way in *My Life as a Man*: Roth: "A successful Italian-American poet in his forties entered into therapy because of anxiety states experienced as a result of his enormous ambivalence about leaving his wife" (239). Castration anxiety is important in the interpretations of both the real and fictional analysts. Kleinschmidt: "It soon became apparent that his main problem was his castration anxiety vis-à-vis a phallic mother figure" (124); Roth: "It soon became clear that the poet's central problem here as elsewhere was his castration anxiety vis-à-vis a phallic mother figure" (240–241). The characterization of the father is identical. Kleinschmidt: "His father was ineffectual and submissive to the mother" (124); Roth: "His father was a harassed man, ineffectual and submissive to his mother" (241).

Nor are these the only similarities. Both analysts view their patients as acting out repressed sexual anger. Kleinschmidt: "His way of avoiding a confrontation with his feelings of anger and his dependency needs toward his wife was to act out sexually with other women. He had been doing this almost from the beginning of his marriage" (125). Roth: "In order to avoid a confrontation with his dependency needs toward his wife the poet acted out sexually with other women almost from the beginning of his marriage" (242). Both the real and fictional analysts offer the identical interpretation of their patients' hostility toward women. Kleinschmidt: "The playwright acted out his anger in his relationships with women, reducing all of them to masturbatory sexual objects and by using his hostile masturbatory fantasies in his literary output" (125). Roth: "The poet acted out his anger in his relationships with women, reducing all women to masturbatory sexual objects" (242).

Jeff's discovery demonstrated conclusively that *Portnoy's Complaint* unexpectedly owes a great deal to "The Angry Act." Roth playfully attributes "Portnoy's Complaint" syndrome to Spielvogel's learned essay, "The Puzzled Penis," published in an authentic-sounding German journal, *Internationale Zeitschrift für Psychoanalyse*. "The Puzzled Penis" is none other than "The Angry Act" in disguise! Spielvogel's sentence, which Roth uses as the frontispiece of *Portnoy's Complaint*—"Acts of exhibitionism, voyeurism, fetishism, auto-eroticism, and oral coitus are plentiful"—parodies Kleinschmidt's language in "The Angry Act": "Practices of voyeurism, exhibitionism and fetishism abound" (125). Every incident described by Spielvogel in *My Life as a Man* corresponds to a similar incident in Kleinschmidt's discussion of the Southern playwright in "The Angry Act." Kleinschmidt's article is also an indispensable commentary on *Portnoy's Complaint*, helping us understand the extent to which Roth projected onto Alex his own ambivalent feelings toward his parents, childhood, and relationships with women. *Portnoy's Complaint*, *My Life as a Man*, and "The Angry Act" represent an unprecedented example of the cross-fertilization of literature and psychoanalysis.

In choosing the title of the psychoanalytic journal in which Spielvogel publishes "The Puzzled Penis," Roth must have been doing his homework, for in 1920 the journal that Freud founded in 1913, the *Internationale Zeitschrift für Ärztliche Psychoanalyse*, was changed to the *Internationale Zeitschrift für Psychoanalyse*—the same journal in which the fictional "The Puzzled Penis" appears! After the Anschluss of Austria in 1938, the International Psychoanalytical Press was liquidated, and the German journal *Imago*, founded by Freud, Otto Rank, and Hanns Sachs in 1912, was succeeded by *American Imago*, founded by Freud and Sachs in 1939. Publishing "The Angry Act" in *American Imago*, Kleinschmidt must have known about the link between the earliest German psychoanalytic journal and one of the most prominent American journals devoted to psychoanalysis and the human sciences. He may have also felt a connection between the creator of psychoanalysis and himself, a link that Roth goes out of his way to establish.

The Bathing Suit Incident

"The Angry Act" led Jeff to a story Roth published not in the *New Yorker* but in the first issue of *New American Review*, published in 1967, which contains an early chapter of *Portnoy's Complaint* called "The Jewish Blues." Alex discloses in this chapter a childhood incident in which his mother shames him over his awakening sexuality. Kleinschmidt includes this material in "The Angry Act." "He was eleven years old when he went with his mother to a store to buy a bathing suit. When trying on several of them, he voiced his desire for bathing trunks with a jock strap. To his great embarrassment his mother said in the presence of the saleslady: 'You don't need one. You have such a little one that it makes no difference.' He felt ashamed, angry, betrayed and utterly helpless" (124). Here is how Alex describes the incident in *Portnoy's Complaint*:

> "I don't want that kind of suit any more," and oh, I can smell humiliation in the wind, hear it rumbling in the distance—any minute now it is going to crash upon my prepubescent head. "Why not?" my father asks. "Didn't you hear your uncle, this is the best—" "I want one with a jockstrap in it!" Yes, sir, this just breaks my mother up. "For *your* little thing?" she asks, with an amused smile.
> "Yes, Mother, imagine: for my little thing." (51)

Neither Roth nor Kleinschmidt knew the other was writing about the bathing suit incident, and by the time they discovered the simultaneous publication of "The Jewish Blues" and "The Angry Act," it was impossible to do anything about it. Roth could have minimized the possibility of readers making the connection between "The Jewish Blues" and "The Angry Act" by excising the material in "The Jewish Blues" from *Portnoy's Complaint*, but that would have been like undergoing an amputation—or castration. Perhaps for that or a related reason, Roth decided to include "The Jewish Blues" in *Portnoy's Complaint*, thus increasing the risk of public exposure.

Jeff was certain that he had discovered the irrefutable evidence that Spielvogel's breach of confidentiality in *My Life as Man* was modeled on Kleinschmidt's breach of confidentiality in "The Angry Act." More importantly, Jeff was certain that Tarnopol's five-year psychoanalysis with Spielvogel was based on Roth's lengthy psychoanalysis with Kleinschmidt. Of the many novelists who have fictionalized the patient-analyst relationship, no one has written more authoritatively about psychoanalysis than Roth. And no novelist has written more frequently or imaginatively about psychoanalysis than Roth, creating a long line of psychoanalysts, including Dr. Lumin in his first novel, *Letting Go*, and Dr. Klinger, David Kepesh's analyst in the Kafkaesque novella *The Breast* and the novel *The Professor of Desire*. The psychoanalytic descriptions of Spielvogel's analysis of Tarnopol in *My Life as a Man* and Kleinschmidt's analysis of the Southern playwright in "The Angry Act" could not be coincidental. Could Roth have plagiarized the material in Kleinschmidt's case study and used it in *My Life as a Man*? It's conceivable—but if so, why would Tarnopol become so furious at Spielvogel for stealing the *novelist's* material? Why would Tarnopol be enraged at a breach of confidentiality if such a breach did not exist? Kleinschmidt could not have plagiarized from *My Life as a Man*, which was published seven years after "The Angry Act." Kleinschmidt tried to disguise Roth's identity, but he doesn't disguise the bathing suit incident. The simultaneous publication of the bathing suit incident in Roth's "The Jewish Blues" and Kleinschmidt's "The Angry Act" eliminated the possibility of chance or coincidence.

An Ethical Dilemma

Jeff's discovery ensnared him in an ethical dilemma. He had two choices—publish an article or book chapter about his discovery or not publish. If he did not publish his finding, he couldn't establish the fascinating interconnections among Roth's two novels and the analyst's case study. He couldn't establish that Kleinschmidt had appropriated Roth's words in "The Angry Act" or that Roth had reappropriated them in *My Life as a Man*. He couldn't establish that Roth based Spielvogel on Kleinschmidt in *Portnoy's Complaint* or that the argument arising from "The Angry Act" became an important section of *My Life as a Man*. He couldn't establish that Kleinschmidt's published article would turn out to be one of the most famous—or infamous—examples of a psychoanalyst's breach of a patient's confidentiality. He couldn't establish that there is more than a little truth in Kleinschmidt's interpretation of his patient's psychological woes—and more than a little justification in Roth's outrage over his analyst's betrayal of trust. Nor could Jeff establish that aggression "may be the prime mover or the unconscious motivating force for a creative push," as Kleinschmidt asserts in "The Angry Act" (125). Jeff couldn't establish any of these observations without publishing his discovery.

Had Jeff made the connection between Roth and Kleinschmidt on the basis of reading *Portnoy's Complaint* and "The Angry Act," he would *not* have published his discovery because he would have felt that he was exploiting the analyst's breach of confidentiality. But Jeff was led to the discovery by *Roth himself*. The novelist gave Jeff all the clues necessary to make the discovery. Without these clues, Jeff never could have made the connection.

Surprisingly, Jeff felt more protective of Kleinschmidt than of Roth. Jeff could imagine the analyst's shock and dismay upon learning that he and Roth had published the same material at the same time. Jeff could also imagine Roth's shock and dismay, even horror, but Roth had made the critical decision to disclose the breach of confidentiality in a new novel, thus further calling attention to it. Ironically, Roth made fewer disguises in *My Life as a Man* than Kleinschmidt did in "The Angry Act." Jeff knew Roth would be angry with his discovery despite the fact that the novelist ambivalently led him to it.

A frequent theme throughout Roth's novels is his displeasure with—even contempt for—literary critics who read biography into his art. Tarnopol expresses his creator's exasperation when he admits, about the harsh reception of his early novel, "My life was coming to resemble one of those texts upon which certain literary critics of that era used to enjoy venting their ingenuity" (*My Life as a Man*, 72). In Roth's 1981 novel *Zuckerman Unbound*, Nathan Zuckerman, Roth's alter ego, is the author of a wildly erotic confessional novel, *Carnovsky*, which resembles *Portnoy's Complaint*. Zuckerman watches in outraged disbelief one Sunday morning "three therapists sitting in lounge chairs on Channel 5 analyzing his castration complex with the program host. They all agreed that Zuckerman had a lulu" (128). In *Exit Ghost*, the last Nathan Zuckerman novel, Roth's counterself expresses only contempt for biographical criticism, or what he calls biographical inquisition: "the dirt-seeking snooping calling itself research is just about the lowest of literary rackets" (102). Jeff had no desire to be pilloried in Roth's stories.

And so Jeff did not send his chapter, "Philip Roth's Psychoanalysts," to the novelist himself, but he did send it to Dr. Kleinschmidt, inviting him to respond to it before he submitted his book for publication. As Jeff reports in "Revisiting Philip Roth's Psychoanalysts," Kleinschmidt's response was not what he naively expected:

> The author of "The Angry Act" was furious at me for linking his case study to Roth's novels. Far from wishing to comment on the chapter, he threatened to sue me if I submitted it for publication, claiming that I was

using "privileged" material, that I hadn't "proven anything" in my argument, and that the publication of the chapter would irreparably damage his reputation. Whatever protectiveness I had felt toward him soon vanished! I told him that I had made the discovery on the basis of published rather than confidential material and that, moreover, I was not interested in assigning blame for the simultaneous publication of his article and Roth's story. Nor did I believe that the publication of my book would damage his reputation. In fact, I did not accuse Dr. Kleinschmidt of a breach of confidentiality or plagiarizing his patient's words. We argued back and forth over the phone, and finally he withdrew his threat and asked me to make minor revisions, which I did, including deleting the reference to Golding's characterization of Spielvogel's behavior. ("Revisiting Philip Roth's Psychoanalysts," 98)

Dr. Kleinschmidt's insistence that Jeff delete all references to "Dr. Golding" suggests that Roth was not the only one suffering from "wounded narcissism." Dr. Golding's condemnation of Spielvogel's unprofessional behavior is unequivocal: "Dr. Golding said that was as reprehensible as anything he had ever heard of between a doctor and his patient" (*My Life as a Man*, 258). Jeff was never able to determine whether Roth actually based Dr. Golding on a real analyst, but there is little doubt that the fictional analyst's judgment, which functions in the novel like a Greek chorus, represented that of the psychoanalytic community.

In a follow-up letter, Dr. Kleinschmidt once again denied Jeff had proven he was Roth's analyst. "Since in my article I introduce the brief case history of a Southern playwright, I in no way allude to or reveal the identity of the patient. Your assumption that Roth's use of lines from my paper is proof positive that my case history relates to him, is inconclusive since authors habitually take material from a wide variety of sources for their fiction" (personal correspondence, December 6, 1980). The analyst also claimed that Jeff had quoted his words in "The Angry Act" without permission, a strange criticism in light of *his* use of Roth's words without permission. Dr. Kleinschmidt ended his letter with a warning that unless Jeff deleted all references to him in his otherwise "excellent and extremely well-written chapter," he would be faced with an "onerous lawsuit." Jeff immediately wrote him back and reminded him that he had every legal right to publish his discovery.

A few weeks later, Jeff brought a revised copy of the manuscript to Dr. Kleinschmidt's office. The psychoanalyst had suggested the meeting, perhaps to dissuade Jeff in person from publishing the Roth chapter. Once again he threatened Jeff with a lawsuit if he attempted to publish the article—and once again Jeff insisted, as he had done during his telephone call, that he was discussing two published novels and a published case study, drawing inferences with which any reasonable person would agree. Jeff said he had no interest in sensationalizing the discovery but that if the analyst attempted to block publication of the book, he would immediately contact the *New York Times*. Neither man wanted that kind of publicity, Jeff reminded him. He explained that he was writing *The Talking Cure* for a scholarly audience, the same type of audience that read *American Imago*.

Though not Dr. Kleinschmidt's patient, Jeff found the transference and countertransference dynamics of their exchange fascinating—and bedeviling. When the analyst called Jeff "narcissistic" and "aggressive," he felt like he had become Philip Roth—and that Dr. Kleinschmidt was *his* analyst! Yet if Jeff felt like a patient in search of an analyst, he also felt like an analyst (and a lawyer), for he knew he had pieced together a story that Dr. Kleinschmidt was vainly attempting to deny. Not once did the analyst acknowledge regret or wrongdoing about publishing "The Angry Act." Jeff understood for the first time Tarnopol's observation that Spielvogel's immunity to criticism was "dazzling." Indeed, Tarnopol adds, "the imperviousness of this pallid doctor with the limping gait seemed to me, in those days of uncertainty and self-doubt, a condition to aspire to: *I am right and you are wrong, and even if I'm not, I'll just hold out and hold out and not give a single inch, and that will make it so*" (*My Life as a Man*, 259).

Dr. Kleinschmidt finally backed off his threat of a lawsuit, begrudgingly praised Jeff's chapter, and told him that he had declined Roth's invitation to read the as-yet-unpublished manuscript of *Portnoy's Complaint* while the writer was still in analysis. He added, in another breach of confidentiality, that Roth was furious when he read "The Angry Act"—no surprise there—and then disclosed that Roth said that he could write anything he wanted about the psychoanalyst but that the analyst could write nothing about him without permission. Dr. Kleinschmidt then asked if Jeff had sent a copy of his chapter to "Philip"—to which Jeff said, without hesitation, "no." Jeff had no intention of provoking Roth's wrath, at least not then. Near the end of their meeting Dr. Kleinschmidt asked Jeff not to publish the chapter for five years, at which time, he hinted portentously, he would not be around to be troubled by it—a prophecy that underestimated his longevity: he did not die until 1997 at the age of eighty-three. As it turned out,

The Talking Cure was not published until 1985. As Jeff was leaving his office, Dr. Kleinschmidt exclaimed, in an unapologetically defiant and proud tone, "Incidentally, I'm Klinger too!"

Jeff wondered how the publication of *The Talking Cure* would affect Roth, if at all. Jeff had no way of knowing apart from asking him, which he was reluctant to do. Roth would, almost certainly, be displeased, to put it mildly, over the discovery of the connection between "The Angry Act" and his two novels, *Portnoy's Complaint* and *My Life as a Man*. It's not likely that Roth would have cheerfully nodded in agreement if Jeff thanked him for supplying all the clues necessary for a literary scholar like himself to uncover the link between biography and art. Jeff feared, realistically or not, that Roth would threaten him with a lawsuit if he attempted publication, as Kleinschmidt had done, and although Jeff knew that he would ultimately prevail in court if Roth decided to sue, such a lawsuit would be financially and emotionally draining.

Invasion of an Author's Privacy

Jeff's dilemma reminded him of Philip Young's situation three decades earlier. Young wrote his NYU doctoral dissertation on Hemingway, probably the most famous writer of the twentieth century, one whose larger-than-life image continues to captivate readers. Young made two crucial observations in his early study of Hemingway: first, that the Hemingway hero is essentially Hemingway himself; and second, that the novelist's major heroes are all wounded, physically and psychologically, wounds that Young traced back to Hemingway's injury during World War I, when he was nearly killed from mortar fire. Young finished his dissertation in 1948, completing one of the first book-length studies of Hemingway, and he then revised the manuscript for book publication.

Young sent a copy of his manuscript to Hemingway, hoping the novelist would grant him permission to quote from his stories, but instead, the writer was distressed, fearing that such books would destroy his creativity, if not his sanity. Young was shocked when he learned that Hemingway was determined to block the publication of any biography of him. The two men warily corresponded with each other, with Hemingway insisting that books like Young's could cause enormous psychological damage to living writers. "To tell a writer he has a neurosis, Hemingway did write me, is as bad as telling him he has cancer: you can put a writer permanently out of business this way" (3). Hemingway finally relented and generously allowed

Young to quote from the fiction, but the novelist continued to believe to the end of his life that critics could permanently injure a writer.

Young's book came out in 1958, and a revised version appeared in 1966, five years after Hemingway's death. In the "Foreword: Author and Critic—A Rather Long Story," Young records his anguish on receiving telephone calls, telegrams, and letters in early July 1961 "congratulating" him on predicting Hemingway's suicide. Writing about himself in the third person, Young denied he had made such a prediction. "He described a situation, a pattern, a process in Hemingway's life and work in which the act of suicide would not be altogether inconsistent" (3).

Literary scholarship has changed profoundly in the fifty years since Young published *Ernest Hemingway: A Reconsideration*, but the questions raised in the foreword still deserve attention:

> The real questions were, first: what are the rights of living authors, and what are the rights of critics? Second: at what point can criticism become an invasion of an author's privacy? And how much privacy can a writer expect when he has allowed himself to become an internationally public figure, or even, according to some, had worked very hard at promoting the image? Lastly, and this was much the toughest of the lot: given a respect for the author, which grew as the struggle progressed, how far is the critic willing to venture, even in defense of the author against many other critics, in violation of the author's deepest wishes that certain theories about him not be published? (10–11)

All of these questions weighed on Jeff's mind. Roth was, like Hemingway, an internationally known public figure, one who was known to nearly every serious American reader as a result of the spectacularly successful *Portnoy's Complaint*, which has become a household expression. Like Hemingway, Roth has gone out of his way in his fictional and nonfictional books to deride and disagree with those critics who have attacked him, fairly or not, in print. To be sure, Roth is more reclusive than Hemingway, but he appears to be no less thin skinned. Perhaps the greatest similarity is that both novelists are highly autobiographical while at the same time seeming to share a contempt for biographical and psychological criticism.

Unlike Philip Young, Jeff did not offer a new interpretation of a novelist's work but rather established an indisputable link between the novelist's life and art. There was nothing in Jeff's chapter that was hearsay, gossip, or speculation. Jeff did not put the novelist on the analytical couch but instead showed the similarities among the couches in "The Angry Act," *Portnoy's*

Complaint, and *My Life as a Man*. Roth had placed his largely autobiographical characters on the analytic couch in two of his major novels; Jeff filled in many of the details of their therapy.

Did Roth consider suing his analyst for breach of confidentiality? He almost certainly would have prevailed in court had he done so. Jeff didn't ask Dr. Kleinschmidt this question, but it's unlikely the novelist would have been inclined to seek legal redress. Tarnopol never raises this question in *My Life as a Man*, and for a good reason. He is so traumatized by the nasty lawsuit in which he is embroiled with his wife over alimony that the last thing he wants is another court case. One of Tarnopol's worst fears is that Maureen will find out about the case study, discover the extent of his sexual acting out with other women, and use the information against him in court. If the psychoanalytic article came to Maureen's attention, he tells Spielvogel, "well, just imagine how happy she would be to read those pages about me to the judge in the courtroom. Just imagine a New York municipal judge taking that stuff in" (253–254). Tarnopol's fear of adverse publicity thus prevents him from suing his analyst. Whatever legal damages he won might have only increased his alimony payments to his ex-wife.

Jeff never had any contact with Dr. Kleinschmidt again. Nor did Jeff send Roth a copy of the chapter. *The Talking Cure* passed largely unnoticed in the scholarly world, enjoying fewer than fifteen minutes of academic fame. Jeff has no reason to believe that the publication of his book damaged Dr. Kleinschmidt's reputation—or that the analyst lost much sleep over the book. Jeff doubts that many people in the 1980s knew that Dr. Kleinschmidt was Roth's analyst. A paid death notice published in the *New York Times* on February 22, 1997, describes Dr. Kleinschmidt as a psychiatrist and art collector who was born in East Prussia in 1913, grew up in Berlin, fled the Nazis in 1933, completed medical school at the University of Padua, and traveled to the United States in 1946, where he became an associate clinical professor of psychiatry at Mount Sinai Hospital and practiced for five decades. His wife of more than fifty years had died in 1990. "Dr. Kleinschmidt retired last year from active clinical practice and throughout his career, wrote on the intersections between creativity and psychoanalysis." The death notice is discreetly silent about his famous writer-patients. After his death, "The Angry Act" became a cautionary tale of the dangers of analysts publishing case studies without their patients' knowledge or permission.

The Kleinschmidt-Roth relationship is not the only example of a psychoanalyst writing without permission about a patient who discovers to his dismay the breach of confidentiality and then writes about his own analysis in disguised form. After we finished a draft of our manuscript, we learned from the psychoanalyst Henry Schwartz (personal communication, December 30, 2013) about the French novelist Georges Perec (1936–1982). After his Polish-Jewish parents died during World War II, Perec began therapy in his twenties with Michel de M'Uzan and analysis in his thirties with J.-B. Pontalis, the coauthor of *The Language of Psycho-Analysis*. According to Perec's close friend Claude Burgelin, Pontalis wrote several papers about his treatment of Perec, but despite the analyst's use of disguises and pseudonyms, Perec's identity was apparent. Perec's termination of analysis was possibly because of the discovery of the breach. In 1978 Perec published a celebrated novel, *La Vie mode d'emploi* (*Life: A User's Manual*), at the center of which are two characters, Bartlebooth and Winkler. Burgelin, the author of a book on Perec, believes that Winkler's elaborately disguised murder of Bartlebooth represents the novelist's revenge on Pontalis. We are grateful to Henry Schwartz, who is writing about Perec, for bringing this case to our attention.

The Analyst's Dilemma

Roth's encounter with Kleinschmidt, observes Peter L. Rudnytsky, is "the most famous example in the annals of psychoanalysis of a patient who discovered that his analyst had published an account of his treatment without his permission, and later wrote about this experience of betrayal from his own perspective" ("Book Review," 1407). Kleinschmidt's breach of confidentiality is often cited in articles and books on privacy issues, and the case has called attention to analysts' ethical responsibilities to their patients. Glen O. Gabbard, the former chair of the American Psychoanalytic Association Joint Committee on Confidentiality, begins his thoughtful article "Disguise or Consent" by discussing the inability of that organization's Ethics Subcommittee on Revision to reach consensus on issues of confidentiality and disguise. The Ethics Subcommittee finally "agreed to disagree" and came up with the following statement, one that Gabbard finds helpful but problematic:

> If the psychoanalyst uses confidential case material in clinical presentations or in scientific or educational exchanges with colleagues, either the case material must be disguised sufficiently to prevent identification of the patient, or the patient's informed consent must first be obtained. If the latter, the psychoanalyst should discuss the purpose(s) of such

presentations, the possible risks and benefits to the patient's treatment, and the patient's right to withhold or withdraw consent. (1071)

Gabbard then explores the various ways psychoanalysts can use confidential case material in clinical presentations and scholarly publications, including the use of thick disguise, patient consent, the process approach, the use of composites, and the use of a colleague as author. Each approach, Gabbard admits, is fraught with difficulty. For example, using thick disguise results in the loss of accuracy and may reflect the analyst's unconscious distortions of the presented material. Another problem with thick disguise is that it forces the analyst to be deliberately deceptive "in the service of a higher ethical standard, namely, protection of the patient's identity" (1073). Gabbard draws several conclusions from the breach of confidentiality in Kleinschmidt's case study about Roth, including the possibility that the analyst's unresolved aggression toward his patient may have been an unconscious factor in the decision to write about him. "Many of us write in an effort to master complex and difficult countertransference situations in our clinical work. Adverse consequences from publishing case material may in some cases reflect our own unanalyzed hostility towards the patient we choose to use as a clinical example" (1075).

The psychoanalyst's narcissism, whether conscious or not, acknowledged or not, also comes into play when writing about a patient. "Publications bring acclaim to analysts and advance their careers and reputations" (1076), Gabbard points out, and then generalizes that many analysts "feel guilty" about the "obvious self-interest" inherent in publishing clinical material. "While they attempt to convince themselves that they are being ethical by obtaining patient consent, they are secretly ashamed of exploiting their patient's trust in them by their writing" (1076). Analysts' guilt is likely to be more intense when they do *not* ask their patients for permission to write about them. Tarnopol mentions in *My Life as a Man* seeing the journal containing Spielvogel's case study about him "displayed conspicuously" on the analyst's desk (232), suggesting Spielvogel's pride over authorship. Dr. Kleinschmidt never discusses in "The Angry Act" the analyst's motives in writing about patients, but the desire for fame and success is a motivating force for analysts and novelists alike. The subtitle of Spielvogel's article, "The Narcissism of the Artist," could have also been called "The Narcissism of the Psychoanalyst."

Human research at colleges and universities is supervised by an Institutional Review Board (IRB), which almost always requires investigators to receive the informed consent of research subjects, but as Gabbard points out, informed consent is always problematic in psychoanalysis because it arouses intense transference and countertransference issues in patients and analysts. Patients may feel pressured by their analysts into giving informed consent, believing that analysts will be hurt or angry if they decline. Patients may give informed consent but later regret it. They may also feel devastated when they read what their analysts wrote about them. A patient's informed consent inevitably influences the course of the analysis. Gabbard's conclusion is that analysts can never know in advance if and when it is appropriate to ask for a patient's informed consent or how the patient will react.

Gabbard ends his overview of "Disguise or Consent" by concluding that the "patient's right to privacy, the profession's requirement to publish advances and new knowledge in the field, and the analyst's need for recognition are inevitably in conflict" (1082). This does not absolve the analyst from the need to act ethically, and Gabbard offers several useful recommendations to analysts, including writing about patients after the termination of their analysis and using vignettes instead of extended case histories.

Judy Leopold Kantrowitz's 2006 book *Writing About Patients* continues to explore the analyst's dilemma of writing about patients without betraying confidentiality. Curiously, she concludes that disguise without consent is acceptable—a position that seems to rationalize the further loss of privacy in psychoanalysis. In his book review of *Writing About Patients*, Peter J. Rudnytsky, an English professor, psychoanalyst, and editor-in-chief of *American Imago* decades after the journal published "The Angry Act," has criticized the "current mainstream position," advocated by Kantrowitz and others, of allowing analysts the freedom to choose whether to ask their patients' permission to write about them.

> The uncomfortable irony of this situation is that whereas psychoanalysts have traditionally prided themselves on their respect for the dignity and worth of the individual, by making the obtaining of informed consent optional rather than mandatory, analysts are arguably showing *less* respect for patients and their rights than is the norm for all other health care professionals and scientific researchers. (1408)

Kantrowitz also ignores Gabbard's discussion of the darker motives that come into play when analysts write about their patients.

Significantly, Elyn R. Saks and Shahrokh Golshan report in *Informed Consent to Psychoanalysis* (2013), an empirical study of sixty-two psychoanalysts who participated in their survey, that less than 20 percent disclosed to their patients that they might present material at a conference (19.3%), present material at an institute (19.3%), publish material (17.5%), or write the case up for a progression committee (13.1%). The authors conclude that threats to confidentiality "were endorsed at a much lower rate than expected given concerns about this in the analytic community" (57). Only one item was agreed with "at least frequently" at a rate greater than 50 percent—the disclosure to an insurance company (57.1%). After weighing the benefits and risks of informed consent, Saks and Golshan conclude that the therapist faces a vexing choice. "If she informs, she may possibly hurt the therapy, but if she does not inform, she may possibly anger the patient and subject herself to a lawsuit" (89). The dilemma is bound to evoke angry acts and counteracts.

Take-home lesson: The publication of anecdotes or case reports based on actual patients always runs the risk of harming a patient and therefore, at a minimum, should require the patient's permission before publication. In today's world of electronic publishing, which has made professional publications much more widely available and searchable, the risks have increased exponentially.

5. Angry Acts and Counteracts in Philip Roth's Life and Art

What to look for: Can a privacy breach by a psychotherapist have profound and lasting effects on a patient's life and even on the patient's future artistic creations?

Dr. Kleinschmidt could not have chosen a more self-fulfilling title for his case study, and the implications of "The Angry Act" continue to reverberate throughout Philip Roth's life and art. He must have sensed when writing *My Life as a Man* that he was aiding, even enabling, the discovery of an important link between his biography and fiction. Surely he must have known, on some level, that a psychoanalytically oriented critic would read *Portnoy's Complaint* and *My Life as a Man* and then search for and probably locate the professional journal with Dr. Kleinschmidt's explosive case study. Attuned to the unconscious as he is, and he is perhaps the country's leading psychological novelist, Roth may have felt that by writing *My Life as a Man* he was exacting revenge on the analyst who had betrayed confidentiality. From one point of view, Roth's revenge would not be complete unless and until the public became aware of "The Angry Act." But from another point of view, Roth's self-disclosures in *My Life as a Man* were risky, even reckless, because he was opening—or reopening—his life to public scrutiny. Didn't he envision the possibility that he might be hoist with his own petard? Kleinschmidt breached confidentiality, but then Roth wrote about the breach, having the last word—until a literary critic entered the scene, offering his own commentary.

Five Layers of Disguise

Jeff didn't write again about Dr. Kleinschmidt's relationship with Roth until 2001, when *Psychoanalytic Psychology*, the official journal of the Division of Psychoanalysis of the American Psychological Association, invited him to review Jason Shinder's edited volume *Tales from the Couch: Writers on Therapy*. Jeff couldn't resist commenting on Adam Gopnik's droll essay "Man Goes to See a Doctor," which first appeared in a 1998 issue of the *New Yorker*, one year after Kleinschmidt's death. Gopnik recounts, with ironic self-effacement worthy of Woody Allen, how he was on the "receiving end of what must have been one of the last, and easily one of the most unsuccessful, psychoanalyses that have ever been attempted—one of the last times a German-born analyst, with a direct laying on of hands from Freud, spent forty-five minutes twice a week for six years discussing, in a small room on Park Avenue decorated with Motherwell posters, the problems of a 'creative' New York neurotic" (18–19).

The analyst was eighty when the thirtyish Gopnik began treatment. The analysis, from 1990 to 1996, was a comedy of errors, with the elderly doctor talking mainly about himself and the famous creative writers he had treated. In the beginning of his essay Gopnik wonders whether to refer to his analyst by name or by pseudonym,

> a choice that is more one of decorum than of legal necessity (he's dead). To introduce him by name is, in a sense, to invade his privacy. On the other hand, not to introduce him by name is to allow him to disappear into the braid of literature in which he was caught—his patients liked to write about him, in masks, theirs and his—and from which, at the end, he was struggling to break free. (19)

Gopnik decides, finally, to use a pseudonym, Dr. Max Grosskurth. To heighten his reader's interest in the mysterious analyst, Gopnik adds coyly the following details:

> He had, for instance, written a professional article about a well-known patient, in which the (let's say) playwright who had inspired the article was turned into a painter. He had then seen this article, and the disputes it engendered, transformed into an episode in one of the playwright's plays, with the playwright-painter now turned into a novelist, and then the entire pas de deux had been turned by a colleague into a further psychoanalytic study of the exchange, with the occupations altered yet again—the playwright-painter-novelist now becoming a poet—so that four layers of disguise (five, as I write this) gathered around one episode in his office. "Yes, but I received only one check" was his bland response when I pointed this out to him. (19)

Confusing? Notice how Gopnik's disguise of his analyst's "professional article about a well-known patient" is far less transparent than Roth's disguise of his own analyst. Gopnik's five layers of disguise prevent a reader from locating an article like "The Angry Act" or suggesting how an actual published case study might be linked to novels like *Portnoy's Complaint* and *My Life as a Man*. But a reader familiar with "The Angry Act" or *My Life as a Man* might infer Kleinschmidt's identity from Gopnik's portrait. Gopnik is aware of his analyst's relationship with this well-known patient, for he cunningly tells us that one way he revived his sleeping analyst, who increasingly drifted off during his patient's ruminations, was to mention Philip Roth's divorce from Claire Bloom. "Instantly, his head would jerk straight up, his eyes open, and he would shake himself all over like a Lab coming out of the water. 'Yes, what are they saying about this divorce?' he would demand" (32).

Dr. Grosskurth comes across as distinctly unfriendly, unempathic, and humorless in "Man Goes to See a Doctor." He is not at all like the "small, wisecracking, scared Mitteleuropean Jews" with whom Gopnik was familiar. Incurably patriarchal, Grosskurth would talk about famous artists he treated or knew, offering monologues that involved, in part, "broken confidences of the confessional" (23). His sagacious clinical advice takes the form of statements like "No one cares" (25), an expression he uses to signify that people have their own problems and don't want to be troubled by others' woes. He describes celebrated writers such as Susan Sontag as "well defended," his formula for psychological health. Gopnik can never figure out why he remained six years with an analyst who offered so little therapeutic help. "He was touchy, prejudiced, opinionated, impatient, often bored, usually highhanded, brutally bigoted. I could never decide whether to sue for malpractice or fall to my knees in gratitude for such an original healer" (23).

Grosskurth's psychoanalytic theory, Gopnik writes, using a memorable metaphor, is that creative people were

> inherently in a rage, and that this rage came from their disappointed
> narcissism. The narcissism could take a negative, paranoid form or a
> positive, defiant, arrogant form. The analyst's job was not to cure the
> narcissism (which was inseparable from the creativity) but, instead, to
> fortify it—to get the drawbridge up and the gate down and leave the
> Indians circling outside, with nothing to do but shoot flaming arrows
> harmlessly over the stockade. (24)

Tellingly, Grosskurth's thesis resembles Kleinschmidt's in "The Angry Act," where the analyst discusses how the creative artist uses aggression in various ways. "The successful channeling of aggression into creativity restores equilibrium within the ego. Whether attempts at channeling aggression are successful or not depends largely on the ability of the ego to tolerate aggression" (125).

Ironically, Grosskurth not only fails to fortify Gopnik's narcissism, but he also appears to injure it further with his blunt statements. Grosskurth maintains, for example, that sometimes people are damaged by their burdensome names, but when Gopnik objects to the length of time the analyst pursues this misguided thesis, he is told that "Gopnik" is a "very ugly name" (26), which gives his frustrated patient something else to ruminate over.

Gopnik's decision to end analysis, partly because he wanted to move to Europe to write, infuriated Grosskurth, a response that, in turn, infuriated Gopnik. In their final sessions, he recalls, with more piquant sadness than anger, their "nonaggression pact." For the first time in six years Grosskurth affectionately calls his patient "Adam." Seeking to impart one last piece of analytic wisdom, the aging doctor muses, "In retrospect, life has many worthwhile aspects" (35). Gopnik reflects on the gnomic statement long after the analysis ends, trying to uncover the hidden meaning in his doctor's words. "Could there have been a more fatuous and arrhythmic and unmemorable conclusion to what had been, after all, *my* analysis, my only analysis?" (36). A year later Gopnik visits the ailing analyst, who is now confined to a hospital bed, and, after voicing the fear that he might "linger indefinitely," Grosskurth recalls reading something controversial Gopnik had written. "You showed independence of mind," Grosskurth declares, and then, turning away in pain, utters his final words: "And, as always, very poor judgment" (36). For all of Grosskurth's many failings, prejudices, and crude oversimplifications, Gopnik admits near the end of the essay that the analyst is "inside me," concluding, with more affection than many readers might be able to understand, that "On the whole, I would say that my years in analysis had many worthwhile aspects" (37).

Grosskurth's patients were drawn mainly from what he called "creative people," especially writers, painters, and composers. According to Gopnik, "he talked about them so freely that I sometimes half expected him to put up autographed glossies around the office, like the ones on the wall at the Stage Deli. ('Max—Thanks for the most terrific transference in Gotham! Lenny.')" (20).

In Chapter 1, we referred to Stanley J. Olinick's article about gossiping psychoanalysts, but Kleinschmidt's behavior appears to transform gossip into a treatment strategy. The analyst's behavior would seem to be a caricature of what Olinick describes—except that Kleinschmidt's betrayals of confidence were shockingly real.

Further confirmation of Kleinschmidt's failure to protect his patients' privacy comes from Jane Statlander-Slote, the editor of a volume of essays on Roth as well as the maker of a documentary on him. Statlander-Slote told us about "someone very close to her" who was in analysis with Kleinschmidt for several years, around the same time that Roth was in analysis, and who was the recipient of another broken confidence of the confessional. This person heard Kleinschmidt describe Roth as a "major modern narcissist" (personal communication, September 3, 2013). According to Statlander-Slote, the former patient wishes to remain anonymous out of loyalty to Kleinschmidt, toward whom he still feels gratitude despite the breach of confidentiality.

Roth would probably agree with Gopnik and Statlander-Slote's friend about Kleinschmidt's many worthwhile aspects, for when Tarnopol sends two "postanalytic stories" to Spielvogel, encouraging him to "feel free to speculate all you want, of course, but please, nothing in print without my permission," he receives a warm letter back praising his novelistic talents, compelling the writer to think, "This is the doctor whose ministrations I have renounced? Even if the letter is just a contrivance to woo me back onto his couch, what a lovely and clever contrivance!" (*My Life as a Man*, 223–224).

Roth has remained evasive and flippant about his actual analysis, despite frequent questions from literary critics who have interviewed him. He has given several interviews in which he refers wryly if guardedly to his long psychoanalysis:

> If he had not been psychoanalyzed, he told Hermione Lee in 1984,
> *Portnoy's Complaint*, *My Life as a Man*, and *The Breast* would not resemble
> the stories that he published. "Nor would I resemble myself. The experience
> of psychoanalysis was probably more useful to me as a writer than as a
> neurotic, although there may be a false distinction there" (*Conversations
> with Philip Roth*, 170). He was able to condense the "eight hundred or so
> hours that it took to be psychoanalyzed" into the "eight or so hours" that it
> takes to read *Portnoy's Complaint* (170). When asked in a 1988 interview,
> following the publication of *The Facts*, whether his treatment was a

"traditional five-days-a-week analysis," Roth replied, "I think there were eight days a week." (Berman, "Revisiting Philip Roth's Psychoanalysts," 101)

Roth Speaks About Dr. Kleinschmidt

New information about Roth's relationship to Dr. Kleinschmidt appears in Claudia Roth Pierpont's 2013 critical study *Roth Unbound: A Writer and His Books*. Pierpont's book is based on extensive conversations with Roth (to whom she is not related), who offers for the first time details about his five-year psychoanalysis. Pierpont captures Roth's dark humor about the experience. Kleinschmidt was known for treating artists and writers, she reminds us; "Roth half jokes that his National Book Award [for his 1959 volume *Goodbye, Columbus, and Five Short Stories*] got him a break on the rates." In talking about his psychoanalysis, Roth sometimes sounds like an Alex Portnoy disillusioned with Dr. Spielvogel, who is no longer mute. "'Why do you resist me?' Roth croons in a heavy German accent, mimicking his psychiatrist. Then, playing himself, several decibels louder: 'Why do you resist *me!*'" (Pierpont, 50).

Humor aside, Roth has few positive words about his psychoanalyst, though he credits Kleinschmidt with helping him end his destructive marriage. "Following a particularly horrendous fight, Roth recalls, he called Kleinschmidt and said that he had to see him that evening. After the session, Roth checked into a New York hotel—a room on an air shaft, at an 'academic rate'—and never went home again except to collect his clothes" (Pierpont, 49). In general, however, Roth is highly critical of his psychoanalysis. He implies that the anger Kleinschmidt awakened in him toward his mother, which he later wrote about in *Portnoy's Complaint* and *My Life as a Man*, may have served a purpose at the time; in retrospect, the anger was "unwarranted and wholly unfair" (Pierpont, 50). Roth now regards his psychoanalysis as a form of "brainwashing" (80).

Kleinschmidt could be "stunningly wrong." In the fall of 1967, he diagnosed as "envy" Roth's strange sense of weakness, which grew worse after he attended a party celebrating the publication of William Styron's novel *The Confessions of Nat Turner*. When Roth asked Kleinschmidt why he began to feel ill *before* the book party, the latter replied, "Because you were anticipating the envy" (Pierpont, 50). Later that day Roth rushed to a hospital, where he was told that he had a ruptured appendix that required immediate surgery. "Two of Roth's uncles, his father's brothers, had died of peritonitis from a burst appendix" (50).

Pierpont devotes only one paragraph to Kleinschmidt's breach of confidentiality. Roth's name was omitted in the psychoanalytic journal, and details about his life were slightly altered, she writes, "but there he was, psychically naked, in his psychiatrist's baleful view" (Pierpont, 50). What was "baleful," we should point out, was not necessarily Kleinschmidt's interpretation of his patient but the fact that the analyst published the case study without permission. Curiously, Pierpont does not ask Roth why he had included the clues in *My Life as a Man* that lead to the discovery of "The Angry Act" nor how he felt when Kleinschmidt's breach of confidentiality became known to the public.

Bernard Avishai's 2013 book *Promiscuous: Portnoy's Complaint and Our Doomed Pursuit of Happiness* also casts further light on Roth's relationship with Kleinschmidt. A long-time friend of Roth, Avishai, an adjunct professor of business at Hebrew University, characterizes his book as a "kind of tribute" (218) to the novelist. Like *Roth Unbound, Promiscuous* is based on many hours of conversation with Roth. Avishai offers what might be described as authorized interpretations of *Portnoy's Complaint*. Avishai adds a few more details about Dr. Kleinschmidt, telling us that his patients included Richard Avedon and Leonard Bernstein. Avishai never challenges Roth's assertion that he told Kleinschmidt largely what the analyst wanted to hear. Additionally, Avishai insists that Alex's relationship with Jack and Sophie Portnoy has virtually nothing in common with Roth's relationship with his own parents. *Portnoy's Complaint*, in Avishai's view, is a parody of psychoanalysis and has nothing to do with Roth's own analysis. "Roth remembers his analyst kindly," Avishai writes, "but the high comic moments in *Portnoy's Complaint* turn psychoanalytic authority against itself by taking its projections to extremes" (180).

Avishai interviewed several writers for his book, including Adam Gopnik, who, in a change of heart, decided not to allow his analyst to disappear into the braid of literature in which he is caught. "Kleinschmidt never offered me," Gopnik reveals, "any kind of classical Freudian account to organize my experience. I saw him in the 1990s, and his approach may well have been different in the 1960s. Instead, and that was part of the comedy of it, he was immensely practical and opinionated" (Avishai, 193).

To judge from his statements to Pierpont and Avishai, Roth looks back at his psychoanalysis with Kleinschmidt as more helpful than not. Roth rejects the psychoanalyst's interpretations, which he regards as reductive, but not the experience of talking to another person who was trying to help him figure out his life. "By the end of the analysis," Avishai observes,

"Kleinschmidt's specific interpretations of Roth's blues often seemed to his patient more useful as material than as a springboard to self-discovery" (177). Roth's outrage in the late 1960s and 1970s seems to have diminished; neither Pierpont nor Avishai suggests that the novelist is still furious over the "breach," a word, significantly, that Avishai never uses. One would never know from reading Avishai's book that there was a betrayal of privacy. Did Avishai and Pierpont believe that it would be inappropriate to ask Roth how he felt about having his case history held up to public scrutiny? Avishai is, in fact, reluctant to criticize the analyst's professional behavior. "I do not mean to judge Hans Kleinschmidt, whom I never met except in print, and who is reputed to have helped a great many people. Nor do I presume to dismiss the virtues of psychoanalysis, of which I am a grateful beneficiary" (192). Avishai points out the sobering changes that have occurred in psychoanalysis as the talking cure has morphed into psychopharmacology; "Nowadays," Gopnik quips, "Portnoy would go to Spielvogel and the doctor would have to give him Prozac and Viagra and send him home" (192).

"Do You Ask Permission of the People You Write About?"

There are radically different conventions, practices, laws, and ethics for mental health professionals and creative writers surrounding privacy. Tarnopol and Dr. Spielvogel battle over this issue in *My Life as a Man*, and we have no doubt inferring Roth's position. "Do you ask permission of the people you write about?" Spielvogel asks Tarnopol, to which the novelist responds: "But I am not a psychoanalyst! The comparison won't work. I write fiction—or did, once upon a time." Tarnopol's fictional stories may arise from autobiography, he concedes, but they are shaped and transmuted into art. "I do not write 'about' people in a strict factual or historical sense." Analysts are bound by ethical considerations, Tarnopol continues, which do not apply to novelists. "Nobody comes to me with confidences the way they do to you, and if they tell me stories, it's not so that I can cure what ails them. That's obvious enough. It's in the nature of being a novelist to make private life public—that's a part of what a novelist is up to" (*My Life as a Man*, 250).

Roth asserts that he is *not* an autobiographical writer and, therefore, not guilty of appropriating others' words and deeds. "You are not an autobiographer, you're a personificator," Nathan Zuckerman tells "Roth" in *The Facts* (1988), which purports to be, as the subtitle suggests, *A Novelist's*

Autobiography (162). "You make a fictional world that is far more exciting than the world it comes out of. My guess is that you've written metamorphoses of yourself so many times, you no longer have any idea what *you* are or ever were. By now what you are is a walking text" (162). Without explicitly referring to "The Angry Act," Zuckerman reminds "Roth" that texts, no matter how transparent, always have other texts that remain hidden or secret: "With autobiography there's always another text, a countertext, if you will, to the one presented. It's probably the most manipulative of all literary forms" (172). "The Angry Act" turns out to be Roth's most famous countertext, one whose psychoanalytic thesis he accepts in *Portnoy's Complaint* but rejects in *My Life as a Man*.

Roth's insistence that he is not an autobiographical novelist includes his rejection of Kleinschmidt's theory of the narcissistic novelist. Tarnopol argues that the artist's success "depends as much as anything on his powers of detachment, on *denarcissizing* himself" (*My Life as a Man*, 240). Roth returns to this theme in *The Counterlife* (1986), where an editor delivers a eulogy of Nathan Zuckerman that implies Roth's rejection of the Kleinschmidt-Spielvogel theory of artistic creativity: "The exhibitionism of the superior artist is connected to his imagination; fiction is for him at once playful hypothesis and serious supposition, an imaginative form of inquiry—everything that exhibitionism is not. It is, if anything, closet exhibitionism, exhibitionism in hiding" (210). Zuckerman claims that he can exhibit himself only in disguise: "All my audacity derives from masks" (275). The American writer "Philip" makes a similar observation in *Deception* (1990): "I write fiction and I'm told it's autobiography, I write autobiography and I'm told it's fiction, so since I'm so dim and they're so smart, let *them* decide what it is or it isn't" (190).

"Invasion of Privacy"

Apart from "The Angry Act," Roth is ingenious enough to make it difficult if not impossible to locate the "real" or "essential" person behind the fictional disguises. He is a master of ventriloquizing himself, whether in the form of Peter Tarnopol, David Kepesh, or Nathan Zuckerman, the three novelists who most resemble their creator—or puppeteer. The novelist's fictional masks run like a leitmotif from his earliest to latest novels. So persistent is this theme that one senses the novelist protests too much.

These privacy issues are evident in Roth's 1993 novel *Operation Shylock*. The early part of the story, the writer's discussion of his near-catastrophic medical crisis around 1988, when he was in his mid-fifties, turned out to be autobiographical. Following "botched" surgery for a minor knee injury, Roth began to experience a severe clinical depression accompanied by alarming suicidal ideation. The crisis lasted for "one hundred days and one hundred nights" (22). What prevented him from committing suicide, he discloses, is that had he done so, his eighty-six-year-old father "would smash to smithereens" (23). During his breakdown Roth discovered that his misery was caused by the sleeping medication Halcion (Triazolam), the pill that had been charged with "driving people crazy all over the globe" (24). Eighteen months after his breakdown, he learned from reading an article that all of the symptoms of his crisis were associated with "Halcion madness," including "severe malaise; depersonalization and derealization; paranoid reactions; acute and chronic anxiety; continuous fear of going insane; . . . patients often feel desperate and have to fight an almost irresistible impulse to commit suicide" (25; ellipsis in original).

Autobiography in *Operation Shylock* soon gives way to fiction when Roth encounters in Israel a preposterous double, an impersonator, who looks and speaks like Roth, pretends he's Roth, and gives interviews in Roth's name. The "other" Roth even meets with the Polish Solidarity leader Lech Walesa and proposes a Diaspora plan to resettle Israeli Jews in Poland. The "other" Roth allows the "real" one to list his many grievances over the critical receptions of his novels. "I know your books inside out," the imposter exclaims. "I know your *life* inside out. I could be your biographer. I *am* your biographer" (*Operation Shylock*, 73). The imposter turns out to be one of Roth's strangest characters, possessing surely the oddest moniker, Moishe Pipik, the "derogatory, joking nonsense name that translates literally to Moses Bellybutton and that probably connoted something slightly different to every Jewish family on our block" (116).

The "real" Roth is so unnerved by the imposter's appropriation of his identity that he threatens to take legal action, but then, unexpectedly, the imposter cites William L. Prosser, the coauthor of *Handbook of the Law of Torts*, to settle the question of the right to be left alone. "In 1960," the imposter declares, sounding like a law professor,

> in the *California Law Review*, Prosser published a long article, a reconsideration of the original 1890 Warren and Brandeis *Harvard Law Review* article in which they'd borrowed Judge Cooley's phrase "the general right to be left alone" and staked out the dimensions of the privacy interest. Prosser discusses privacy cases as having four separate branches

and causes of action—one, intrusion upon seclusion; two, public disclosure of private facts; three, false light in the public eye; and four, appropriation of identity. The prima facie case is defined as follows: "One who appropriates to his own use or benefit the name or likeness of another is subject to liability to the other for invasion of his privacy." (75–76)

Roth's reference to William Prosser's forty-page article, a classic in the area of privacy rights, is fascinating for many reasons. Prosser, the dean of the University of California School of Law, at Berkeley, when he published the article, was a gifted writer, and his statements resonate with eerie relevance more than half a century later, as when he states that "All of us, to some extent, lead lives exposed to the public gaze or to public inquiry, and complete privacy does not exist in this world except for the eremite in the desert' (396). Prosser traces the historical clash between two basic principles, the right of privacy and the constitutional guarantee of freedom of the press. The result of this clash, he suggests, has been a "slow evolution of a compromise between the two" (410).

Quoting Prosser allows Roth to stake out the legal dimensions of his *own* interest in privacy. The issue for Roth is not the creation of an impersonator like Moishe Pipik, who derives from the long literary tradition of the double, but from the way in which novelists like Roth appropriate "real" characters for thinly disguised fictions. In Roth's view, the public's right to be left alone does not limit the novelist's freedom to write about "real" people in his stories. But what happens if a person experiences "mental distress" as a result of a writer's characterization of him or her in a novel or memoir? Prosser observes that although the biographies of celebrities can be written and their life stories be set before the public in unflattering detail, some limits nevertheless exist on the disclosure of private facts. Prosser's example? The "private sex relations of actresses" (417).

Leaving a Doll's House

Nowhere is the right to privacy in Roth's world a more contentious issue than in *Leaving a Doll's House*, Claire Bloom's unsettling 1996 memoir. Bloom describes her marriage first to the actor Rod Steiger, with whom she had a daughter, Anna, and then to the producer Hillard Elkins, but she focuses on her twenty-year relationship with Roth, whom she portrays as capable of great kindness and great cruelty, a man who is at once psychologically astute and emotionally unstable. Their relationship began in

1974, when the British actress wrote a letter expressing her admiration for *My Life as a Man*. Bloom's response to the novel became an ominous prophecy of her long relationship with the novelist.

> Under Roth's brilliant inventiveness, beneath his diamond-sharp obser-
> vation, was a deep and irrepressible rage: anger at being trapped in
> marriage; fear of giving up autonomy; and a profound distrust of the
> sexual power of women. I noted the warning signals; but of course the
> situation would be different with me. So most women imagine it will be
> with them, as they enter a new and challenging relationship. (145–146)

Bloom and Roth lived together for many years before marrying in 1990, and the novelist dedicated *Operation Shylock* to her. The marriage ended bitterly, however, in 1995.

One of the most disturbing aspects of Bloom's memoir is the intense hostility she describes between Roth and her teenage daughter, who was studying to become an opera singer. "Emotionally fragile as a result of my second marriage, she was deeply distrustful of the strange new man in my life and felt full of anxiety, lest the damage done to us both by Elkins should be repeated in yet another painful relationship" (150). Bloom readily admits that Roth was right when he accused her of having an unhealthy relationship with her daughter. "I cannot recall my actions during that time without experiencing an unrelenting feeling of guilt" (159). According to Bloom, Roth exasperated the situation by trying to "take revenge on me for his perceived exclusion from my 'symbiotic relationship with Anna'" (225). Bloom also began to feel that Roth was demonizing her, as he had done with his first wife. "Whatever self-fulfilling prophecy made him marry her in the first place, she became only the first in a long parade of women over whom he exercised his considerable power to transform through his writing. The dead woman has gone into the realm of Roth's fiction; the truth about her real self has been left far behind" (170).

There is no evidence that Bloom knew about Kleinschmidt's "The Angry Act," but her portrait of Roth uncannily resembles the psychoanalyst's characterization of the "successful Southern playwright," Kleinschmidt's "disguised" version of Roth in "The Angry Act." Kleinschmidt writes that the price of his patient's love for his mother and wife was submission to them, resulting in castration anxiety (124–125); Bloom writes about Roth's "need to escape from a woman at the moment when he realizes his affection makes him vulnerable to her. Implicit in this notion is the sense that, through a woman's dangerous, clandestine power, she is bearer of his physical and mental castration—possibly, even, his death" (*Leaving a Doll's House*, 169). Roth had once observed to her that, in her own words, "it was a strange coincidence that his most consequential and far-reaching relationships had always been with fatherless women." She finds the comment "hardly a coincidence" insofar as "fatherless women gravitate toward emotionally unavailable men." What she does find strange is that "any of the women who shared even a morsel of his life—and here I must include myself—could have hoped to find a paternal figure in Philip Roth, so austere, so conditional, so far removed from the warm and protective figure of our childhood imagination. Instead, he was the fleeting shadow of the one who disappeared" (236).

As described by Kleinschmidt and Bloom, a man's fear of women goes beyond castration fear and may include other fears as well: fear of the loss of boundaries, fear of merger, and fear of the loss of self. Many psychologists have offered theories of the male fear of the feminine, including Karen Horney in "The Dread of Women," Erich Neumann in *The Fear of the Feminine*, and Wolfgang Lederer in *The Fear of Women*. Novelists, too, have explored this fear along with male insecurity; a more accurate title for D. H. Lawrence's *Women in Love* would be *Men in Love*.

Bloom acknowledges in *Leaving a Doll's House* that the suicidal depression Roth writes about in *Operation Shylock*, which was the most severe psychological crisis in his life, is "neither inaccurate nor overblown; it was just as he recorded it, with the added factor, unmentioned in the book, that I was dangerously close to going down with him" (178). Part of the reason for her own psychological crisis is her discovery, when reading the manuscript of Roth's novel *Deception*, of his merciless characterization of her relatives. She is distressed by his description of her "self-hating, Anglo-Jewish family," with whom he had lived in England. "Oh well, I thought, he doesn't like my family. There was a description of his working studio in London, letter-perfect and precise" (183). Bloom reads about her husband's sexually explicit descriptions of encounters with several women who may or may not have been fictional, but then she comes across something that horrifies her.

> Finally, I arrived at the chapter about his remarkably uninteresting, middle-aged wife, who, as described, is nothing better than an ever-spouting fountain of tears bemoaning the fact that his other women are so young. She is an actress by profession, and—as if hazarding a guess would spoil the incipient surprise lying in store—her name is Claire.

It is not her husband's erotic imagination that mortifies Bloom but his mean-spirited portrayal of her and the invasion of her privacy. "What left me speechless—though not for long—was that he would paint a picture of me as a jealous wife who is betrayed over and over again. I found the portrait nasty and insulting, and his use of my name completely unacceptable" (183).

"You Cannot Stop Me from Writing"

Furious, Bloom demands that Roth delete her name from the novel. She remains adamant about this even when he tells her that because he calls his protagonist "Philip" he wants to call the wife "Claire" to add to the richness of the novel's texture. "I replied I didn't care whether it did or not. I reminded him that, like him, I was a public figure also and would seek any means at my disposal—even legal means—to have my name removed. For once, confronted by my opposition, Philip agreed to remove it from the novel" (184). He eliminates his wife's name in the penultimate chapter of *Deception* but not the portrait of her as jealous, possessive, and narrow minded, and he insists on his right to publish anything he wants regardless of who gets hurt in the process. "You cannot stop me from writing what I write for a simple and ridiculous pathological reason—because I cannot stop myself! I write what I write the way I write it, and if and when it should ever happen, I will publish what I publish however I want to publish and I'm not going to start worrying at this late date what people misunderstand or get wrong!"—to which the now unnamed wife replies, sardonically, "Or get right" (*Deception* 191).

Roth's response is troubling for many reasons. The admission that he cannot prevent himself from writing the way he does suggests the obsessional nature of his writing, implying he cannot write differently even if he wanted to. Rationalizing his lack of respect for others' privacy as stemming from a self-confessed weakness of character absolves him of taking responsibility for his actions. Would the same "innocence by reason of insanity" satisfy him if his psychoanalyst breached confidentiality and used the same kind of defense: "I just gotta write up my case study!"

Roth's insistence that he can write about anyone he pleases, without permission, suggests, curiously, the position that Dr. Kleinschmidt took in breaching Roth's own confidentiality. It's as if Roth has become his own psychoanalyst here, able to write about anyone or anything with impunity. What may be Roth's identification with Kleinschmidt's imperious disregard for confidentiality could be seen as by some as startling. Anna Freud, in her classic study *The Ego and the Mechanisms of Defense*, writes about "identification with the aggressor," a defense mechanism often observed both in children and in adults that has far-reaching implications. "By impersonating the aggressor, assuming his attributes or imitating his aggression, the child transforms himself from the person threatened into the person who makes the threat" (113). Just as Kleinschmidt justified violating Roth's privacy and confidentiality by invoking the claim of advancing science, one might speculate that so does Roth justify violating others' privacy and confidentiality by invoking the claim of advancing art. In that view, Roth's violation of privacy could be seen as part of his continuing attachment to Kleinschmidt. By writing about each other without permission, both Kleinschmidt and Roth become involved in an odd pas de deux in which each aggresses on the other, supporting Kleinschmidt's thesis that art is an angry act. It is possible that Roth may not be conscious of his possible identification with his "breaching" psychoanalyst, but it is not impossible to imagine that it could be one of the most disturbing consequences of his psychoanalysis. Identification with the aggressor is not, of course, a positive outcome of psychoanalysis, nor is any action that remains beyond a patient's—or writer's—conscious control.

Roth walks a tightrope when writing about the abyss, and the consequences of losing artistic control extend to his life as well. Solange Leibovici has observed insightfully that central to Roth's fictional and nonfictional writings is the mise-en-abyme, a French term, derived from the language of heraldry and popularized by Andre Gide in his 1926 novel *The Counterfeiters*, which literally means "placed into [an] abyss" and figuratively refers to an endless series of internal duplications, such as a story-within-a-story. "Roth needs the alter ego and the mise-en-abyme, the specular self," Leibovici argues, "not to avoid the boring genre of autobiography, but to keep some distance from the chaos and destructivity of the hysterical ghost." The mise-en-abyme structures psychoanalysis as well as writing, but whereas patients can work through their conflicts in a successful analysis, Roth's protagonists cannot. "The mise-en-abyme goes on and on and on, the traumatic ghost keeps on reappearing again and again and is always acted out, never worked through." Without using the term mise-en-abyme, Peter Rudnytsky has noted that one of the recurrent dynamics in Roth's art and life is the character who "sets himself up for failure in the form of rejection and then feels justified in his anger once the rejection or disappointment has been meted out to him" ("Interview," 154).

Bloom writes about her shock, several months after their wedding, when she asked Roth for a blank tape to prepare for a recital. "I pressed the play button by mistake, and was stunned to hear our telephone conversation in full. It had been recorded by Philip for reasons I can only guess at. As far as I knew, this accidental revelation remained undiscovered by him. I certainly never mentioned it" (191). Bloom cites this as an example of the "bouts of paranoia" seizing her husband, but she never comments about the loss of privacy implications here. Prosser cites eavesdropping upon private conversations as an example of "intrusion upon a person's seclusion or solitude" (390).

The publication of *Leaving A Doll's House* provoked Roth's own counterstory in the form of his 1998 novel *I Married a Communist*. The radio actor Iron Rinn—born Ira Ringold—is a man who is passionately committed to social justice, but he is destroyed by a catastrophic marriage to the celebrated silent-film actress Eve Frame, who has much in common with Claire Bloom: beautiful looks, precise locution, English nationality, and an "ultracivilized, ladylike" role (53). Married several times, like Bloom, including once to a famous actor, enmeshed with her musical daughter who despises her stepfather, Eve Frame writes a sensationalistic memoir, *I Married a Communist*, to denounce and destroy her husband, who has ended their marriage. Roth portrays Eve Frame as a vindictive, insecure, self-deceived woman, depriving her of any dignity or complexity. Roth imputes only the worst motives behind the actress's book and, in a final, mocking irony, suggests that it was ghostwritten.

I Married a Communist dramatizes the destructive historical forces unleashed during the McCarthy era, but it also reveals, on a more personal level, the devastation that occurs when a marriage dissolves into angry recriminations and when private lives become lurid public stories. Roth's novel can be read in many ways, including as a roman à clef. For Roth, *Leaving a Doll's House* becomes another version of "The Angry Act," the violation of his privacy, resulting in the novelist's immediate counterattack, illustrating, in Dr. Kleinschmidt's clinical judgment, the interplay of narcissism and aggression.

"Ira is Roth's stick to beat the old Stalinist Left, while Eve Frame is his stick to beat his ex-wife," exclaims Mark Shechner, one of Roth's most sympathetic critics, but even Shechner is disturbed by Roth's portrayal of Bloom's daughter.

If the ex-wife is fair game, especially after Claire Bloom had struck first with her book about Roth, *Leaving a Doll's House*, is the daughter a fair target as well? It is bootless to ask this of Roth, for whom turning aside wrath is a crime worse than McCarthyism. Privacy? What privacy? In tabloid America, where all is material, closed doors are open invitations, and what is the envious novelist to do but press on with the illicit pleasures of exposure and revenge, with all the bile he can muster. (149–150)

Do we hear in Roth's rage toward his ex-wife and ex-stepdaughter in *I Married a Communist* echoes of his anger toward his ex-analyst in *My Life as a Man*? Surely the details of the two novels are vastly different. The earlier novel dramatizes his understandable anger toward an ex-analyst who was guilty of a stunning breach of confidentiality; the later novel dramatizes Roth's fury over his ex-wife's portrayal of him in a memoir. But there are several similarities between the two novels. Both focus on invasions of privacy that have a traumatic effect on the Rothian protagonist. As a patient, Roth was the innocent victim of his analyst's breach of confidentiality. Reversing roles, he struck back with all of his imaginative power in *My Life as a Man*. As a husband, Roth's role in the breakup of his marriage to Bloom is far less clear: he was hardly an innocent victim. Nevertheless, *I Married a Communist* depicts Ira Rinn as his wife's victim. His public exposure is nothing less than a public humiliation, and his life is destroyed as a result. In both novels, the violation of the Rothian protagonist's privacy becomes an "angry act" that results in a more furious counteract.

Roth had no objection to other writers offering disguised portrayals of him, as can be seen in Janet Hobhouse's posthumously published novel, *The Furies*. Raised in New York City and educated at Oxford, Hobhouse died in 1991 from ovarian cancer at the age of forty-two. *The Furies* includes a brief amorous relationship with a famous writer who, though not identified, is recognizably Roth. One of the characteristics of their affair, the autobiographical protagonist, Helen, admits, was his "obsession with secrecy. There wasn't the slightest gallantry in any of this; it was not my reputation he was trying to protect, but his own comfort" (195). She finds herself "seduced by the dullest thing about him, which was how he actually lived, the monkish habits of his solitude, the grim, even depressive minimalism of his life" (197). Helen respects the way he organizes his life around writing two pages a day and fiercely protects his privacy. Both agree to end their relationship, and the celebrated author "withdrew himself from my life as politely and agilely as he knew how" (204).

"Don't Tell the Children"

One of the most curious details in *Leaving a Doll's House* occurs when Bloom describes Roth's pent-up rage while recovering in a sanitarium from a breakdown. For two hours he pours out his fury over her words and deeds, including his wrath over a remark she had made when his father was dying from a massive tumor. "When his father lost control of his bowels, I said it was a pity he couldn't control himself" (201). He "spewed out every mistake I had ever made," Bloom reports. "Everything that had made him angry over our seventeen years together. Nothing was omitted. Some were ridiculous; some were petty; some, unfortunately were true, like my comment about Herman's bowels" (201–202).

Roth may have objected to Bloom's use of "pity," an ambiguous word that sometimes connotes condescension or disapproval, but he couldn't have objected to her calling attention to this incident because he uses it in *Patrimony*, his 1991 memoir about the last months of his father's life. Indeed, Roth's description of his father's loss of control over his bowels is the most poignant—and problematic—moment in the memoir, one that raises many questions about the ethics of privacy. The memoir opens in 1987 when Herman Roth, eighty-six years old, widowed for six years, begins losing his eyesight, a problem caused by a massive brain tumor that, though benign, is life threatening. Roth mythologizes his father throughout the story: "He could never understand that a capacity for renunciation and iron self-discipline like his own was extraordinary and not an endowment shared by all" (*Patrimony*, 79). The father appears as "all-powerful" (16), the embodiment of "survivorship, survivorhood, survivalism" (125).

Partly as a result of anesthesia taken for a biopsy and lying in bed, Herman Roth is constipated for four days and then suddenly has an explosive evacuation in the bathroom of his son's house. "In a voice as forlorn as any I had ever heard, from him or anyone, he told me what it hadn't been difficult to surmise: 'I beshat myself,' he said" (172). Humiliated, he pleads with his son not to tell "the children" or "Claire," to which his son promises, "Nobody." But of course Roth tells everyone who reads his book. On the penultimate page of *Patrimony* Roth confesses that "in keeping with the unseemliness of my profession, I had been writing all the while he was ill and dying" (237).

Roth's readers won't find it unseemly for him to write a loving memoir of his father, one that verbally brings Herman Roth back to life, where he can be honored and memorialized, but they may find it unseemly to break a solemn promise to a dying man. Roth never offers an explanation for the breach of promise except to say that this is what writers do, betray confidences. He has long insisted that privacy is the "domain of the novelist," as he told Mervyn Rothstein in an interview in the *New York Times* on December 17, 1986. "The serious, merciless invasion of privacy is at the heart of the fiction we value most highly"—and, presumably, at the heart of memoir, too. Janet Malcolm reflects a similar point of view when she argues that the writer "is like the professional burglar, breaking into a house, rifling through certain drawers that he has good reason to think contain the jewelry and money, and triumphantly bearing his loot away" (9). Roth is honest enough to tell us about his broken promise—he could have omitted this detail from the story—but his honesty doesn't make the betrayal less transgressive. In an interview published in *Conversations with Philip Roth*, the writer responded evasively to the question of how his father would have reacted to the broken promise. "'I think about [that question] too,' he says. 'One doesn't want to be sentimental answering it. Well, I don't know what things he might not have liked so much. I don't know, he might not have liked some things. Who could?' he asks. 'Who could? But he's dead. So we needn't speculate'" (272).

The Ethics of Life Writing

Is a writer released from the moral and ethical requirements of honoring a promise to a dying parent after the parent's death? Professional ethics require physicians and psychotherapists to respect confidentiality, but what are the ethics of life writing? What are a writer's moral and ethical responsibilities? These remain haunting questions in all privacy stories, especially those like *Patrimony*. And they are questions without easy answers—or perhaps *any* answers.

G. Thomas Couser observes in *Vulnerable Subjects* that whereas physicians and mental health professionals must protect the confidentiality of their patients and clients, no such regulations constrain creative writers. Nor does Couser call for regulations. He points out that "whether a biographical subject is living or dead would seem to change the ethical standards, as it does the legal ones: one cannot libel a dead person, and the right to privacy is also held to terminate with death" (6). He argues, however, that legality does not always imply ethical propriety; sometimes autobiographical authors must balance the benefits and risks of life writing the way scientists do in their research with vulnerable subjects. "If life writing entails potential harms, such as violation of privacy, are they, can they, be offset by countervailing benefits? What good does life writing do, and whose interests does it finally serve—the subjects', the writers', or the readers'?" (10).

Couser's work reminds us that a deliberate consideration of the ethics of life writing entails weighing competing values: "the desire to tell one's story and the need to protect others, the obligation to truth and the obligations of trust" (198). Couser also reminds us that people become most vulnerable when they are in a state of dependency, such as infancy, old age, or illness. In an intriguing parenthesis, Couser suggests that, from one point of view, death does not qualify as a state of dependency, representing complete invulnerability, but from another point of view, death represents a state of "maximum vulnerability to posthumous misrepresentations because it precludes self-defense" (16).

Couser doesn't discuss Philip Roth, but many of the vexing ethical questions he raises apply to *Patrimony*. Throughout the story Roth insists on the importance of remembering, honoring, and bearing witness. "You mustn't forget anything—that's the inscription on his coat of arms," Roth says about his father. "To be alive, to him, is to be made of memory—to him if a man's not made of memory, he's made of nothing" (*Patrimony*, 124). Like father, like son, seems to be the writer's credo. Later he reminds himself to remember everything about his father: "'I must remember accurately,' I told myself, 'remember everything accurately so that when he is gone I can re-create the father who created me.' *You must not forget anything*" (177). He repeats the injunction in the last sentence of the memoir: "You must not forget anything" (238). For Roth, remembering trumps promising: the writer's challenge is to remember everything and to write everything, good and bad, noble and ignoble. For writers like Roth, writing the truth requires the betrayal of promises.

Remembering and Forgetting

Yet Roth does not convey *all* of the truth, for although he insists that everything in *Patrimony* is true, he omits crucial information about his parents and his childhood that appears in "The Angry Act," information that calls into question the idyllic portrait of his parents and childhood that he dramatizes in the memoir. The passive, ineffectual father and controlling, intrusive, castrating mother who appear in *Portnoy's Complaint* and "The Angry Act" are nowhere to be found in *The Facts* or *Patrimony*. Nor is there any hint of the humiliation arising from the bathing suit incident. "The Angry Act" is the dark countertext of *Patrimony*, as Nathan Zuckerman, the authorial interlocutor, seems to hint at in Roth's earlier memoir, *The Facts*. As in *Patrimony*, Roth presents in *The Facts* a highly edited portrait of his parents. He doesn't omit entirely the son's Oedipal battlefield with his father, but he gives us an idealized portrait of his mother and father, one that is consistent with Tarnopol's version of the truth in *My Life as a Man*.

Indeed, Roth replays in *The Facts* many of the incidents that appear in *My Life as a Man*, including the relationship with his duplicitous first wife, but he never mentions Tarnopol's anger over Spielvogel's breach of confidentiality. Roth's autobiographical account of his relationship with his parents seems too good to be true, Zuckerman implies near the end of *The Facts* when he advises "Roth" not to publish the memoir. "There's an awful lot of loving gentleness in those opening chapters of yours," Zuckerman observes dryly, "a tone of reconciliation that strikes me as suspiciously unsubstantiated and so unlike what you usually do. . . . Are we to believe that this warm, comforting home portrayed there is the home that nurtured the author of *Portnoy's Complaint*?" (165). A few pages later Zuckerman asks "Roth," tauntingly, "Your psychoanalysis you present in barely more than a sentence. I wonder why. Don't you remember, or are the themes too embarrassing?" (169).

Zuckerman is right: Roth does idealize his parents and childhood in both *The Facts* and *Patrimony*, but his question is vague enough—he never refers to Kleinschmidt's case study—so that most readers will not give it a second thought. Zuckerman allows Roth to hint at hidden secrets without actually disclosing them. But our knowledge of "The Angry Act" allows us to see that despite Roth's promise that everything in both memoirs is true, he has omitted the dark countertext of his life. And for a writer who promises, as Roth does in the subtitle of *Patrimony: A True Story*, that he is presenting and re-presenting the whole truth of his family life, or at least enough of the truth that can be conveyed in two memoirs that total 433 pages, readers may conclude that he has committed a breach of narrative trust.

"Nothing is never ironic," Alex Portnoy complains when he reflects on the "literature of my childhood" (93), but there are more ironies in Roth's life and art than he could have imagined when he wrote *Portnoy's Complaint*. The novelist who was victimized by his psychoanalyst's breach of confidentiality in "The Angry Act" later wrote a novel, *My Life as a Man*, that takes full advantage of this violation of trust. It is an obvious irony that "The Angry Act" provoked Roth's furious anger—anyone would have reacted as he did—but it is a more compelling irony that Roth keeps returning ambivalently to this dark countertext, in the guise of his fictional counterself, Nathan Zuckerman, without disclosing the actual case study. The writer who becomes infamous for creating a rebellious, transgressive son in *Portnoy's Complaint*, one of the most shocking American novels of the twentieth century, portrays himself as a loving and devoted son in two of the most acclaimed memoirs of the 1980s and 1990s, *The Facts* and *Patrimony*. *Portnoy's Complaint* demonstrates Kleinschmidt's thesis that art is an aggressive act; *The Facts* and *Patrimony* reveal art as a reparative act.

Biography as Calamity

In an interview with Roger Catlin published in the *Hartford Courant*, Roth, who turned eighty on March 19, 2013, commented on *Philip Roth: Unmasked*, a ninety-minute PBS "American Masters" special adapted from a French television interview. Roth had announced, shortly before the interview, that he was no longer writing fiction and that he had chosen a new biographer, Blake Bailey, to replace Ross Miller, whom the novelist had selected previously to write his biography. Not entirely unmasking himself to Catlin or, for that matter, to anyone else, Roth wryly referred to the major challenges still confronting him. "'In the coming years, I have two great calamities to face, death and biography,' Roth says at the outset of 'Unmasked.' 'Let's hope the first comes first.'"

Roth may not have been joking when he expressed his belief that biography is calamity for the writer. Biography is calamity when a psychoanalyst has breached confidentiality and exposed the writer's innermost wishes and fears to public scrutiny. Biography is calamity when a literary critic, following the writer's leads, further exposes his life to public scrutiny. Biography is calamity when the writer's ex-wife exposes his life to public scrutiny. And biography is calamity when the writer exposes former lovers and himself to public scrutiny.

Despite its title, *Philip Roth Unmasked* is an "authorized" documentary. Roth exerted strong editorial control over every aspect of the documentary by giving a list of approved questions and then his answers to Livia Manera, one of the film's cowriters and codirectors. There is no unmasking in this sanitized documentary, which omits any mention of the many controversies and crises in Roth's life. By contrast, Jane Statlander-Slote's competing, unauthorized documentary, *Out of Newark: The Life and Work of Philip Roth*, is a candid and balanced study that focuses on many of the troubling aspects of Roth's life, including the significance of "The Angry Act." Susan Lacey, of WNET, Channel 13, interviewed Statlander-Slote about her documentary, which was shown in Newark during Roth's eightieth birthday. "Philip asked me," Lacey told Statlander-Slote, "if I knew anything about this 'other' film," referring to *Out of Newark*. Lacey told him that she hadn't yet seen it. "Do you think the film will be defamatory?" Roth then asked her. The question startled Lacey, who recounted it to Statlander-Slote. "It was at this point," the latter reports, that she and Lacey burst out laughing when they reflected on Roth's question. The two women "both enjoyed the unbelievable irony of that comment-of-the-century remark" (personal communication, August 14, 2013). One can only speculate on how the young Roth, who elevated character defamation into high art in *Portnoy's Complaint*, would comment on his older self.

What shall we say, finally, about the impact of Dr. Kleinschmidt's breach of confidentiality on Philip Roth's life and art? Acutely sensitive to having his story appropriated without permission by his psychoanalyst, Roth remains equally insensitive or indifferent to the anger and indignation of those who find their own words and deeds appropriated, without permission, in his fictional and nonfictional writings. Like the unapologetic Kleinschmidt, he claims that nothing is off limits to the creative writer in service of art, including a promise of confidentiality to a dying father. Roth may believe that a novelist's success depends on *de*narcissizing himself, yet his experience with psychoanalysis appears to have heightened and hardened his narcissistic defenses, as Kleinschmidt hoped would happen. In the process of becoming "well defended," as Adam Gopnik reports his psychoanalyst saying about Susan Sontag, Roth has kept a tight rein on his empathy. Throughout his long and distinguished career, Roth has insisted that his ultimate dedication is to art, crafting stories of abiding truth, yet even when he claims full transparency in his memoiristic writings, "The Angry Act" remains in the shadowy background, a demon voice that cannot be silenced.

Take-home lesson: Like a course of extended psychotherapy itself, a privacy breach, however inadvertent, can have a lasting and profound effect on a patient's future life and artistic productions.

6. The Case of *Jane Doe v. Joan Roe and Peter Poe:* The Most Extensive Violation Ever of a Psychotherapy Patient's Privacy

What to look for: Is it permissible for a psychotherapist to publish a book containing confidential information about a patient's psychotherapy sessions without the patient's permission?

A reader encountering the 1973 book *In Search of a Response* for the first time cannot help being impressed by its imposing physical appearance. A huge tome, much larger than a typical hardcover book in its weight and dimensions, *In Search of a Response* weighs in at about 4.5 pounds and is as thick as a large city's phone book. Overall, it resembles a ponderous medical or technical textbook of that era. The book's heft is reinforced by a cover illustration consisting entirely of an enormous photomicrograph of a nerve synapse, giving the book the aura of a scientific publication. *In Search of a Response* constitutes what is almost certainly the most egregious violation of patients' privacy in the history of psychotherapy.

In retrospect, the title of the book could not be more ironic, for the case study provoked plenty of responses, none of which the authors imagined or desired. Like "The Angry Act," the publication of *In Search of a Response* triggered a furious counteract, in the form of a lawsuit involving appeals and cross-appeals. The psychiatrist Leida Berg and her psychologist husband, Harold Steinberg, claimed to have received permission to publish a case study about their two patients, "Helena" and "Henry," who were married to each other when they began treatment. According to Helena, however, neither she nor her ex-husband, who died two years before the book was published, granted such permission. When Helena discovered that the book was published without permission, she immediately sought an injunction to block distribution, mainly because she feared the authors did not disguise her identity sufficiently.

In Search of a Response represents a different kind of breach of confidentiality from the one we saw in "The Angry Act." Neither Philip Roth nor Dr. Kleinschmidt knew each was writing about the other. It's true that Kleinschmidt never apologized for the breach, never took responsibility for the violation of medical responsibility, and never tried to understand his patient's dismay, but despite his insensitivity to Roth's justifiable outrage, the psychiatrist never would have published the case study if he suspected the novelist was writing about the same material. By contrast, not only did Dr. Berg publish *In Search of a Response*, but, in light of the patient's subsequent protests, she also used every legal maneuver to allow the book to be sold. There is never any recognition, rueful or otherwise, that the psychiatrist is betraying her responsibility to her patients and profession.

Unlike Kleinschmidt's breach of confidentiality, which many contemporary analysts have heard of, there are few references to *In Search of a Response* in the psychoanalytic literature—though used copies of the book can be easily bought on Amazon. Glen Gabbard, for example, does not mention the book in "Disguise or Consent," but the case study certainly illustrates his warnings about the host of potential problems arising from a psychiatrist's decision to write about a patient. Nor does Judy Leopold Kantrowitz refer to *In Search of a Response* in her book *Writing About Patients: Responsibilities, Risks, and Ramifications.*

In Search of a Response is longer than nearly all of Freud's case studies *combined*, yet it breaks no new ground. Only 220 copies of the book were sold before a judge issued a temporary restraining order. The book engendered little or no interest as a psychiatric case study, but the case of *Jane Doe v. Joan Roe and Peter Poe* aroused extraordinary legal interest in state and federal courts. The case focused on competing claims: a patient's right to privacy versus a therapist's right to publish a scientific work. *In Search of a Response* may well be the longest psychiatric case study ever published, and it raises many troubling privacy issues.

Our chapter focuses on the way in which this violation of patient privacy came about and on one woman's legal struggle, eventually reaching the U.S. Supreme Court, to find a rationale to redress this shocking betrayal of her trust. First, however, we need to understand the cast of characters in the story: the patients whose confidences were violated; the psychiatrist who ignored their privacy concerns; the details of the therapy, including what was revealed and concealed; and the bitter aftermath of the lawsuit. The legal and psychiatric dimensions of the case are fascinating, but it's the human story that compels our interest.

The Affidavit of Harriet G. Werner

With the help of an article published in the *New York Times* on March 16, 1973, as well as the affidavit of Harriet G. Werner ("Helena") and other documents, we can now piece together the patients' identities. From 1956 through 1963, Harriet Werner "underwent continual psychoanalysis and psychotherapy" with Dr. Berg, seeing her usually three times a week. She shared with her psychiatrist "virtually every intimate detail" of her life and those of her children and her then husband, Steven L. Werner, called "Henry" in the case study. Beginning in 1956 or early 1957, her husband, who later became an attorney, also went into therapy with Dr. Berg. Their son, Frank ("Jan" in the case study), went into therapy, mainly with Harold Steinberg. The Werners divorced in 1963, and in the same year Steven Werner married Edith Lond Fisch, also a lawyer (called "Fay" in the case study). The major characters of this story are now deceased. Steven Werner died in 1971, at age sixty-two; Dr. Berg died in 1985, at age eighty-six; Edith Lond Fisch died in 2006, at age eighty-three; and Harriet Werner died in 2010, at age eighty-seven. The only character still alive is Frank Werner.

The main problem with *In Search of a Response* is that Dr. Berg failed to disguise her patients' identities, doing little more than changing their names. Many identifying details remain: Helena was a therapist; Henry went to Harvard Law School, dropped out to become a writer, returned to law school to complete his degree, and later married a lawyer after divorcing Helena. The book discloses the patients' ages and Jewish backgrounds. Nor is this all. We learn that each had been previously married and that their "autistic" son had an unusual talent for music. The book also discloses that the patients saw Dr. Berg (who used her real name) in her Manhattan office and also at her country home in Katonah, New York.

A Prodigious Case Study

In Search of a Response consists of several sections, all narrated from the first-person-singular point of view. The introduction, five pages long, is the only section written by Harold Steinberg, who refers repeatedly to his wife and coauthor as "my Leida." The first reference to "my Leida" sounds endearing, but the next seven references (including an eighth, when he refers to her as "my beautiful Leida") are cloying. Next comes "Diary of Jan," ten pages long, in which Leida Berg tries to imagine her patients' child from the moment he is born until he arrives home from the hospital. (There is no mention of the Werners' other child in the case study.) This is the only section of the book that is entirely fictional. The two major sections of *In Search of a Response* are "Helena," 354 pages long, and "Henry," a staggering 613 pages long. The five-page conclusion, "Exegesis," gives an all-too-brief and misleading history of how the book came into existence. Curiously, the pagination is not consecutive: after finishing "Helena" on page 354, the reader begins "Henry" on page 1. All together, *In Search of a Response* is just short of a thousand pages long, with a mind-numbing 1,716 footnotes. (Amazon incorrectly describes the book as 613 pages long.)

How can a case study be so long? Neither Leida Berg nor Harold Steinberg addresses this question. Instead, we are told in the introduction that in January 1960 Dr. Berg was asked to present a paper to parents whose children attended a school for schizophrenic children. She initially wrote a paper entitled "The Relationships of a Disturbed Child," but following her husband's suggestion, she changed it to "Diary of Jan," a "composite of one specific little boy's experiences together with the accounts of experiences of infinite numbers of patients" (introduction, vii). The presentation understandably provoked outrage among the parents in the audience, who believed that the authors were unfairly blaming them for their children's illnesses.

"A Hornet's Nest"

That response distressed Steinberg but not his unflappable wife.

> I, myself, felt quite confused and discouraged by this experience. We had stirred up a hornet's nest. I slept badly that night—I felt tired and disillusioned. Next morning at breakfast my Leida, however, was bright, fresh and stimulated. "Time to do a diary of the parents," she announced. In response to this I groaned. But my Leida did produce a diary of the parents. *Helena* and *Henry* (these, of course, are not their real names) are the notes, reproduced as closely as possible, of hundreds of therapeutic sessions with a husband and wife who were the parents of that one special little boy to whom we shall refer as Jan. (introduction, viii)

Steinberg never explains how they chose the title of their book, but it's noteworthy that his first "response" to his wife's book proposal was a "groan."

Dr. Berg began seeing Jan when he was seven months old, when his mother brought him to the therapists' country home. The infant struck

Dr. Berg and Steinberg as disconnected from everyone, including his mother. Hence, we infer, the meaning of the title, *In Search of a Response*: Jan's need to evoke emotional responses from a family who cannot easily express their feelings and who remain disengaged from one another and his own need to respond to and engage with the world. Though Dr. Berg had never worked with children before, she began treating Jan and his troubled parents. Dr. Berg must have taken a special interest in Helena, who was herself a therapist. (Harriet Werner was a university social worker who later became a psychoanalyst.) Her therapy with Dr. Berg lasted for seven years, though this detail is omitted from the case study. Dr. Berg states cryptically in "Exegesis" that the book "was written 12 to 13 years ago but, because of certain personal problems, we didn't publish it at that time. Now that we come upon it again we decided to publish this wealth of material and offer it to the public as a contribution and enrichment of understanding." What were the "personal problems" that delayed publication for more than a decade? Dr. Berg doesn't say. Nor does she explain why they finally decided to publish the book.

Unlike Freud's case studies, where we hear him talking with his patients, engaging in conversations that appear lively, spontaneous, and interactive, "Helena" and "Henry" are long monologues interrupted only by Dr. Berg's brief questions that are placed in parentheses. The therapeutic sessions have no beginnings or endings, and we don't know how much time passes from one session to another. To understand how Dr. Berg feels during the therapy sessions, we must read the footnotes. After hundreds of pages chronicling the patients' anger, mistrust, sadness, and disillusionment, the reader is tempted simply to read the footnotes, which offer a distillation of the patients' and analyst's feelings.

The footnotes themselves are uneven. Some are compelling; others are either obvious or unnecessary; still others reveal analytic interpretations that are not always persuasive. Helena and Henry occasionally comment on how they feel about their therapy sessions being published as a book, but, maddeningly, Dr. Berg does not elaborate on the significance of these statements. Clinical boundaries did not exist for Leida Berg and Harold Steinberg, who invite Helena, Henry, and Jan to visit them in their country home even when the therapists are on vacation. In addition, the therapists encounter their patients at various social gatherings. To complicate the situation, Helena and Henry see Dr. Berg in individual therapy, but they also participate in group therapy led by Steinberg. These boundary transgressions make the therapies even more problematic.

Leida Berg

Leida Berg reveals little about herself in the book. She was not exactly the "blank screen" that Freud urged psychoanalysts to be for their patients, but she didn't believe that psychiatrists should share aspects of their lives with patients—or readers. We know little about her apart from the fact that she received her medical degree from the University of Tartu, Estonia, in 1925 and that she was licensed to practice medicine in New York in 1943. According to the information she supplied to the New York Court of Appeals, she worked "under the direction of Dr. A. A. Brill" from 1933 to 1940 and was connected with Bellevue Hospital in the mid-1940s, New York University from 1945 to 1947, and Jewish Family Services beginning in 1944. She also claimed to have been connected with the Columbia University Psychoanalytic Clinic from 1944 to 1946, but the records and archive staff of the clinic are unable to find any record of her having been at the Columbia Institute. Finally, she did claim to at least one patient that she herself had been a psychoanalytic patient of Dr. Abram Kardiner, one of the four founders of the Columbia Institute. She evidently had some differences with Dr. Kardiner, and she eventually abandoned any interest in psychoanalysis. Kardiner himself had been psychoanalyzed by Freud, a relationship that would have allowed Berg, as Kardiner's patient, to trace her psychoanalytic lineage to a mere two removes from the Great Man himself. *In Search of a Response* was her only book.

Unlike most psychoanalysts, then and now, Dr. Berg apparently did not believe it was useful for patients to understand their transference toward analysts or for analysts to understand their countertransference toward patients. Freud insisted that both transference and countertransference were central to psychoanalysis, distinguishing psychoanalysis from other forms of psychotherapy. The patient sees in the analyst, Freud writes in *An Outline of Psycho-Analysis*, "the return, the reincarnation, of some important figure out of his childhood or past, and consequently transfers on to him feelings and reactions which undoubtedly applied to this prototype" (174). The psychic mechanism behind transference is projection, in which an anxiety-provoking emotion or idea is displaced onto another person. Countertransference is the term Freud used to describe the analyst's largely unconscious projections onto the patient.

What's striking about Dr. Berg's treatment of Helena and Henry is her insistence that they do not view her as a transference figure. As a therapist herself, Helena recognizes her tendency to project her feelings toward her mother onto Dr. Berg, but the analyst discourages her from doing so. In Helena's words: "You used to say it clearly, 'No transference. You must divorce yourself from the image of mother.' But the fact of the matter is that you are mother, but you don't let me have a distorted image of mother. That's the way you stay m–o–t–h–e–r. You don't stand for the pathology." The analyst elaborates on this in a footnote. "Whenever she started to act toward me as if I were her mother, I would say to her, 'Who am I? Out of what do you feel this? What is reality?' In this way I discouraged all transference and demanded that she relate to me out of what was real. This is the connection with the world that I helped her to establish" ("Helena," 144).

It was not common in the 1970s for psychoanalysts to write articles about countertransference or to discuss it with their patients, but Dr. Berg never analyzes why she believed readers would be interested in a nearly thousand-page case study. The authors' claim that their intention was "to further the state of knowledge concerning the treatment and cure of schizophrenia" (Affidavit of Harriet G. Werner, 64) touches only the surface of their motivation.

Helena: A Sympathetic Character

Unlike most case studies, *In Search of a Response* lacks a coherent theoretical or clinical thesis. It also lacks a statement of the clinician's approach, traditional or innovative, to individual and family psychotherapy. The book offers, instead, a therapeutic "slice of life," or a slice of therapeutic life, that captures many of the prevailing psychiatric assumptions of the early 1960s. Helena, despite her many problems, including being traumatized by her father's early suicide (in real life Harriet Werner's father had attempted suicide) and verbally and physically abused by her husband, remains a sympathetic character even when she loses self-control, as she describes in the following family crisis. "Made dinner for Henry and myself—I yelled at Jan. Henry came in and said, 'What are you doing to the child?' I became furious—threw the whole supper on the floor. I was so mean to Henry. 'Do you want me to kill you?' he asked. I said, 'I don't care.' He started to choke me—I bit his finger. I said, 'You take care of Jan.' Jan came out and said, 'No, Mommy, you take care of me.' Then I wanted to go to sleep. If only I could vanish! I am sick and tired of it! Everything is on me" ("Helena" 121).

Dr. Berg diagnoses Helena as suffering from "postpartum psychosis" ("Helena," 18, 43). There is much evidence to suggest that she suffered from postpartum depression but not from psychosis. She appears severely depressed after the birth of her son, and she becomes more attuned to her feelings as a result of therapy. "Gradually I am aware of my hostility to Jan. I experience myself as a mean mother" (129). There are many sources of her depression. She dates the end of her childhood to her father's suicide. Dr. Berg informs us in a footnote about the traumatic details of the suicide, and we have no way of knowing how much she embellished this event. "Father lay there for half an hour and mother waited for her to come home as if she herself were paralyzed, incapable of action" (58). One can begin to understand the depth of Helena's horror when her husband threatens to kill himself almost on a daily basis. Less convincing is Dr. Berg's diagnosis of Helena as a schizophrenic. The analyst uses the word not as contemporary therapists would, to signify a serious thought disorder, but as a metaphor for disconnection. There is no doubt that Helena feels disconnected—from her mother, husband, son, herself, and life itself—but she never displays the symptoms of schizophrenia.

Henry: "Infinite Wrath, and Infinite Despair"

Dr. Berg uses a quotation from John Milton's poem *Paradise Lost* as the epigraph to her case study on Henry: "Infinite wrath, and infinite despair? / Which way I fly is Hell; myself am Hell." Henry is, like Milton's Satan, filled with wrath along with intense mistrust. He feels massive rage toward everyone in his life, including himself. Midway through the case study he refers to his extensive history of psychiatric treatment.

> I saw a psychiatrist long ago; he told me to go to a psychiatric hospital. There was no extreme of emergency of any symptoms in me. I was very unsophisticated about clinics and psychiatry. I entered a psychiatric hospital. I never saw their diagnosis until years later when my [first] wife sued me for more alimony—paranoid, schizophrenia, unimproved. The recommendation was that I get a job and get treatment, not analysis. I'd rather be sick and schizophrenic and everything else and have money and do nothing. ("Henry," 357)

Reading *In Search of a Response*, Helena calls Dr. Berg the "the Boswell of Schizophrenia" ("Helena," 211), implying that she meticulously documents every detail of her patients' lives. Though he may not be schizophrenic, Henry strikes us as deeply disturbed and violent throughout the book. He is not lacking in insight—in many ways he is brilliant. He mentions studying philosophy at an Ivy League university. He quotes philosophers, psychoanalysts (he had therapy with a disciple of Wilhelm Reich), and novelists like Henry James and James Joyce (he cites Molly Bloom's monologue at the end of *Ulysses*). He also takes pleasure in paradox, as when he opines, sounding like Oscar Wilde, that "the only thing that is real is my phoniness" ("Henry," 508). Rarely can Henry apply intellectual insights to his own life, however. He observes at the beginning of his case study that "Most of the things I don't like in Helena might turn out to be my own difficulties," adding, ominously, "I will kill myself when I have reached a certain level financially—because I am trapped in a miserable situation. It is a frightening situation—I have passed the point of no return" (2). He knows that his suicide threats are a form of manipulation, but that doesn't prevent him from making additional threats. "When I talk about killing myself, I am testing everybody and myself too—the strength of my own negativism—how bad everything really is" (49). He dares others to encourage him to kill himself, just as he taunts his analyst to order him to harm his wife. "If you told me to murder Helena, I would be hopeful."

There must have been times when Dr. Berg wondered whether Henry would act out his anger toward her. During one session he acknowledges his murderous rage. "I feel I would like to destroy you—I feel you are destructive to me. I don't know whether I feel it or think it. I am impressed by the intensity of my desire to kill you." His violence here may seem mainly rhetorical, but Dr. Berg's footnote indicates she takes it seriously. "At this point his body was poised as if he were ready to pounce on me—his eyes were ferocious, his face reddened. He was responding from the depths of his very core—the intensity of feeling was transmitted into his nervous system and his motor apparatus. I felt he could kill me. There was a pause and he slumped back in his seat and continued talking" ("Henry," 183–184).

How was Dr. Berg able to treat a patient as violent as Henry? He not only threatens violence but also acts on it. On one occasion, he refers obliquely, with mordant understatement, to an actual attempt to murder his wife at a party attended by the two therapists. "Too bad we had to go home so early from K's. We got a million dollars worth of psychiatry from you and Harold. There was such clarity." Without Dr. Berg's footnote, the clarity would be lost to the reader. "He is referring to an incident at the home of a friend. He became very hostile and abusive. He acted out murder, with a knife at Helena's throat, and tried to jump out of the window" ("Henry," 345).

Nor is this Henry's only act of violence. During group therapy he hits one of Steinberg's patients. "We all knew that he hit Ralph because of his jealousy," Dr. Berg informs us, and when she asks Henry whether he was acting "in full awareness" of himself, he replies in the negative ("Henry," 418). During a therapy session with Dr. Berg, he "knocks over the chair, throws the ashtray at the radiator, smashing it to bits, continues to bang, all this accompanied by menacing grimaces and noises." Dr. Berg's reaction? "This demonstration did not frighten me at all. He was manifesting power by smashing objects. He no longer wants to kill or commit suicide" (492).

How can Dr. Berg be so confident that Henry no longer wants to kill or commit suicide? His violence continues, as we soon discover. "I started Sunday morning by beating up Helena, I felt that she was doing something terrible to me, that she could annihilate me and I had to kill her. She is out to kill. I don't give her love and sex and then this female destructiveness comes out" ("Henry," 538). Dr. Berg doesn't footnote this moment in the case study, nor does she comment on his most shocking murderous act, one he committed before entering therapy with her. "Many years back when I got a job, I had a dog that was my constant companion for many years. The only creature that loved me and whom I loved. I virtually murdered this dog. I left the dog alone—I couldn't be bothered with it. One very hot day it suffocated in my car" (205).

A Jewish anti-Semite, Henry realizes that his prejudice is a "mixture of truth and fantasy," yet he maintains that most of his "negative accusations" against Jews are accurate ("Henry," 80). He identifies with "superior" German Jews and counteridentifies with "inferior" Eastern European Jews. "Germans had to protect themselves from Jews by putting them into gas chambers" (63). He knows his anti-Semitism is a projection of self-hatred. "So many things I say about Jews I am saying about myself" (170). Nevertheless, like his recognition that what he despises in Helena is self-projection, he cannot let go of his anti-Semitism. What value is self-awareness if one cannot act upon it? Henry cannot stop spewing anti-Semitic venom, believing that "Yiddish is a corruption of German" (361), that the "only nutty thing about Eichmann was that he went out to find Jews" (363), and that Helena was "making a Jew" out of Jan by "encouraging him to maintain his infantile feelings" (363).

An attorney, Henry finds the mental health and legal professions incompatible. "Psychiatry and law can never find a common ground" ("Henry," 95). Emotions are dangerous to him, mainly because his own emotions are (self-)destructive. Believing that rage has characterized his entire life, he asserts that psychiatry represents a false hope. "What happened ten or twelve years ago, I got a lot of money and I got into psychiatry—I thought I could evade life by psychiatry. I was much better before I started having those hopes" (355). He mistrusts all psychiatrists, including Dr. Berg, for several reasons: they demand all of his money, leaving him destitute and powerless; they talk in bewildering abstractions; they make promises they cannot fulfill; and they always lie. "I hate you," he tells Berg. "If you gave me everything, I would still wait for the hook in the bait. Everybody ought to be allowed to plead the Fifth Amendment about their feelings. Not a damn thing is real, big laugh! That's where I can die. Everything is a lie. Nothing exists. All the Jews know it. I am still schizophrenic" (373).

Helena: "The Book Hurts Me Too Much"

After informing Henry and Helena of her intention to publish a book about them, Dr. Berg gave each a copy of the manuscript. She describes her patients' initial objections to the publication of *In Search of a Response*, but we are never sure whether she adequately conveys the intensity of their opposition. Nor do we know from the case study whether their objections decreased over time, as she implies. Dr. Berg's first comment about the manuscript appears in a footnote when she explains that before she and her husband published the book, they "felt it was imperative that Helena and Henry read it through. They read their own and each other's productions" ("Helena," 211). She tells us about their reactions to reading their own and their spouse's case studies, but she doesn't tell us about the promises she made to them prior to giving them a copy of the manuscript. Harriet Werner states in her affidavit that she and her husband had been in therapy for five years when Dr. Berg informed them of her wish to write about them, but she obviously had begun writing about them near the beginning of therapy. What made Dr. Berg think that Helena and Henry would agree to her request?

There are other questions. Does a therapist's request to write about her patients while they are still in therapy represent coercion? Did Helena and Henry have the right to demand additional disguises? Were they told how the book would be marketed and distributed? Would they share in the book's profits? Did Helena worry that she would be recognized by other therapists and perhaps even by her own patients? Did Dr. Berg fear that Jan would be harmed when he eventually read the case study of his parents and himself? Dr. Berg discusses none of these questions.

Instead, we are given only a few of the patients' reactions to the manu-script, none of which Dr. Berg interprets in depth. Helena reports being "scared of the depth and extent to Henry's disorganization." When she asks the analyst to explain "what this core is that you talk about in your foot-notes," a reference to the "core of schizophrenia," Dr. Berg replies, "It is the beauty, the warmth, the love that is covered up with the distrust, the rage, the ugliness" ("Helena," 211). We see none of this beauty, warmth, or love, however. A page later Dr. Berg writes that "Helena learned a great deal from reading about her own treatment." Helena then reveals her fan-tasy of "writing a book like Schechehaye's [sic] Renee did," a reference to *Autobiography of a Schizophrenic Girl: The True Story of Renee*, first published in 1951 by the French analyst Marguerite Sechehaye (1887–1964).

Reading Dr. Berg's manuscript is troubling for Helena, heightening her feelings of competition and jealousy. It is painful reading, she tells Dr. Berg, and then explains why: "you are not committed to me, but to Henry. Maybe I should go to somebody else" (214). She believes that Dr. Berg favors Henry over her, a fear that may arise from the fact that his case study is twice as long as her own. In her final reference to the book, Helena states that she became "terribly upset after Sunday. I felt I was left out. Harold and you concentrated on Henry. I was just an accessory. I came away feeling that it was a struggle between you and Henry and that I was excluded. Am I really gifted or am I a figment of your omnipotence where Henry is concerned. Then I thought, why this mistrust? You are all for Henry." Dr. Berg doesn't respond to any of these questions, telling us only that "On this Sunday night Henry and Helena were over to discuss publication of this book" ("Helena," 229).

Henry: "I Am Angered by the Book"

Unlike Helena, who never expresses any privacy concerns about the pub-lication of *In Search of a Response*, Henry immediately raises an objection upon reading the manuscript. "Helena is very upset. I am upset too. I am angered by the book. It adds a new dimension to our relationship. There is some element of privacy in the relationship that has been destroyed. A cold-blooded objectivity enters into it, an evaluation at a distance. At the beginning I felt like being watched through a microscope" ("Henry," 421). Tellingly, Dr. Berg never responds to any of his accusations. Does she be-lieve that his arguments have no merit and, therefore, need not elicit a response from her? Or does she believe that Henry's arguments have too much merit and, therefore, cannot be countered?

Later we hear Henry's angry outburst following a meeting the night before when Dr. Berg and Steinberg tried to obtain a "release" for the publication of the case studies.

> Who do you think you are, writing a book? You think that by your artful, wiley [*sic*], dishonest, political intriguing—you get anything done. Your schizophrenia empire, 1,000 jet powered, will work free for you. My life has been wrecked by you, you lousy stinking analyst! The treachery of it, to make such a book. All these years people have been struggling to write a book and you come along and think that the book writes itself. You are a mad scientist. The mad scientist doesn't know he is mad. The way he cuts people up in a bland way. "And now to the torture machine," you say. ("Henry," 438)

Dr. Berg makes no attempt to disentangle Henry's legitimate objections from the "paranoid" ones. The legitimate criticisms are that, without adequate disguises, readers will discover Henry's and Helena's real identities, a discovery that would then seriously harm their personal and professional lives. Anyone who is deeply mistrustful by nature, as Henry is, will have this fear. But others will have this fear as well! The "paranoid" objections, or perhaps, more accurately, the "grandiose" objections, are that through the publication of her book Dr. Berg will expand her "schizophrenic empire."

Henry's implacable opposition is clear. "I don't really feel that I ever will let you publish this book. The feeling is that the only hold I have on you is the book. I feel the magic or compulsion of holding on to treatment. When I say that I will not let you publish the book—it means that I will never be secure" ("Henry," 440). A few pages later he adds, "When the world crushes me you can publish your stinking book" (443).

One of Henry's most intriguing dreams follows his attack upon Dr. Berg's book. He dreams he is making love to a woman with three eyes—one in the center of her forehead.

> I said, "You are very beautiful." She didn't believe that anyone could love a woman with three eyes. I looked at her again and she wasn't so beautiful. She had a veil to cover it up. I was kissing her and had sex with her—this was in a room like the room my brother and I occupied when

we were young. Someone else—a child in the room. I was somewhat self-conscious because of this presence, and then other people too. I was also aware that Helena would come home and I would be accused of adultery. I went ahead just the same. She did come home—the dream ended. ("Henry," 447)

It's surprising that Dr. Berg doesn't question Henry about the significance of this dream or interpret it in a footnote, especially since he admits to her that the woman in the dream is his psychiatrist. The woman with three eyes may symbolize the analyst who is watching him with the eye of a writer. If so, her veil may suggest the hidden motivation behind her need to write about Helena and Henry.

Henry's last comment about the book occurs when he finds himself in dire economic straits, unable to pay any of his bills: "As soon as I am financially independent I'll let you publish the book" (489). He never explains why he has overcome his many objections to publication.

The End of Therapy

"Henry" continues for another one hundred and fifty pages, but there are no further references to the publication of the book. The rest of the case study describes his divorce from Helena, his growing relationship with Fay, whom he eventually marries, and the end of treatment.

Were Helena and Henry helped by therapy? Helena recognizes, as her case study draws to a close, that she can survive without Henry. She misses sleeping with someone, but she is hopeful about the future. We learn, in Dr. Berg's final footnote, that she sees Helena only occasionally to discuss her work. "Each time, however, she brings me up to date about her own emotional state, especially if and when it has an effect on her work" ("Helena," 354).

Henry's recovery is more problematic than Helena's. Sometimes he suggests that the price of recovery is the loss of creativity. "I no longer feel schizophrenic. But I feel that there is no fire. I am integrated—but what for? I'd much rather have a crazy drive even if the fire doesn't break through. The schizophrenia was the fuel that kept me going" ("Henry," 474). Other times he makes characteristically negative statements: "The only way life can be for me is to the extent that I could enforce it upon the world—if the world doesn't acquiesce I can kill the world. I can have an effect in the world by not being in it." Dr. Berg then claims, unconvincingly, that these comments represent his old point of view in which he no longer believes. She tells us in a footnote at the end of the case study that she has seen Henry on several occasions outside of therapeutic sessions and that he "seems integrated and doing well in all areas of his life" (613). According to court records, Henry continued to see Dr. Berg until his death, suggesting that he never believed his treatment was over.

In what reads like a postscript to the case study, Harold Steinberg reports that "some time ago Fay called to inform us that Henry had died of a coronary thrombosis." The widow then visits Dr. Berg and Steinberg to bring them up to date about Henry. "Fay wanted us to know how happy she had been with Henry and how much Henry had contributed to and enriched the lives of everybody who knew him—especially Fay." It's hard to believe that Henry was able to overcome the intense anger, mistrust, and cynicism that pervade the case study. Fay claims that she and her husband were "very successful in their mutual endeavors." This, too, is difficult to believe. Fay brings Henry's copy of the manuscript, which, she says, he read from time to time. "Fay knew how important it had been to him—how lovingly he handled it." When Steinberg asks her, "What shall I do with it?" she replies, "Publish it!"

Are we to assume that Henry would have agreed with Fay's decision to publish the book? If so, what made an act of "treachery" acceptable? Why didn't he communicate his change of mind directly to Dr. Berg? What were Fay's motives for allowing the book to be published? Henry makes many hateful comments about Fay in the case study; why, then, wasn't she worried about the nasty characterization and potential loss of privacy?

Fay

The historical Fay adds another element of complexity to the story. Born in 1923, Edith Lond Fisch was stricken with polio at the age of twelve, and despite three years of treatment, she was confined to a wheelchair for the rest of her life. Illness did not prevent her from leading a noteworthy life. According to Dorothy Thomas, Fisch was the first physically disabled woman in New York City to obtain a high school "academic" diploma. She was the first person to earn all three degrees offered by Columbia University Law School, and she became the first female professor of law in New York. "Lawyers and judges who know only her writings and then meet this lively wheel-chair bound woman for the first time are stunned," adds Thomas. "Inevitably, they shake their heads and say, 'Remarkable woman.'"

Edith Lond Fisch was married to Steven Werner for nine years until his death. Most of that time, however, took place after he stopped seeing Dr. Berg on a regular basis. The author of several books on law, Edith may have felt that appearing as Henry's new wife in *In Search of a Response* represented one more notable achievement. Yet questions remain about her marriage to Steven. If Henry was physically abusive to Helena, as Dr. Berg suggests repeatedly in the case study, was he abusive to the real Fay? And if he was virulently anti-Semitic, as Dr. Berg reports, how did he feel being married to a woman who, as Thomas points out, was "ardently proud of being a Jew"?

Leida Berg's Other Patients—and the Real Jan

Three reviews of *In Search of a Response* posted on Amazon give us more information about Leida Berg and an update on the "real" Jan. In a February 19, 2003, review, Neil Friedman identifies himself as one of Dr. Berg's patients in the 1970s and calls her the "most unforgettable" person in his life. "This book shows a little of her at work. You can sometimes feel her fierce independence and honesty, her stripping away of conventional realities. She was an existential analyst originally from Estonia who developed her own way of doing therapy. It is to get a glimpse, a taste, a smell of her that I recommend this book."

Friedman expands on his portrait of Leida Berg in his 2005 book *Remarkable Psychotherapeutic Experiences*, stating that she had been analyzed by Abram Kardiner and that she began her career as a psychoanalytically oriented psychiatrist. "She was not effusive about analysis. 'Kardiner liked me more than I liked him,' she told me. At some point, she said, she had thrown out her psychoanalytic text books and become what I would call an existential analyst" (15). Friedman calls the husband-and-wife team of Berg and Steinberg "Harold and Maude," a reference to the 1971 film directed by Hal Ashby. Friedman portrays Berg as a "handsome older woman, diminutive in size, a real powerhouse. She dressed always in what to my lower-middle-class eyes looked like 'opera' clothes. She was elegant" (15). Dr. Berg was always blunt and outspoken, demanding Friedman to express his bottled-up rage. "I expected a warm, supportive, sympathetic ear. I was expecting to gain insight into the 'whys' of my misery. I was in for a very rude shock. Leida broke every implicit expectation about psychotherapy that I held" (18). Friedman's therapy lasted for three years, and he often asked himself why he continued to see her. The answer, he admits, is because he found her fascinating. She comes across as a maverick analyst who was less interested in the past than in the present and who insisted that her patients "live intensely!" (35). Friedman now believes she was, "for her own reasons, focused on anger-interpretations a bit too much" (98), an observation with which readers of *In Search of a Response* would probably agree.

Friedman's description of Dr. Berg's penthouse Park Avenue apartment on the corner of Eighty-Fifth Street evokes old-world affluence and comfort. "The building had a doorman and a special elevator to her floor. Of her apartment I remember Oriental rugs and modern paintings" (15). She seemed to take her own advice and lived intensely—and elegantly. Money was apparently not a motive behind her decision to publish *In Search of a Response*. Nor did she lack self-confidence. Two years into therapy, Friedman found himself becoming increasingly assertive, ready to exchange roles and analyze the analyst. To his question whether she was satisfied with anyone in her life, she pointed to herself, suggesting the extent of her self-esteem.

In a review posted on May 13, 2011, Janet Holloway also portrays Dr. Berg as a bold, iconoclastic therapist. "I was a patient of Leida Berg in the mid-70's," Holloway writes. "An extraordinary person in every way, Dr. Berg cut to the bone of truth in every comment. One example is when I was lamenting the end of my marriage, wallowing in despair that my husband admitted he was gay but I still loved him, she practically shouted, 'Garbage! How can you love someone who lied to you and maybe himself for so many years? There was no truth from the beginning! Now, talk about something else.'" Holloway admits that Dr. Berg's observation changed her life. "She opened my eyes to truths I might never have discovered on my own. This book shows how Leida connected with her patients on the deepest, most truthful level, challenging you to find the courage to take the next step."

The third review, posted on February 16, 2012, is written by Frank Werner, who begins by stating, "I am the little boy known as 'Jan.'" After briefly summarizing the court case, he offers his opinion of *In Search of a Response*. "I really feel this book is a piece of trash. It is sensationally written and seems to have been fictionalized. Certainly 'The Diary of Jan' [is] total fiction." His most interesting revelation is that the woman his father married after he divorced Helena, Fay, "once called me a monster to other adults." He concludes by stating he would like to speak to Friedman and Holloway and, if possible, with Steinberg "if he's alive."

Frank Werner's Point of View

We were able to contact Frank Werner through that Amazon posting, subsequently edited by him, and he was kind enough to read an early draft of our chapter and respond to our questions. He offered much new information about *In Search of a Response*. In the introduction, Harold Steinberg misleadingly refers to the "notes" of the case study, "reproduced as closely as possible" (viii). The truth was that Leida Berg and Harold Steinberg tape recorded all the therapy sessions, a detail never mentioned in the book or court documents. "They had a reel-to-reel tape in their office, like the type Nixon used, which recorded all sessions."

Frank recalls his mother telling him that she felt the book was a form of "stealing" of her and her husband's words for the authors' use and exploitation. He listed the minor fictional changes the therapists made, including fabricating some details and omitting other details, but these changes were insufficient to disguise the patients' identities. He speculates that Dr. Berg never would have published the book if his father were still alive, partly because she feared his violence. Frank also believes that the therapist made little effort to disguise her favoritism. "My mother always felt Dr. Berg preferred my father to her. Certainly she tolerated his abuse of my mother."

The therapists made a major change that distorted the accuracy of the case study. "One thing I was always taken aback by was the negative and depressing tone of the book. I talked to my mother about that, and she said that Leida deliberately excluded positive events in my parents' life and included everything that was negative." Consequently, the therapists portrayed his parents' life together as much bleaker than it was.

Frank agrees that his mother suffered from postpartum depression but disagrees that she was psychotic. He believes his father would today be diagnosed as having bipolar disorder. Frank disputes Dr. Berg's assertion that his mother was traumatized by her father's suicide attempt. "My grandfather made his suicide attempt when my mother was around twenty-five years old and living independently. As far as I know, it wasn't all that traumatic for my mother, just to be clear about what really happened." He remains unsure whether Dr. Berg believed his parents were "schizophrenic" and he "autistic" or whether she simply made the diagnoses for "dramatic effect, to sell books."

Frank has several memories of traveling with his parents when he was a young child to Leida Berg and Harold Steinberg's summer home in Katonah. He told Dr. Berg that he preferred his mother to her, a statement that elicited contradictory responses from the two women. "I remember Leida replying, 'sons should love their mother better than anyone else,' or words to that effect. My mother probably mistrusted Leida by that point. I told my mother what Leida said, and she said Leida didn't want to tell you she would want it if I liked Leida better than mom. My mother told me Leida dislikes all mothers."

"Ah, Little Jan, will you ever have a self?" Frank recalls reading Steinberg's strange question in the introduction to *In Search of a Response* (xi), though he has no recollection of the psychologist asking him the question. Frank now believes that Steinberg's rhetorical question was addressed to the reader. "It's a weird thing to say in a book, let alone to ask a child. Why did he say it? It's almost like psychobabble. It sounds like an impression he got from a child doing something or saying something childish." The statement attributed to him at the end of "Diary of Jan"—"My solution to everything is to withdraw and my life gets emptier and emptier until I have nothing to strive for" (10)—also strikes him as psychological jargon. He points out that "Diary of Jan" was the section of the book his mother hated the most because it was all "lies."

Frank has nothing good to say about his psychotherapy experience, which he now likens to a form of brainwashing. "My mother had to be reassured by clinicians that I was not autistic." If he could now speak to Steinberg, he would tell him that he has indeed achieved a self, along with a master's degree in social work. His mother's experience as a psychotherapist and her ordeal with the lawsuit influenced his own career choice. His life, he adds, has had many successes and a few hardships.

The last time Frank saw Berg and Steinberg was during the trial in 1976, at which time Frank was an undergraduate at Clark University in Worcester, Massachusetts. He attended only one day of the trial, though afterward he regretted not being there more frequently. Judge Martin B. Stecher stared curiously at him in the courtroom, perhaps because the jurist had read about him in the case study. Frank recalls an awkward encounter with Harold Steinberg. "Harold came up to me and asked me some questions. I told him I was a psychology major, and he grimaced. He then asked me if I would be willing to go up to and speak with Leida. I then said, 'let Leida know if she wants to approach me to speak to me she can.' And that was the end of our exchange. She never did."

The publication of *In Search of a Response* changed his mother's life. Before the book appeared, she was "kind of sweet" and perhaps "naive"; afterward, "she became more acerbic, I think, almost 'brittle.'" The publication of the book "hardened" his mother and perhaps made her more "critical" of him. Nevertheless, he believes she was approving of him most of the time. His mother lost "a lot of her innocence" as a result of the ordeal, but she became more resilient, confirming the truth of Nietzsche's wry observation that anything that doesn't kill you makes you stronger.

After the book's publication, Harriet Werner was extremely worried that her practice would be destroyed, but that did not happen, according to her son:

> As it ended up, my mother's practice did very well from the time in early 1973 when the book was published, before going off the shelves, until the time Judge Stecher made the final decision in my mother's favor, sometime during the first half of 1977. As far as I know, no patient ever found out about the book or if they did, they did not leave treatment with my mother. And she continued to get referrals (actually her main source of referrals was from either former or current patients, not fellow clinicians). During this time my mother also stopped teaching a course at NYU's Graduate Psychology Department in Family Therapy. She was an adjunct professor at NYU. She did tell me one or two students in her course or ex-students found out about the book and that it was about her. And of course she took time off from her practice for court appearances.

Did Frank's mother "indoctrinate" him with hatred for Leida Berg and Edith Fisch? Frank raised this question and concluded that the answer is complicated.

> My mother severed her relationship with Leida and Harold in 1965. My dad continued to be a loyal patient until his death in 1971 while Edith was his wife from 1963 until 1971. My mom told me that when she told Leida Steve was abusive to her, Leida said, "you provoke him." It was only really after the book publication that Leida became an enemy. I continued to see Edith on visits with my dad after the divorce. She was my stepmother. But she in no way, shape, or form helped to raise me.

The enmity between his mother and stepmother did not begin until the lawsuits started, after his father's death. Elaborating on a statement he had made in his Amazon posting, Frank said that Edith called him a "monster" during a pretrial deposition for the lawsuit over his father's estate.

How did his mother's lawsuit with Edith, contesting Steven's trust fund for his two children, affect Frank? "I myself was not greatly disturbed at being 'disinherited' in Edith's favor in my dad's will, at least not consciously."

Frank believes Edith characterized his father as doing well as "evidence he was fully capable of leaving his money to her in his will." Frank offered a similar interpretation of her claim that the case study was fictionalized. "It could have been she was saying that to counter any contest of my father's will on the basis that he was mentally incompetent. My sister had conspiracy theories about the will; she even thought maybe Edith had my father murdered."

Frank told us that he began to break away from his mother's influence and point of view while he was in college, an act of independence that complicated their relationship:

> When I went to college I changed from someone who always took my mom's side in all disputes to someone who questioned everything. It was in 1977 around the time I graduated when I kind of played the devil's advocate about Leida (not so much Edith, who my mom sometimes conflated with Leida on her enemies' list) and said maybe she has a side too. My mom was enraged and disturbed I would see things from Leida's point of view. I think it was specifically about her right, or the value in, having the book published. In college I became a bit of a rebel, at least intellectually speaking, and questioned everything I was told. The best professors in college (unlike graduate school) mostly encouraged a questioning stance toward all accepted beliefs of all kinds so maybe this was not so rebellious.

Judge Stecher said in his final decision that the authors of *In Search of a Response* "were stupid but not malicious," a statement with which Harriet Werner disagreed. "She felt it was a malicious act," Frank said. The twenty-thousand-dollar judgment she was awarded was not enough to pay her extensive legal fees. She did not regret filing the lawsuit, however.

Did Dr. Berg's other patients worry that she might write a case study about them without permission? Perhaps. Frank recalls that a female psychologist who was one of Dr. Berg's patients was told by someone close to the psychiatrist's circle, possibly by Harold Steinberg, that if she was "not careful," Dr. Berg might also write a book about her. This psychologist had earlier informed Frank's mother about the planned publication of *In Search of a Response*. Even if this story is apocryphal, it reveals a patient's understandable fear arising from a therapist's breach of confidentiality. Frank doesn't own a copy of *In Search of a Response*. His mother kept a copy until her death, and then it was thrown out. "Nobody wanted it."

"A Minefield"

"Writing about patients is like walking in a minefield," Judy Leopold Kantrowitz states, adding, "There are no good solutions" (40). Leida Berg and Harold Steinberg must have known, when Harriet Werner's lawyer contacted them after Harriet became aware of their plan to publish *In Search of a Response*, that a landmine loomed perilously near, yet they persisted in their efforts to publish their explosive book. What were they thinking? They used the thinnest of disguises in their case study, avoiding the thick disguise that is necessary but not always sufficient to protect confidentiality. The therapists may have thought that Harriet Werner had given them permission to write about her life, but her implacable opposition became apparent when she hired a lawyer to block the distribution of *In Search of a Response*. Didn't the therapists realize that whatever help they provided "Helena" over a period of several years would be destroyed by their efforts to publish a book that not only invaded her privacy but also caused her incalculable anguish? Didn't they realize that they were elevating their own needs over their patients' by publishing the book? Harriet Werner certainly felt this way, according to her son. "My mother feared that if she did not take action [by filing a lawsuit], Leida Berg might fulfill what my mother viewed as Leida's dream and ambition of having a bestseller."

Leida Berg never received her patients' written permission for publication, and the explanation she offered at the trial—consent "was there one day and not there another day. That was the nature of the illness I was treating, unreliable"—is hardly credible from a psychiatric or legal point of view. Had the Werners given Dr. Berg and Steinberg their written permission, *In Search of a Response* would have been published, but it's not likely it would have received many responses from psychiatrists and psychologists, few of whom would have been willing to read such a ponderous tome simply to learn about the "flow" of therapy.

Steven's Death: The Two Wives Battle Over Money

Steven Werner's death left Edith, his third wife, a wheelchair-bound widow. Shortly thereafter, she filed an application to collect Steven's pension benefits, but since he had designated Harriet, his second wife, as the beneficiary of his retirement account, Edith's application was denied. At the end of July 1971, Harriet Werner filed an application for the same benefits. Edith, a law professor, immediately requested a formal hearing, and at the same time she filed a lawsuit against Harriet Werner and the state controller. It took one year for the case to be decided in favor of Harriet (*Werner v. Werner*).

A second battle took place over Steven's trust fund, which at the time was worth $62,000, roughly $350,000 in present-day funds. In Steven and Harriet's separation agreement, which was officially recognized as part of their Mexican divorce, he had promised to designate the two children from his marriage with Harriet as the ultimate beneficiaries of the trust proceeds he was entitled to receive from the trust established by his father. However, four months later, after his marriage to Edith, he revised the will and left all his property including the trust to his new wife. This led to another legal battle between the two wives that was not resolved until 1975, at which time a judge ruled that Edith, not the children, was entitled to the funds (*Seidel v. Werner*, 81 Misc.2d 220 [1975]). That judge's ruling was appealed to the next higher court (the Appellate Division) and was upheld in a four-to-one decision (*Seidel v. Werner*, 50 A.D.2d 743 [1975]). Finally, the lawyer for Frank and Anna, the two children who were no longer beneficiaries of the trust, appealed to the court for payment *from the trust* of his legal fees involved in the case, fees that, presumably, Harriet would have had to pay out of pocket. The court agreed that the legal costs could be paid from the trust even though the children had lost the case (*Seidel v. Werner* 81 Misc.2d 1064 [1975]).

We recount these extensive legal battles because they lead us to infer intensely hostile feelings between Steven Werner's second and third wives. It is not clear whether Tiresias Press, which published *In Search of a Response*, knew about this hostility; if so, it did not prevent its publication. Edith, now bereft of a husband yet a highly accomplished law professor, used the most potent weapon available to her, her knowledge of the law, to wage war against Harriet and her two children with Steven.

In addition, Edith's hostility toward Harriet raises the question of a possible vindictive motive in seeking out Dr. Berg, bringing with her Steven's copy of the book manuscript, and urging the psychiatrist to publish the book. One wonders whether Edith used the prestige of being a law professor to assure Dr. Berg that publication of the book would not be improper. Of course, even if she had the power, as Steven's executrix, to authorize the publication of the material about him, she certainly could not authorize the publication of hundreds of pages of transcripts of Harriet's sessions.

Banning a Book

Harriet Werner states unambiguously in her affidavit that both she and her husband were adamantly opposed to the publication of *In Search of a Response*:

> In 1961, after we had been patients of Dr. Berg for five years, she told Steven and me that she wanted to publish a record of our sessions together. After reading the manuscript, Steven—speaking for both of us—told Dr. Berg that we would not permit her to publish the manuscript, because we would be easy to identify even though our names (but nothing else) were changed, and because the manuscript contained highly confidential material that would be damaging to us if disclosed. Thereafter I always kept my copy of the manuscript under lock and key so that my children and other persons would not see it. I believed that Dr. Berg had abandoned the idea of publication, for I did not hear of the manuscript again until 1972, after Steven had died.

When she did hear that the book might be published, Harriet immediately contacted a lawyer who, in turn, wrote a letter to Dr. Berg on October 9, 1972, and requested a copy of the book for review. The letter contains a thinly veiled threat of legal action against Dr. Berg:

> Dear Dr. Berg,
>
> I am the attorney for Harriet G. Werner who was formerly married to the late Steven L. Werner.
>
> My client informs me that it has come to her attention that you are in the process of publishing a book containing privileged information arising out of your treatment of both Steven L. Werner and Harriet G. Werner.
>
> Before distribution of your manuscript, may I ask that you submit to me a copy thereof so that I can determine whether there has been a violation of your obligation not to reveal communications between a doctor and patient, and further, whether your manuscript contains any libelous material or references.
>
> I trust that you will avail yourself of this opportunity to avoid any further [. . . ?] or litigation in connection with the aforementioned obligation.
>
> Please guide yourself accordingly.
>
> Very truly yours,
> [. . .] Katzman

A week later, Dr. Berg responded to this letter (Affidavit of Harriet G. Werner, 30) with a handwritten note. Parts of the letter are illegible in the handwritten version in the court record but have been supplemented by an almost identical version included within a longer letter by Dr. Berg published in *American Psychiatric News*, January 1, 1975:

> Dear Mr. Katzman:
>
> I am rather surprised by your letter. Mrs. Werner apparently has heard rumors that I am publishing a book & has assumed that it is about her.
>
> 10–12 years ago I was preparing a manuscript. I talked with Mrs. Werner about it & she agreed to the publication. I gave her the manuscript in which is even the statement of agreement. Mrs. Werner was pleased and flattered and thought the book would be a contribution to all therapists. Mrs. Werner still has *that* manuscript in her possession. I am surprised to hear that she has contacted an attorney instead of calling me & talking with me & straightening out whatever bothered her. She knows that I have never rejected her and that I do not want to harm her in any way.
>
> Sincerely,
> Leida Berg, M.D.

Dr. Berg's response was profoundly misleading, so deliberately ambiguous that one suspects she consulted with Edith before responding. She neither confirms nor denies the "rumors" and then suggests, infuriatingly, that Harriet should have contacted her instead of an attorney when, in fact, *she* should have contacted the patient herself. Didn't the psychiatrist realize that she was betraying the Hippocratic Oath to do no harm? Didn't she realize she was destroying the patient's trust in her, a trust developed over a seven-year analysis? Dr. Berg may not have "rejected" her patient; she merely betrayed her.

No further communication apparently took place until the publication of the book in February 1973. Advertisements appeared in professional publications, newspapers, and magazines, including the *New York Times* and the *New Republic*. According to court documents, one of Harriet Werner's friends, who knew nothing about a plan to publish a book about her, saw the advertisement in the *New Republic*, leading the friend to examine the book in a New York bookstore. She immediately recognized Harriet and her family from the book despite the altered names used in the book. She informed Harriet of her discovery.

In its advertising of the book, Tiresias Press, which is no longer in existence, presented the book with two faces, depending on the place the advertisements appeared. In advertisements in publications intended for professional readers, the book was described this way (*American Journal of Nursing* 73, no. 4 [April 1973]):

in search of a response

Leida Berg, M.D. Harold Steinberg

A documentary account of the psychiatric treatment of a man and his wife, both diagnosed as schizophrenic, showing how they were reached and treated successfully without drugs or shock therapy.

Includes extensive notes on the philosophy of treatment.

1008 pages, 1973, $20.00

On the other hand, advertisements in the public press, such as the *New York Times* (April 8, 1973), took on a more lurid form (see accompanying illustrations):

in search
of a
response
Leida Berg, M.D.
Harold Steinberg

You will be stirred and
agitated, but you are left
with an affirmation of life and living.
A Symphony of feeling—
read it for the sheer
pleasure of an intense
experience.

In Search of a Response is about the psychiatric treatment of a husband and his wife—both diagnosed as schizophrenic—how they were reached and treated successfully without drugs or shock by a therapist who was a sensitive human being. As you read on, these individuals no longer frighten you and the label with which they have been stamped disappears. They emerge as people.

1008 pages, 1973, $20.00

Advertisements of this second type, which also appeared in the *New Republic*, cast doubt on Dr. Berg's later claims that the book had been published because of its scientific value for other psychotherapists.

On March 15, 1973, Harriet's attorney filed a petition with the New York Supreme Court (the lowest-level state court in New York) to halt sales of the book. The following brief account then appeared in the *New York Times* on March 16, 1973, under the headline "Psychiatrist's Book Subject of Suit":

> A psychiatric social worker said that her life story formed the major case history in a psychiatrist's new book, and she has begun a suit to prevent further distribution and sale of the 1,008-page book, *In Search of a Response.*
>
> Harriet G. Werner, the social worker, asked the State Supreme Court for an order restraining the Tiresias Publishing Company from selling

in search
of a
response

Leida Berg, M.D.
Harold Steinberg

You will be stirred and
agitated, but you are left
with an affirmation of
life and of living.
A symphony of feeling –
read it for the sheer
pleasure of an intense
experience.

In Search Of A Response is about the psychiatric treatment of a husband and wife—both diagnosed as schizophrenic—how they were reached and treated successfully without drugs or shock by a therapist who was a sensitive human being—responding to another human being. As you read on, these individuals no longer frighten you and the label with which they have been stamped disappears. They emerge as people.

1008 pages   1973 $20.00

THE TIRESIAS PRESS, INC., NEW YORK
116 Pinehurst Avenue, New York City 10033

and advertising the book, which is by Dr. Leida Berg and her husband, Harold Steinberg, of 1020 Park Avenue.

"I shared with Dr. Berg virtually every intimate detail of my life and of the lives of my children and my then husband," Mrs. Werner said. She said that her privacy rights had been violated and that she and her family were identifiable in the book's presentation.

According to advertisements, the book is "about the psychiatric treatment of a man and wife both diagnosed as schizophrenic."

The problem faced by the court was that although there seemed to be a clear invasion of the patients' privacy (assuming that the patients actually

were recognizable), no one was able to cite a specific law that would allow
a court to interfere with Dr. Berg's First Amendment right to publish her
book! In addition, the judge in the case accepted the argument of Dr. Berg's
attorney that changing the patients' names was a form of disguise consistent with previously published medical case histories. Harriet responded
by listing many undisguised characteristics of her family that, taken together, could clearly identify her to anyone who was aware that she was
Dr. Berg's patient. In his ruling, the judge, who noted, offhandedly, that
he himself had not read the book, decided to issue a temporary *limited* injunction that had the effect of halting sales of the book in some but not all
stores.

Titicut Follies

This Solomonic compromise might have been patterned on the similar resolution in the notorious *Titicut Follies* case that had taken place six years
earlier. A 1967 American documentary directed by Frederick Wiseman,
Titicut Follies focuses on the patient-inmates of Bridgewater State Hospital, a state facility for the criminally insane located in southeastern Massachusetts. The documentary exposes the horrific conditions in which
patient-inmates live in subhuman conditions and are routinely humiliated and taunted by callous guards, social workers, and psychologists.
Many patient-inmates are catatonic. Nearly all are drugged into a stupor.
Nothing in the documentary is staged or fictional. Wiseman spent a hundred hours filming in the facility, but he does not editorialize, instead allowing viewers to reach their own conclusions about the harrowing
treatment of the patient-inmates, some of whom are brutally force-fed
while being kept naked. The title of the film, which comes from the yearly
stage show put on by patient-inmates and guards, is grimly ironic, for the
psychiatric correctional institution appears more dehumanized and sinister
than the nightmarish hospital in Ken Kesey's *One Flew Over the Cuckoo's
Nest*. In fact, the cast and crew prepared for the filming of *Cuckoo's Nest* by
watching *Titicut Follies*. Called "despairing" by the film critic Roger Eb-
ert, *Titicut Follies* received awards in Italy and Germany. It is now considered one of the greatest documentaries of all time.

Titicut Follies also provoked one of the greatest legal furors in the United
States. It has the distinction of being the only American film banned for
release for reasons other than obscenity or national security. As he reported in interviews with Jesse Pearson in *Vice* and Jesse Walker in *Reason*,

Wiseman, a former law professor, had received the permission of the patient-inmates, the superintendent of Bridgewater State Hospital, and the lieutenant governor of Massachusetts, Elliot Richardson. The documentary received positive reviews when it was shown at the New York Film Festival in 1967, but when a person who had not seen the film complained, Richardson, who had become by then Massachusetts' attorney general, reversed his opinion and expressed opposition to its showing.

The state brought a suit against Wiseman, charging that he was exploiting the patient-inmates because they had not signed release forms. A judge ruled against Wiseman, finding for the first time a right of privacy in the Commonwealth of Massachusetts. The judge ordered that all negatives of the film be destroyed. He called *Titicut Follies* a "nightmare of ghoulish obscenities," but Pearson describes it more accurately as a "documentation of ghoulish obscenities." "The next thing I did," Wiseman told Pearson,

> was appeal to the Massachusetts Supreme Court. They decided that the film had value but could only be seen by limited audiences: doctors, lawyers, judges, health-care professionals, social workers, and students in these and related fields, but not the "merely curious general public." And this was on condition that I give the attorney general's office a week's notice before any screening and that I file an affidavit after that everyone who attended was, of my personal knowledge, a member of the class of people allowed to see the film. Those were the conditions under which I could screen *Titicut Follies*.

The conditions were, Wiseman remarks, impossible.

The court's real reason, as one reviewer pointed out, had less to do with the law than with politics. "While the issue of the inmates' privacy was used to keep the film from the public eye, after viewing the extraordinary *Titicut Follies*, it's obvious that the court was also out to protect the privacy of state officials, who wanted to run their tax-supported facility without the burden of public scrutiny."

As Wiseman indicated to Walker, "in any number of cases before and after the *Titicut Follies* case, the U.S. Supreme Court found that when the right of privacy and the public's right to know are in conflict, the public's right to know is the dominant value. Even where the common law or statutory right of privacy exists, it falls before the overriding importance of the First Amendment." Wiseman appealed to the U.S. Supreme Court, which refused to accept the case. Years passed, and in the mid-1980s Wiseman filed another suit in Massachusetts. The original judge had died, and a new judge ordered an investigation to determine whether a showing of the film would damage the surviving patient-inmates. In 1991 the court allowed the film to be shown to the public, citing the passing of time and the end of privacy issues, since many of the patient-inmates were no longer alive. The following year it was aired by PBS.

In Search of a Response and *Titicut Follies* both involve the clash between free speech and the right to privacy, and both involve lawsuits that began in state courts and were deemed important enough to be considered in federal court. Both involved judges issuing injunctions, temporary or permanent. Both involved professional organizations filing amicus curiae ("friend of the court") briefs. The ACLU first recommended that the documentary's distribution be highly limited, and then, years later, it reversed itself and fully supported the film.

Despite these similarities, there are two fundamental differences between the cases. First, the director of *Titicut Follies* had done his best to secure the permission of everyone involved with the film, including the patient-inmates and the superintendent of the facility. Everyone in the hospital knew Wiseman was making a documentary. By contrast, the authors of *In Search of a Response* did not have the permission of the two patients. The patients did not even know they were being taped. Second, *Titicut Follies* was recognized from the beginning as a landmark documentary: a classic of social criticism (*Time* compared it to Upton Sinclair's *The Jungle*) and an enduring work of art. No such high praise can be bestowed on *In Search of a Response*. Even if the book were half as long, none would wish it longer.

A Split Decision

Clearly unable to resolve the conflict between the patient's privacy rights and the doctor's First Amendment right to publish *In Search of a Response*, the judge similarly "split the difference" by partly granting and partly denying Harriet Werner's motion for the preliminary injunction. The book could be sold only in stores that had separate sections for books on "medicine, psychiatry, psychology, social work, or works of science," and the book had to be placed in these sections. The judge also ordered the papers in the case to be sealed, thus preventing the publication of his opinion and order. From that point on, the names in all published legal proceedings in the case are Doe, Roe, and Poe—despite the fact that the actual name of one of the patients (Werner), the names of the authors, the book's title, and the name of the publisher had appeared in the *New York Times* on March 16, 1973.

The judge's decision was for all practical purposes a defeat for Harriet Werner, who had hoped to stop all sales of the book. Two hundred and twenty copies had already been sold. Because she was a mental health professional, limiting sales to similar professionals, many of whom she knew, would hardly protect her privacy.

Given this disappointing decision, Harriet Werner's lawyer wrote a letter to the judge to whom the case had been assigned and asked for a rapid resolution of the complaint through a trial in ten days. Dr. Berg's lawyer responded by opposing a quick trial and instead claimed that a prolonged process would be needed to interview all the witnesses before a trial could be scheduled. This would mean, of course, that the book could continue to be sold, at least in some stores.

In addition, Dr. Berg's attorney filed an appeal of the limited injunction in the next higher level court, the Appellate Division of the New York Supreme Court. In response, Harriet Werner's attorney "cross-appealed," setting up a major confrontation (*Doe v. Roe*). The case was heard by a four-judge panel, all of whom apparently sympathized with the patient's ongoing complaint about breach of privacy. The panel knew that she would in effect lose the case, through further sales of the book, before she could have a legal resolution of the complaint. In its opinion in the case, the appellate court wrote:

> While we agree with [the lower court] that the granting of a preliminary injunction, under the circumstances here disclosed, would not constitute an invalid prior restraint upon publication, *we find no justification for the distinction attempted to be drawn between "scientific readers" and the general public.* Pending the outcome of this litigation plaintiff is entitled to either full protection or to no protection at all. We are of the view that, upon the record before us, she is entitled to full effective interim relief. (*Doe v. Roe*, 42 A.D.2d 559 [1973]; emphasis added)

The court took the view that because the patient had a sufficiently plausible complaint, she had a legal right to a judicial resolution of it. She was therefore entitled to a complete ban on publication of the book while the case was pending. The Appellate Division thus expanded the temporary limited injunction of the lower court by extending the injunction to cover all sales of the book until the dispute was resolved by a trial. Harriet

Werner, however, was required to post $5,000 (about $27,000 in present-day funds) as a fund to cover possible losses of the publisher in the event that she eventually lost the case. The order was issued at the end of June, several months after the book had first gone on sale in stores all over New York City.

What had begun as a privacy dispute was now a major legal battle between a patient and not only her former psychotherapist but also the therapist's husband and the publisher of the book. The case involved a complex array of legal questions, some of which had hardly been explored in courts before and therefore lacked any precedents upon which the issues could be decided.

Following their loss in the Appellate Division, which prohibited all sales of the book until a trial, Dr. Berg and her husband appealed to the highest court in the state, the New York Court of Appeals. The Court of Appeals decided the case on December 28, simply affirming the injunction of the Appellate Division but without a written opinion (33 N.Y.2d 902 [1973]). Of interest is that one of the Court of Appeals judges who decided the case, Sol Wachtler, later to become the chief judge, is the subject of another chapter in this book.

Unable to sell their book at least until a trial and having exhausted all legal possibilities in New York State, Dr. Berg, her husband, and Tiresias Press decided to appeal to the U.S. Supreme Court to seek relief from the injunction.

An appeal to the U.S. Supreme Court is a kind of "Hail Mary pass" because the odds of a case's being accepted are very small. Thousands of cases are brought to the Supreme Court each year, but unlike most appellate courts, which must consider all cases brought to them, the nation's highest court selects relatively few cases each year, between one hundred and two hundred. In 1973 the number of appeal petitions was in excess of two thousand, so probably the court only accepted between 5 and 10 percent of all cases that year.

However, the federal courts, unlike some state courts, take a hard line in the protection of the First Amendment constitutional right of free press. A decision by the Supreme Court can have a profound effect on the way the entire legal system views an issue. This case was accepted by the court, presumably because it raised the novel and important question of the conflict between the First Amendment guarantee of freedom to publish whatever one wants with the obligations that certain people, such as psychotherapists, have to protect the privacy of others, including psychotherapy patients. Expressed differently, can the government step in to prevent the publication of a book that egregiously violates a person's privacy?

Eighteen months after the publication of *In Search of a Response*, after the sale of only 220 copies, with further sales halted on December 18, 1973, by the New York Court, lawyers for both Harriet Werner and Dr. Berg appeared before the nine justices of the U.S. Supreme Court, the highest legal authority in the country, to argue the case. In addition, an influential amici curiae brief had been filed by a group of mental health organizations, including the American Psychiatric Association and the American Psychoanalytic Association, urging the court to reverse its decision to accept the case.

Within the American Psychiatric Association, the decision to ask the court to reverse itself by refusing to rule on the case was widely supported, but this decision caused Dr. Berg to write an angry letter to the association's newspaper, *Psychiatric News*, on January 1, 1975. Feeling betrayed by her colleagues and fellow APA members, she expressed outrage and indignation—but not a word about the psychological damage done to her patient. Her long letter was printed in full and without editing, and ended with this revealing paragraph:

> In their decision to file an amicus brief against me, the Executive
> Committee of APA has condemned me without a hearing. In publishing
> this article, they further malign and injure me—because of their insecu
> rity, their confusion, and self-consciousness about their public image.
> They fail to recognize their full responsibility to APA membership—to
> me as a member of APA. Does it not occur to them that they have an
> obligation to me—to stand behind me and to affirm my right to publish
> a worthwhile book, ten years after treatment was discontinued—a book
> in which the identities were carefully disguised? Does it not occur to
> them that they have an obligation to APA membership to encourage, not
> stifle, the spirit of free inquiry?

In reply, Dr. Berg's letter was followed by an equally long letter from Dr. Alan Stone, chairman of APA's Commission of Judicial Action and a highly regarded professor of law and psychiatry at Harvard University, in which he explained the APA's concern about the possibility of a Supreme Court decision in the case either way. A number of amici briefs were also submitted by publisher groups and author groups siding with Dr. Berg and her publisher. In addition, the American Civil Liberties Union, in its brief, argued strongly for the Supreme Court to declare the injunction from the state court unconstitutional.

In the one-hour oral argument the court devoted to the case, much of the questioning focused on the issue of whether the court should have taken the case in the first place. In the written opinions of the New York courts and the briefs supporting the injunction, several legal theories were proposed to justify the government's interference with Dr. Berg's free speech, but no single explanation seemed compelling. (For a detailed account of the myriad legal issues presented by the case, see "*Roe v. Doe:* A Remedy for Disclosure of Psychiatric Confidences.") The professional organizations feared that the case might establish a precedent that would make it nearly impossible to publish any descriptions of clinical cases in the professional literature. Indeed, during the oral argument, both sides seemed to agree hypothetically that if the material in the book had been published without any disguise of the patients' identities, and if the patients had not given their consent to the publications, then there would be a legitimate role for the government to step in to block publication of the book despite First Amendment guarantees. In that respect, then, the disagreement about the facts in the case that were in dispute (Had the patients actually been disguised? Had they given consent?) would need to be settled *before* a decision on banning the book could be seriously addressed.

In fact, during the oral argument, Justice Potter Stewart, confronting the patient's lawyer, pointed out that if changing a patient's name and making a few other changes of identifying information were not a sufficient disguise, then Freud himself could not have published his case reports, since he had done little more to protect his patients' identities than Dr. Berg had done. Justice Stewart commented, "This means if New York State had wanted to, back in the era where Freud was writing, [it] could have enjoined the publication of everything he wrote." Justice Stewart was correct with respect to Freud's case studies. As we noted in the Introduction, Freud was never able to resolve the conflict between preserving patient confidentiality and publishing psychiatric case studies. The names of all the patients in Freud's major case studies have been identified by historians despite his attempt to conceal their identities by means of the same devices Dr. Berg used in *In Search of a Response.*

An actual audio recording of the one-hour argument before the Supreme Court can be accessed on the Oyez Project's webpage devoted to this case.

By the end of the hour of argument, it appeared that the court had wandered into territory too uncharted to support a precedent that would be created by *any* decision. And so, two months after the oral argument, the court took the unusual step of reversing its own decision to accept the case and issued a terse 9–0 decision that simply read, *"The writ of certiorari is dismissed as improvidently granted."* In plain English, "We shouldn't have accepted this case in the first place, and we are washing our hands of it."

Dismissing a writ of certiorari has occurred in only about fifty cases in the court's history. Therefore, after all the excitement, angst, and expense of both sides' preparation for the case to be decided at the highest legal level in the country, and the cost of lawyers for both sides traveling to Washington to argue the case before the court, possibly the biggest professional event of their lives, the situation in effect turned out to be the same as if the Supreme Court had not accepted the case in the first place.

Clearly relieved by the court's having accepted the arguments of the American Psychiatric Association's brief in the court's decision to avoid ruling in the case, Dr. Stone commented in *Psychiatric News:* "If the Court had said there was no Constitutional right to publish the material, we would have been unhappy. If the Court had said that there was a Constitutional right to publish the material, we would have been unhappy." Consequently, Stone added, "we are delighted" (March 19, 1975). The temporary complete prohibition on the sale of the book therefore continued, and the case was returned to the New York courts for a final decision.

"A Matter of First Impression"

The legal battle finally ended in 1977, when the case was decided by Judge Stecher of the New York State Supreme Court. Acknowledging that he was dealing with something odd and unique in legal history, Judge Stecher began his opinion with these words:

> This action for an injunction and for damages for breach of privacy is a matter of first impression in this State, and so far as I am able to ascertain, a matter of first impression in the United States. It arises out of the publication, verbatim, by a psychiatrist of a patient's disclosures during the course of a lengthy psychoanalysis. (93 Misc.2d201 [1977], 204)

A "matter" or "case of first impression" (*primae impressionis*, in Latin) is, according to uslegal.com, "one which presents a question of legal interpretation that has not yet been before the court. It is a case which presents a new issue never before decided in a reported case." After reviewing the legal situation presented to him, Judge Stecher wrote:

Every patient, and particularly every patient undergoing psychoanalysis, has such a right of privacy. Under what circumstances can a person be expected to reveal sexual fantasies, infantile memories, passions of hate and love, one's most intimate relationship with one's spouse and others except upon the inferential agreement that such confessions will be forever entombed in the psychiatrist's memory, never to be revealed during the psychiatrist's lifetime or thereafter? The very needs of the profession itself require that confidentiality exist and be enforced. (213)

Unable to cite a specific law or case precedent on which to base his opinion that a dreadful wrong had been done, Judge Stecher pointed out that "Despite the fact that in no New York case has such a wrong been remedied due, most likely, to the fact that so few physicians violate this fundamental obligation, it is time that the obligation not only be recognized but that the right of redress be recognized as well" (213). And unable to come up with a name for what Leida Berg had done to Harriet Werner, the judge concluded:

What label we affix to this wrong is unimportant (although the category of wrong could, under certain circumstances—such as determining the applicable Statute of Limitations—be significant). It is generally accepted that "There is no necessity whatever that a tort must have a name. New and nameless torts are being recognized constantly." . . . What is important is that there must be the infliction of intentional harm, resulting in damage, without legal excuses or justification. (213)

Harriet Werner was awarded a judgment of $20,000 (about $78,000 in present-day funds) to compensate her for her expense and emotional distress—though as Frank pointed out to us, the judgment was not enough to cover her legal fees. In addition, Judge Stecher ordered Dr. Berg to refrain from disclosing any confidential information about her former patient.

Damages, of course, do not provide an adequate remedy; for should the book circulate further, beyond the 220 copies already sold, the damage must accrue anew. The plaintiff is entitled to a judgment permanently enjoining the defendants, their heirs, successors and assigns from further violating the plaintiff's right to privacy whether by circulating this book or by otherwise disclosing any of the matters revealed by the plaintiff to Dr. Roe in the course of psychotherapy. (218)

What shall we finally say about *In Search of a Response?* As Barry Landau suggests, the title of the book hints at a series of "unconscious enactments" committed by Berg and Steinberg, part of a pattern of egregious behavior. The book had no effect on the theory or practice of psychiatry. Few psychotherapists have heard about the book, and even fewer have read it. Harriet Werner must have felt that her deceased ex-husband was right when he characterized the book as an act of "treachery." The treachery arose not only from the authors' writing of the book but from its publication over the patient's strenuous objections. *In Search of a Response* abounds in unintentional ironies, including the name of the publisher, Tiresias Press. In Greek mythology, Tiresias was a blind prophet. Dr. Berg was figuratively blind but no prophet.

In Search of a Response remains a cautionary tale. Its publication and the legal aftermath we have described were an important milestone because it established the principle that where the ethical obligation of a psychotherapist to protect a patient's confidences breaks down, the government itself, through action of the judicial system, will step in to punish the therapist and attempt to restore the privacy and dignity of the patient.

Nevertheless, the fact that the participants in this case were plainly identified in the *New York Times* before an effort could be made to shield their identities through a court order shows that psychotherapy patients must rely for the most part on the *ethics of the therapist and not on the law* to protect their identity and privacy. The same might be said about the distribution of *In Search of a Response*, which despite the permanent injunction preventing sales of that book, is to this day widely available (fifty-eight libraries have the book, according to Google, and numerous copies are for sale on various Internet sites, including Amazon). These facts make the legal attempts to protect Harriet Werner's privacy, including changing the names on the case to Doe, Roe, and Coe, sealing the court records, and the permanent injunction, seem almost farcical.

Only about 220 copies of *In Search of a Response* are in circulation as a result of sales before the injunction and the distribution of numerous review copies by the publisher to magazines and newspapers. During the past several decades, the gradual replacement of self- and professionally imposed ethical restraints by feeble external legal restraints on therapists is in itself an example of the erosion of the privacy protection psychotherapy patients need and deserve.

Take-home lessons:

1. Courts can step in to prevent the publication of confidential psychotherapy information when the patient has not consented.

2. By the time courts act to prevent a disclosure, it is often too late to protect the patient's privacy.

3. Patients need to rely on not only their therapists' ethics but also their adherence to professional standards of conduct, which includes maintaining appropriate boundaries.

7.   The Anne Sexton Controversy:
      "There Is Nothing Like This in the
      History of Literary Biography!"

What to look for: Is it permissible for a psychotherapist to release confidential tape recordings of a patient's psychotherapy sessions after the patient has died, and without the patient's earlier consent?

A patient, or his or her executor if the patient is deceased, generally has the right to copies of the patient's medical records. If the patient is a famous poet who commits suicide, should her biographer have access to extensive audiotapes of her therapy sessions without the patient's explicit permission? Should the deceased poet's daughter, who is also the poet's literary executor, be allowed to make a decision that will expose the most intimate aspects of her mother's private life to public scrutiny? Should the deceased poet's psychiatrist cooperate fully with the biographer by supplying sensitive information that is always deemed private and confidential? If so, what are the risks and benefits of such collaboration for the psychotherapeutic and literary communities? These are only a few of the daunting questions surrounding Anne Sexton's life and death.

Sexton was forty-five when she asphyxiated herself in her bright red Mercury Cougar in the closed garage of her home in 1974. Controversial in life, she became more controversial in death. Her older daughter and literary executor, Linda Gray Sexton, who was twenty-one when her mother committed suicide, requested that Martin Orne, the psychiatrist who treated Sexton from 1956 through 1964, turn over to the biographer Diane Wood Middlebrook three hundred hours of audiotapes he had made of his patient's therapy sessions. Linda Gray Sexton, who edited her mother's books, then allowed Middlebrook to use presumably confidential material from the tapes of Sexton's sessions with Martin Orne and from her hospital psychiatric records. With the family's permission, Dr. Orne cooperated fully with Middlebrook, who interviewed him for a total of eighty hours. He even wrote a foreword to Middlebrook's biography, in which he referred to the soul-searching that had gone into his decision to give the audio-tapes to his patient's family.

The publication of *Anne Sexton: A Biography* in 1991 created a literary sensation as well as an unprecedented psychiatric controversy, one that shows no signs of abating. The appearance in 1994 of Linda Gray Sexton's memoir, *Searching for Mercy Street: My Journey Back to My Mother, Anne Sexton*, her 2011 memoir *Half in Love (Surviving the Legacy of Suicide)*, and, in 2012, of Dawn Skorczewski's scholarly book *An Accident of Hope: The Therapy Tapes of Anne Sexton*, continues the debate.

Played out in the *New York Times* on the front page, the op-ed page, and in the letters column, as well as in professional journals and other publications, the controversy over the propriety of Dr. Orne's decision to turn over the therapy tapes to Middlebrook has never reached a resolution. In general, mental health professionals excoriated Dr. Orne, accusing him of acting unethically, while many humanities scholars believed the contribution the tapes make to understanding the sources of Sexton's poetry outweighs concerns about the breach of confidentiality.

Anne Sexton first entered therapy for depression in 1954 at the age of twenty-six, a year after Linda's birth. Her younger daughter, Joyce Ladd, was born in 1955. The following year, after the first of many unsuccessful suicide attempts, Sexton was hospitalized for three weeks at Westwood Lodge, a private sanatorium outside of Boston, after developing a severe mental disorder in which "she became paranoid, depressed, and suicidal. She heard voices, fell into apparent trances, and twirled her hair into knots" (Skorczewski, xv). Her Viennese-born psychiatrist, Martha Brunner-Orne, who had treated Sexton's father for alcoholism years earlier, saw her for several months and then, as she was leaving for a vacation, made a referral to her young physician son, Martin Orne, who continued to work with her thereafter. Sexton thus became Martin Orne's first long-term therapy patient.

The younger Orne began keeping detailed notes of their sessions together. In one of the most revealing paragraphs in the biography, Middlebrook observes that Sexton told Orne, in her first interview with him in 1956, that she thought "her only talent might be for prostitution: she could help men feel sexually powerful. He countered that his diagnostic tests indicated she had a good deal of undeveloped creative potential, and he later proposed that she might try to do some writing about her experiences in treatment. This might help others with similar difficulties to feel less alone, he suggested." Middlebrook adds that Sexton "subsequently singled out that conversation as the first encouragement she had ever received to think of herself as a capable person" (42).

Sexton's belief that her only talent was to make men feel sexually powerful may provide us with an important part of the dynamic underlying her wish to make Orne feel powerful by seducing him. Middlebrook does not comment on this, but Sexton's statement to Orne during their opening interview may have been the moment when his countertransference first became evident. By encouraging Sexton to sublimate, or channel, her sexual desire into poetry, Orne could, in effect, serve as her muse while at the same time maintaining appropriate boundaries with her.

Following Sexton's second suicide attempt in 1957, Sexton recalled Dr. Orne saying to her, "You can't kill yourself, you have something to give. Why, if people read your poems (they were all about how sick I was) they would think, 'There's somebody else like me!' They wouldn't feel alone" (Middlebrook, 42–43).

A Confessional Poet

To understand the psychiatric controversy, one must appreciate the ways in which Anne Sexton's poetry and therapy were inextricably related. Her first published poem, "You, Doctor Martin," focuses on her relationship with Orne while she was recovering from a breakdown. "Your business is people, / you call at the madhouse, an oracular / eye in our nest" (*Complete Poems*, 4). Seeing herself and the other patients as part of the "moving dead," she loves the psychiatrist because of his "third eye" that moves among the patients and "lights the separate boxes / where we sleep or cry." The poem honors the person who had a transformative influence on her life. "You, Doctor Martin" appears in Sexton's first volume of poems, aptly entitled *To Bedlam and Part Way Back*, published in 1960. The epigraph to the volume is a letter written by the German philosopher Arthur Schopenhauer to Johann Wolfgang von Goethe in 1815: "It is the courage to make a clean breast of it in face of every question that makes the philosopher. He must be like Sophocles' Oedipus, who, seeking enlightenment concerning his terrible fate, pursues his indefatigable enquiry, even when he divines that appalling horror awaits him in the answer. But most of us carry in our heart the Jocasta who begs Oedipus for God's sake not to inquire further."

Sexton could not have chosen a better epigraph to reveal the haunting themes that would dominate her life and art. Nor could she have chosen a more ironic epigraph, one leading to the iconic play that forms the cornerstone of psychoanalytic theory. She made a number of statements in therapy suggesting her belief that her father had sexually abused her, though it's unclear, Middlebrook concludes, whether Sexton was reporting a memory or a fantasy. Regardless, she always sought dark knowledge no matter where it led. "Why else keep a journal," she wrote in a 1961 poem intended for posthumous publication, "if not to examine your own filth?" (*Complete Poems*, 564). She used the insights gleaned from therapy in her intensely personal poems, which interrogate, with startling candor, the most private details of her life. She knew that the quest for personal knowledge, whether through solitary introspection or therapy, is fraught with peril. Linda quotes her mother's statement to Dr. Orne in 1961: "Oh, God, therapy is a dirty mirror" (*Searching for Mercy Street*, 229). Anne Sexton was ambivalent about disclosing dark knowledge, but despite her hesitation, she identified more with the fearless Oedipus than with his terrified mother, Jocasta.

Anne Sexton was known famously and infamously as a "confessional" poet, writing openly and graphically about depression, abortion, incest, masturbation, drug and alcohol addiction, and suicide. Self-disclosure remained her poetic credo from the beginning to the end of her career. In her poem "For John, Who Begs Me Not to Enquire Further," dedicated to John Holmes, her first poetry professor whose workshop inspired her dazzling early poems, she defiantly vowed to tap into that "narrow diary of my mind, / in the commonplaces of the asylum / where the cracked mirror / or my own selfish death / outstared me" (*Complete Poems*, 34). Holmes, whose wife had committed suicide, was alarmed by Sexton's poems. "It bothers me that you use poetry this way. It's all a release for you, but what is it for anyone else except a spectacle of someone experiencing release?" Holmes feared that Sexton's poems would later prove deadly to her or her family. "Don't publish it in a book. You'll certainly outgrow it, and become another person, then this record will haunt and hurt you. It will even haunt and hurt your children, years from now" (Middlebrook, 98).

Many literary critics found Sexton's poems objectionable. "These are not poems at all," Charles Gullans complained in a review of *Live or Die*, "and I feel that I have, without right or desire, been made a third party to her conversations with her psychiatrist. It is painful, embarrassing, and irritating" (Colburn, 148). James Dickey's review of *To Bedlam and Part Way Back* was no less condescending. "Anne Sexton's poems so obviously come out of deep, painful sections of the author's life that one's literary opinions scarcely seem to matter; one feels tempted to drop them furtively into the nearest ashcan, rather than be caught with them in the presence of so much naked suffering" (Colburn, 63). After Anne Sexton committed suicide, Linda discovered in her mother's wallet a copy of Dickey's scathing *New York Times* review of *All My Pretty Ones*. The poet kept the review as a reminder of the opposition she had to overcome to achieve critical acceptance of her work. Despite winning the Pulitzer Prize in 1967 for her poetry volume *Live or Die*, she remained insecure about her reputation, particularly since she had never attended college.

The label "confessional poet," Diana Hume George observes, became for Sexton a "trap that prevented readers and critics from interpreting the range of her achievement. Poetic typecasting did not prevent Sexton from writing poems that reached beyond the personal boundaries that ostensibly formed the confessional territory; it merely kept readers from noticing that she had done so" (xvii). Sexton generally accepted the label of confessional poet, acknowledging in an interview that the "writer is stuck with what he can do. Any public poem I have ever written, that wasn't personal, was usually a failure" (*No Evil Star*, 50).

Sexton gave many poetry readings during which she excelled as a performer, but she regarded them as freak shows where she was forced to expose herself to a voyeuristic, predatory audience. "Some people hope you will do something audacious—in other words (and I admit to my greatest fears) that you vomit on the stage or go blind, hysterically blind, or actually blind." No one, of course, was forcing her to make her poetry readings into freak shows; her ambivalence toward her exhibitionism is apparent. Sexton's description, part of an essay called "The Freak Show" published in the *American Poetry Review*, evoked an angry letter from a woman who insisted that her appreciation of the reading was based on compassion, not cruelty. Sexton apologized for implying the audience came as "executioners" and vowed to "read my God-damned heart out" for those who genuinely appreciated her work (*Self-Portrait*, 397–398).

Poetic Healing

Sexton never missed an opportunity to extol the therapeutic benefits of writing poetry. In a letter to John Holmes in 1959, she made explicit what she had long wanted to tell him: "And I didn't say that poetry has saved my life; has given me a life and if I had not wandered in off the street and found you and your class, that I would indeed be lost" (*Self-Portrait*, 59). In 1968 she taught a poetry class at McLean Hospital, the renowned psychiatric institution in Belmont, Massachusetts, where many famous artists have been patients, including Sexton herself. "Poetry led me by the hand out of madness," she confided to a student. "I am hoping I can show others that route" (335). Those who knew Sexton agree that poetry gave her a reason to live and that writing became part of her therapy. "The sheer existence of the task of writing poetry," Dr. Orne declares in the foreword to Middlebrook's biography, "through which she could describe her pain, her confusion, and her observations, provided the basis for a critical sense of self-esteem" (xiv). Orne takes credit here for having sent her along that path at the start of the treatment, unaware of what we can speculate was his countertransference difficulty with her sexuality. Middlebrook notes that early in Sexton's career, writing poetry had "taken over her life, as the only activity outside therapy that gave value to existence" (61). It may be that writing poetry for Sexton was an extension of her therapy rather than an activity outside of it.

Linda Grey Sexton, who witnessed the extent to which her mother's creativity held in check her demons, understood that "while the way she looked when she was writing might remind me unconsciously of her sickness—like a secret tide that tugged at memory—in fact, she was rarely crazy when she was writing" (*Searching for Mercy Street*, 92). Both Anne and Linda Sexton knew about the mysterious relationship between artistic creativity and mood disorders, but they were wise enough not to equate the two or reduce art to illness.

Poetry was one of the major pillars of Sexton's life support system. Without it, she would have almost certainly died years earlier. Sexton needed to write poetry, just as she needed psychotherapy. Writing poetry, with the approval of Orne, was essential for her health.

Martin Orne

The Viennese-born Martin Orne was only a year older than Anne Sexton when she entered into treatment with him, just beginning what turned out to be a noteworthy career as both a clinician and researcher. He received his medical degree from Tufts University Medical School in 1955, one year before he began treating Sexton, and in 1958 he received his Ph.D. in clinical psychology from Harvard. Orne went on to a distinguished career as an experimental psychologist and psychiatrist, holding a professorship at the University of Pennsylvania along with many other academic positions. He served for more than three decades as the highly respected editor-in-chief of the *International Journal of Experimental and Clinical Hypnosis*. On the other hand, he did not complete training as a psychoanalyst and evidently had limited experience in intensive individual psychotherapy, at least early in his career, when he began his work with Sexton. At that time Orne was on the faculty of Harvard and was involved in a number of research projects, including clandestine studies secretly funded by the CIA under the program then referred to as "MKULTRA."

These factors may account for the unconventional techniques Orne eventually introduced into his treatment relationship with Sexton. He had a particular scientific interest in memory, false memory syndromes, hypnotic recovery of buried memories, and trance phenomena. Sexton's "memory problems" became apparent to Orne when she couldn't recall what she had discussed with him during the preceding therapy session. He suggested near the end of 1960 an unusual therapeutic technique. With Sexton's permission, he tape recorded each session. Orne reveals in the foreword to Middlebrook's biography that as part of this unusual psychotherapeutic process, Sexton made extensive notes about everything she remembered immediately after each therapy session. The following day she came to his office an hour or more before her scheduled appointment, and Orne's secretary placed the tape on a recorder and left her alone in a room off his waiting room to listen to the session. "She was asked to note particularly the discrepancies between her memories, her notes from the previous day, and what actually happened on the tape" (xvi). She spent what seemed like an inordinate amount of time in Orne's office: listening to each tape (which for some she did twice) would take as long as the actual session.

The therapy tapes thus offer a unique account of a long and intense therapy in which both patient and therapist remained committed to the talking cure and, given Sexton's ability to transmute the conflicts of everyday life into poetry, the writing cure. The regular audiotaping ended in 1964 when Orne was appointed professor of psychiatry and psychology at the University of Pennsylvania Medical School. He continued to see her intermittently when he returned to Boston once a month to see a number of his patients for follow-up. He audiotaped at least two of these sessions. He tried to help her when a serious problem arose in her relationship with her next psychiatrist, but Orne was no longer her primary therapist.

Sexton's transformation during her therapy with Orne from a nondescript housewife and mother of two into a world-famous poet seemed so miraculous and improbable that to this day the possible errors in Orne's management of this extremely ill woman have not been scrutinized, particularly the handling of the separation with which they both struggled when he moved to a city in another state and she began treatment with another psychiatrist. It appears that Sexton never adequately recovered from Orne's unconventional termination of her treatment.

After Anne Sexton's death, Linda began to read the boxes of unpublished letters and manuscripts the poet had placed in her archive at Boston University. She came across the extensive therapy notebooks her mother had transcribed from the years of tape-recorded sessions with Orne. Linda then discovered three reels of audiotapes in her mother's storage cabinet. Playing them, she realized, to her astonishment, that they were the actual recordings of her mother's therapy. Just as Anne Sexton had placed the therapy notebooks in her poetry archive at Boston University, where they were to be part of her history, so had she intended the three tapes, Linda was convinced, to be deposited in the archive. At the biographer's request, Linda wrote to all the hospitals in which her mother had been treated, asking for copies of her medical and psychiatric records. All complied, though McLean Hospital required her first to obtain a court order, and Massachusetts General Hospital maintained that the records could not be found. The Sexton case created a precedent, which is now Massachusetts law, that patients (along with the executors and administrators of their estates) have the right to copies of their medical records. After receiving the medical and psychiatric records, Linda was stunned to learn that Orne had in his possession the remaining three hundred hours of her mother's therapy tapes. With Linda's approval, Orne gave them to Middlebrook, although not, according to Orne, without a considerable amount of soul-searching.

In choosing Middlebrook as Sexton's biographer, Linda Gray Sexton resolved not to be like Sylvia Plath's mother, Aurelia Schober Plath, and husband, Ted Hughes, both of whom attempted to sanitize the poet's life by concealing their troubled relationships with her. Mrs. Plath at first denied her daughter committed suicide, and Hughes destroyed many of his estranged wife's journals. Linda, by contrast, sought to honor her mother's memory by holding back nothing. Declaring that her mother's "brutally frank poetry spared no one in its effort to bring enlightenment about the most human of experiences" (*Searching for Mercy Street*, 192), the daughter has allowed readers to see the unvarnished truth about Anne Sexton's life. Linda worked on putting together her mother's posthumous poetry books and then co-edited, with Lois Ames, a collection of Anne Sexton's correspondence. She was determined to "make my candor match that found in Mother's poetry" (210). Unlike Ted Hughes, who made it difficult for literary scholars to write about Plath's biography by denying them access to her unpublished writings and by refusing to allow them permission to quote from her published writings, Linda gave Middlebrook complete control over her biography and encouraged the Sexton family to share their memories of the poet with the biographer.

Anne Sexton: A Biography

Middlebrook's biography enjoyed immediate popular and critical success. It spent eight weeks on the *New York Times* bestseller list and was a finalist for both the National Book Award and the National Book Critics Circle Award. At the same time the biography generated enormous controversy even before its publication. In "Poet Told All; Therapist Provides the Record," appearing in the July 15, 1991, issue of the *New York Times*, Alessandra Stanley remarked on the historical importance of Middlebrook's book: "the first known time a biography of a major American figure relies on material taken from the subject's private therapy sessions with a psychiatrist."

Middlebrook's biography was controversial from the beginning. Two reviews appearing the same week in the *New York Times* reached opposite conclusions about the ethics of using the therapy tapes. "Although the reader can understand why a biographer would find it difficult to turn down the opportunity to use such available material," Michiko Kakutani wrote on August 13, 1991, "it's impossible to condone Dr. Orne's decision to violate his former patient's confidentiality, whatever the stance of the estate's executors." Regarding Middlebrook's biography as a pathography, a term coined by Joyce Carol Oates to suggest a life based on dysfunction and disaster, Kakutani saw Sexton as a "deeply disturbed, highly unstable, selfish and self-absorbed woman, who happened to possess the talent to channel her neuroses into the therapeutic channels of art." Erica Jong's review, published in the *New York Times* on August 17, 1991, celebrated Sexton as a "rare and generous" poet and lauded the Middlebrook biography for honoring the spirit of a fearlessly self-disclosing writer. "In general, the world has been more injured by silences, erased tapes, rewritten records, shredded documents than it has ever been damaged by revelations of the heart."

Middlebrook is not the first literary scholar to have used psychiatric typescripts for a biography. In his 1981 book *Some Sort of Epic Grandeur: The Life of F. Scott Fitzgerald*, Matthew J. Bruccoli includes a five-page "Extracts" from a 114-page transcript of a May 28, 1933, meeting or "therapy session" that included Fitzgerald, his wife, Zelda, and her psychiatrist, Dr. Thomas Rennie, of the Henry Phipps Psychiatric Clinic at Johns Hopkins Hospital, where Zelda had been treated. The Fitzgeralds discussed their faltering marriage, Scott's alcoholism, which he angrily denied, and Zelda's fierce desire to be a writer despite the intense opposition of her husband. As the "professional" writer in the family, Scott believed that only he, not Zelda, could write about their marriage and her experience with "insanity." Frances "Scottie" Fitzgerald Smith, the executor of her parents' estate and their only child, allowed Bruccoli to include a portion of the psychiatric typescript in his biography. (He acknowledges her help in the beginning of his book.) She has also allowed other biographers to quote from the typescript. Zelda's most recent biographer, Sally Cline, explores in detail Zelda's problematic psychiatric history and treatment. Scottie's children, after her death, gave Cline access to most of Zelda's medical records, including the "sealed" medical records held at Princeton University Library. Cline twice interviewed Zelda's last psychiatrist, Dr. Irving Pine, who stated that one of her earlier psychiatrists, Dr. Robert Carroll, had been involved in a rape case with a patient. Cline quotes Pine as saying that "Dr. Carroll took advantage of several women patients including Zelda" (375). Readers of the 1933 "Extracts" will become aware of a different kind of psychiatric abuse, in which Dr. Rennie consistently sides with Scott's efforts to silence Zelda.

Psychiatric Controversy

The use of psychiatric material in the biographies of Scott Fitzgerald and Zelda Fitzgerald pales in comparison with Martin Orne's collaboration with Middlebrook. Alessandra Stanley observed that Orne's action "has caused far more consternation in literary and more particularly psychiatric circles than any other revelation in the book." Stanley quoted several psychiatrists who regarded Orne's actions as unprofessional, including Dr. Willard Gaylin, a Columbia University psychiatry professor and authority on ethics: "Doctors have no obligation to history and certainly should not act as a research assistant to a biographer." Gaylin's conclusion, that Orne had betrayed his patient and his profession, was echoed by others in the article. "A patient's right to confidentiality survives death," stated Dr. Jeremy A. Lazarus, the chair of the ethics committee of the American Psychiatric Association. "Our view is that only the patient can give that release. What the family wants does not matter a whit."

Many clinicians worried that the loss of confidentiality arising from the publication of Middlebrook's biography would undermine psychotherapy. "The question is not if Anne Sexton's wishes were correctly perceived," Josef F. Weissberg, a psychoanalyst and president of the American Academy of Psychoanalysis, wrote to the *New York Times* on July 26, 1991, "but whether or not a psychiatrist is ever justified in disregarding confidentiality without the patient's explicit, freely given permission. If Dr. Orne's action were condoned, no patient would have any basis for trusting any psychiatrist, since the decision to disclose could be made unilaterally if it furthered the goals of scholarship or any other ostensibly worthwhile purpose." Sexton's story is fascinating not only because her poetry arose from her therapy, and because of the many parallels between the talking cure and the writing cure, but also because of the contrasting dilemmas her story poses to psychotherapists and literary critics.

Dr. Orne remained convinced that he had made the right decision. "I have no question that she would have jumped at the opportunity to share what we did," he told Stanley in an interview, adding, "I was often more concerned about her privacy than she was." Others faulted Orne but praised Middlebrook's sensitive and empathic use of the psychiatric material. Some of the people quoted lauded both the psychiatrist and biographer. Maxine Kumin found the biography of Sexton to be "very balanced and judicious," described Orne's decision as "gutsy," and dismissed the objections of Orne's colleagues as "pietistic" (Stanley).

Legal Controversy

Beyond those ethical concerns, Middlebrook's use of the psychiatric tapes also generated legal controversy. As Joseph Onek pointed out in a talk presented to the American Academy of Psychoanalysis in 1992 and published in the *Journal of the American Academy of Psychoanalysis and Dynamic Psychiatry*, because the disclosure had been authorized by Sexton's executor, there was little question that Orne had acted legally.

> The Massachusetts law . . . states that a health care provider who maintains records for a patient treated or examined by such provider shall permit inspection [and copying] of such records by such patient or an authorized representative of the patient. It is thus clear that while Ms. Sexton was alive she could have obtained copies of her tapes from Dr. Orne. The only question is whether the term "authorized

representative" includes the heirs of a deceased patient. The answer is almost certainly yes. (655)

The law professor Sharon Carton disputed that conclusion, however. In a thorough review of possible causes of legal action that could have been taken by Sexton's estate, Carton argued that "even if Sexton's views can be ascertained from beyond the grave, and even if those views do not condemn the release of the tapes, it must still be considered whether there are rights belonging to other potential victims of that action, and whether their rights have been violated" (117). In Carton's opinion, the publication of the biography served neither the poet nor society. "Harm was done, not just to Sexton and certain members of her family, but to the legions of mental health patients who rely on psychiatrist-patient confidentiality." Carton believed that the Sexton estate had three common law causes of action arising from the wrongdoing caused by the publication of the biography: invasion of privacy, breach of confidentiality, and the contractual covenant of confidentiality. "While it is unlikely a claim will be brought," Carton conceded, "it is a noteworthy instance of privacy and confidentiality violations which should not be tolerated and perpetrated. It is hoped that the outpouring of community and media outrage in the Sexton case will inhibit, if not prevent, reoccurrence of such a disclosure" (164).

In offering her view, Carton could well be referring to the fact that a tape recording of a session may contain confidential information about other persons, information about confidential business dealings, or any number of other "secrets" a patient holds about others but discloses in a therapy session with the conviction that such information will forever remain confidential. Thus a patient who later consents to the release of such tapes may be inadvertently violating a solemn promise or a legally binding business obligation. The release of tape recordings of sessions is therefore a much greater threat to privacy than the obvious threat entailed by the therapist's session notes, where such secrets of others may not have been recorded or recorded with identifying information removed (such as the use of initials only to signify other persons in the patient's life). What obligation does a therapist have to third parties mentioned in such tapes? How could a family member or executor gain the right to authorize their disclosure without hearing them first, which could, itself, be a violation of another person's privacy?

Contrary to Carton's hope, the Sexton family sought neither an injunction to prevent publication and distribution of Middlebrook's biography nor legal redress for invasion of privacy—despite the fact that the psychiatric tapes contained confidential and embarrassing information about other members of the Sexton family.

How would the Sexton-Orne controversy be played out in the courts today? In recent years, the laws regarding the protection status of confidential information about a deceased person have been changing. In 1998 the Supreme Court ruled in the Vince Foster case (*Swidler & Berlin v. United States*) that the lawyer-client privilege, asserted by a lawyer on behalf of a deceased client, endures indefinitely. The principle probably will also apply to the psychotherapist-patient privilege. The U.S. Department of Health and Human Services similarly created in 2000 an indefinite duration of protection for health information in the original HIPAA Privacy Rule, but as a result of protests from historians, archivists, and researchers, the department changed the rule in 2010 so that information loses all HIPAA protection fifty years beyond a patient's death. This applies to the more stringently protected "psychotherapy notes" as well as to other health data. In addition, in guidance published in September 2013, the department reiterated that during the fifty-year period of protection, the data can still be released on an authorization from the deceased person's executor ("Health Information of Deceased Individuals"). The department nevertheless pointed out that the information may still be protected beyond fifty years by other laws and by ethical requirements incumbent on professionals.

Another Controversy

The contentious debate surrounding Linda Gray Sexton's decision to allow Middlebrook to use Orne's audiotapes of Anne Sexton has overshadowed what may have been a far more problematic chapter in her long psychiatric story. After Orne moved from Boston to Philadelphia, Anne Sexton entered treatment with a psychiatrist with whom she then developed a sexual relationship. The first hint of this appears in a 1964 letter published in *A Self-Portrait in Letters*. "My therapy is degenerating to SEX" (231). Middlebrook supplies many details about Sexton's long affair with "Dr. Zweizung," a pseudonym which means "forked tongue" in German. Sexton never attempted to conceal the affair from her relatives, friends, and former psychiatrist, all of whom were understandably disapproving. The affair did not *appear* to be harmful to Sexton, Middlebrook notes, trying not to be judgmental: "during the years Dr. Zweizung was her therapist, her suicidal thoughts and attempts abated, she underwent fewer hospitalizations, and she seemed far more stable to her family" (314). Middlebrook leaves little doubt, however, that those who knew about the affair regarded it as a serious breach of professional ethics. In a 1985 interview, Dr. Orne told Middlebrook he was deeply troubled by the romantic relationship, which he tried unsuccessfully to end in 1966:

> Anne told me what was happening, and I said to her, "All right, you've got to stop this." She said, "I can't." And I said, "Okay, I'll be in Boston, I want to see you and Dr. Zweizung together in my office."
>
> Dr. Zweizung wasn't eager to come but I asked whether he'd prefer me to go to an ethics committee. He came, and we talked. I told him, very straightforwardly, "Well, therapists are people too, and these things happen. But it isn't something which is good for Anne. And you know you have no right to charge her: you use her, that's destructive." I said, "Look, if you want to continue to treat her, you go back to your analysis, and if you don't want to go back into analysis, then you can't treat her." (Middlebrook, 315)

Alessandra Stanley named the psychiatrist in question as Dr. Frederick J. Duhl, who refused to comment on the affair for reasons he made clear to Stanley in a telephone interview: "You are dealing with an explosive subject: basically any doctor who has an affair with a patient loses his license in Massachusetts." Orne decided not to bring up ethics charges against Duhl. "I didn't want to ruin his career," Orne told Stanley. "Today, I might have done it differently."

Curiously, Stanley devoted less space to the second controversy, Anne Sexton's sexual relationship with one of her psychiatrists, than to the first one, the use of her therapy tapes in a biography. Few people who have written on the Anne Sexton controversy have speculated on why a biographer's use of a deceased poet's therapy tapes is a more serious transgression than the boundary violation arising from a therapist's sexual relationship with a patient. Indeed, the professional mental health organizations that were so vociferous in their criticisms of the use of Sexton's therapy tapes were much more muted in their disapproval of the psychiatrist who committed what is probably the most serious boundary violation in therapy.

Linda Gray Sexton referred to this psychiatrist in *A Self-Portrait in Letters* as "Dr. Samuel Deitz" (225) to protect his privacy, but she used his real name in *Searching for Mercy Street*, telling us in a footnote that "since the *New York Times* named Duhl in their front-page story covering the variety of explosive issues raised by the biography, I no longer feel such compunction"

(136). Only by listening to the therapy tapes and reading Middlebrook's biography did Linda begin to realize the extent to which her mother was hurt by her romantic involvement with Duhl. He promised repeatedly to leave his wife and marry Sexton, but he broke the promise, broke off therapy with her, and, soon afterward, she fell down a flight of stairs and broke her hip—three breaks that appear causally connected.

It took Middlebrook ten years to research and write her biography. She notes in the preface that everything she learned about Sexton indicates the poet would not have held back anything from the biographer, including the therapy tapes. "Sexton was not a person with a strong sense of privacy. She was open and impulsive: many people found her exhibitionistic, and some of the people who lived with her found her outrageously, immorally invasive. But her lack of reserve had a generous side as well, which was, I think, connected to her spirituality" (xxii–xxiii).

Hearing the audiotapes compelled Middlebrook to rewrite the biography, though she doesn't explain why. The explanation appears in *Searching for Mercy Street*, where Linda quotes Middlebrook's letters reflecting her shock upon hearing the tapes, when she learned for the first time the "depth of her sickness in its earliest forms." The tapes are for Middlebrook a "setback" because they added years to the biographical project. The tapes are also a "true goldmine of the incidental stuff that makes biographies so good." The result is a new kind of biography. *"I mean, there is nothing like this in the history of literary biography!* Braiding together the doctor's notes gives glimpses of her development as a writer and as a person in those years, as she emerges from deeper despair than I ever realized before" (238).

Reaching Different Conclusions About the Controversy: "Privacy, Professionalism, Psychiatry"

The journal *Society* devoted a special issue in 1992 to "Privacy, Professionalism, Psychiatry," and the seven psychotherapists and academics writing for that issue reached different conclusions about Orne's collaboration with Middlebrook. Barbara L. Lewin, a clinical psychologist, condemned everyone associated with the case. "Therapeutic mismanagement seems to have plagued every aspect of Anne Sexton's psychotherapy. The Middlebrook biography provides a chilling narrative of compromised treatment at all levels, including clinical self-aggrandizement, poor observance of boundaries, and outright violations of professional ethics" (9). Lewin was no less censorious of Anne Sexton, whom she saw as both a victim and victimizer and for whom she had little sympathy. "She was a victimizer in that she tyrannized her family with her abusive mood lability and seduced her therapists into dealing with her as someone quite extraordinary—a literary *enfant terrible*" (10). Lewin concluded that "it is essential to hold Martin Orne accountable for his flagrant disregard of boundaries and for his professional exhibitionism" (11).

Jerome Kroll, a professor of psychiatry at the University of Minnesota School of Medicine, opined that Dr. Orne should have destroyed the tapes or handed them over to Sexton as soon as her therapy ended. "Orne was absolutely incorrect in violating confidentiality—absolutely because the tapes were generated within the privacy of a special relationship that recognizes the inviolability of confidentiality" (18). Kroll described as "spurious and self-serving" the claim made by some psychiatrists that Orne's actions would make patients distrust their therapists—only because the trust had already been lost. Kroll predicted that psychotherapy would be inevitably compromised in the future: "in the final analysis, every time confidentiality is breached, however good the cause or urgent the demand, the principle of confidentiality is weakened and eroded" (20).

Moisy Shopper, a clinical professor of child psychiatry and pediatrics at St. Louis University School of Medicine, also denounced Orne's decision to release the tapes. "The claim that a deceased patient no longer deserves autonomy and integrity of the self is not tenable. Death only precludes the individual from knowing about the violation; it does not negate the violation itself" (25). Rejecting Orne's belief that the use of a patient's psychiatric tapes in a biography may be valuable, Shopper argued that, for the physician, "an even greater ethical precept than 'doing good' is that of 'doing no harm'" (24). Shopper concluded that Orne's actions would make his work with some patients harder. "No longer will the therapist and patient deal with 'mere' fantasies and fears of disclosure; now, one must deal with the possibility and precedent of actual disclosure" (26).

Two therapists took more moderate positions. Francis Degen Horowitz, a developmental psychologist and president of the Graduate School and University Center of City University of New York, asserted that the "issues related to confidentiality and privacy are never absolutes. Societies develop conventions that govern when, where and how confidences and privacy are kept and when, where and how they are breached" (28). Horowitz suggested that the private letters, diaries, and, presumably, therapy tapes saved by individuals "were meant (consciously or unconsciously) by their authors for someday serving public purposes" (28). Mary Gergen, a professor of psychology and associate of the women's studies program at Penn State University, Delaware County Campus, concluded that despite the furor over the controversy, one might view Orne's actions as helpful rather than harmful to therapy. "From the tapes, one can see in slow motion how psychiatry heals the sick" (23).

The two academics who contributed to the special issue of *Society* reached different conclusions. Paul Roazen, a professor of social and political science at York University of Toronto, Canada, and one of the preeminent historians of psychoanalysis, faulted Orne for breaking the rule of confidentiality. Roazen speculated that most therapists would now diagnose Sexton as a "schizophrenic psychotic." She did have a number of schizophrenic symptoms and was treated with antipsychotic medication at Mass General, but later therapists have not offered this diagnosis. Roazen suggested that even if she had authorized the use of the tapes in her biography, "the question remains whether she had the mental competence to do so" (14). Roazen did not fault Middlebrook for using the tapes: he found her biography engrossing "partly because it is a tale of transgression" (14). Vern L. Bullough, a distinguished professor of history at the State University of New York College at Buffalo, commended both Orne and Middlebrook. Anne Sexton's medical records belong to the family, Bullough asserted, and once the family requested them from Orne, he had only two choices: he could either destroy the tapes or take the "honorable and honest course of action" in turning them over (12). Praising the integrity of Orne, Linda Gray Sexton, and Middlebrook, Bullough believed that as a result of all three, we gain a better understanding of Anne Sexton.

Had the seven contributors to "Privacy, Professionalism, Psychiatry" commented on one another's positions, they would have realized the conspicuous absence of a clear consensus among them. They disagreed over every aspect of the controversy: whether Sexton was helped or hurt by her years of psychotherapy, whether the release of the tapes would help or hurt present and future psychotherapy patients, whether the biographer should have used the tapes, and whether the ethics of the case rested upon absolute or relative guidelines. They also differed over the motivation of all those involved in the controversy. Even those who censured Orne for his decision to turn over the tapes disagreed over why he should not have done so or the long-term implications of his actions. Some interpreted the controversy as a cautionary tale about the growing dangers of the loss of confidentiality and privacy in psychotherapy; others saw the controversy as a welcome opportunity for a greater understanding of the process of psychotherapy and its influence in shaping the life and art of a major American writer.

Linda Gray Sexton

No one was more familiar with the Anne Sexton controversy than Linda Gray Sexton. She was the one who, after all, made all the decisions that led to the strident debate. From one point of view she was the best person to be her mother's executor, for she was the person closest to her mother, knew her perhaps best of all, and was herself a developing author. But from another point of view Linda was so enmeshed with her mother, struggling herself with many of her mother's psychological conflicts, that she may not have had the objectivity to think through all the implications of the "Anne Sexton controversy." To understand Linda's decisions, we must turn to her two memoirs, which prove to be as confessional as her mother's poetry.

Growing up with a suicidal mother was understandably difficult for Linda Gray Sexton, as she admits throughout her memoirs. As her mother's older daughter, confidante, caregiver, editor, and literary executor, Linda has worked tirelessly, if at times ambivalently, to promote Anne Sexton's literary reputation. The story of their tangled mother-daughter relationship forms a major part of *Searching for Mercy Street*, which was named a *New York Times* Notable Book of the Year. The memoir dramatizes Linda's anger over her mother's repeated invasion of her privacy as a child, teenager, and adult. Writing the memoir both required and resulted in a further loss of the daughter's privacy, a disrobing in public worthy of Anne Sexton herself.

"I am determined not to be like my mother, spilling secrets like water from a cup, writing out a family's shame. I will lock our dirt up tight" (*Searching for Mercy Street*, 41). The memoirist is keenly aware of the multiple ironies of her statement. How could she not be? Throughout the story she never lets us forget the extent to which her mother's extended family felt hurt, angered, and betrayed by the poet's disclosures. Throughout the memoir one senses that Linda is torn between the need to protect her family's privacy, on the one hand, and the quest for truth, on the other.

Searching for Mercy Street reveals not only the reversal of roles in the mother-daughter relationship but also the blurred boundaries between them. Gender theorists such as Carol Gilligan and Nancy Chodorow have noted the remarkably porous boundaries in the mother-daughter bond, resulting in a closer relationship than usually exists in the mother-son relationship.

Sometimes this closeness becomes problematic, as Linda demonstrates repeatedly. At first she welcomed closeness with her mother—they often felt like they were two sisters—but she came to realize the psychological harm caused by physical and emotional intimacy. "Gradually I stopped confiding in her and stopped listening to her secrets in exchange. I began to resist her intrusive questions" (131). A crisis arose when Linda decided to see a psychiatrist to treat her own depression. Anne Sexton responded "nearly joyfully" to her daughter's decision, regarding it as an opportunity for "reunion," and she helped her find a psychiatrist, Dr. Adele Shambaugh. The mother even drove her daughter to therapy. The problem was that Anne Sexton, clearly demonstrating her distorted view of appropriate boundaries, demanded her daughter tell her everything discussed in therapy. "'What did you talk about?' she would inquire. 'I'm paying for this, so I have a right to ask'" (133). Anne Sexton's resentment of Dr. Shambaugh turned to hatred when the psychiatrist called her to complain about the extent to which she was threatening to breach her daughter's privacy.

As difficult as it was for Linda to disclose these boundary transgressions, it was even more wrenching to reveal her mother's sexual transgressions, which became traumatic memories almost impossible to verbalize. "Living with Mother, I had seen enough sexual acting out to last me a lifetime" (105). Her mother would sometimes crawl into bed with Linda, press herself against her body, and masturbate. The frightened teenager could only close her eyes and pretend to be asleep. Only near the end of the memoir does Linda disclose the most horrific memory, which occurred in 1961, when she was nine and her sister was seven. One night when her father was out of town for business, her mother joined her in bed, lying on top of her and pressing her tongue deep into her mouth, an act that Linda Gray Sexton's psychiatrist likened to incest. The repressed memory was both triggered and corroborated when Linda Sexton listened to one of her mother's therapy tapes with Dr. Orne:

> "You said [talk about] sex.—Could talk about my children.—I don't want to say what comes to mind. . . . The children's relation to my body: six months ago taking a bath, touching [my] breasts, later in bedroom with Linda says did milk really come out? Can I find out? I said yes. I didn't know at what point to back out. . . . They make me feel more womanly. Joy kissing the other breast, I wonder if I'm taking it a little too far. . . . Linda got sand in her bottom at the beach. I gave her the washcloth and told her she could wash it off herself. She pointed to her clitoris and said

her friend Nancy said this was just like a boy, that it will grow. I said that's your clitoris and it won't grow, but when you have intercourse, mating, it will feel good there. She said you don't have one of those all you have is hair. So I showed her. So I told her she should masturbate then she'll know that it feels good there. But maybe this is too intimate. This is usually what you do with your friends, not your mother." Orne: "Why do you want her to masturbate?" Anne: "I'm just making it a little more right if she does." (*Searching for Mercy Street*, 269–270; brackets and ellipses in original)

This was Linda's most traumatic memory of her mother's abuse, but there were many other distressing recollections as well, some of which resurfaced when the daughter listened to the therapy tapes or read her mother's transcripts of them. In the late 1950s, when she first went into therapy, Anne Sexton alternated between depression and rage. She became so violent that once she slapped her daughter and tried to choke her. Linda bore the brunt of the violence because her mother favored Joy. Sometimes Anne Sexton's violence was internalized. Linda tells us, without using irony or an exclamation point, that her mother attempted suicide "only nine times" (33). Anne Sexton would sometimes dissociate, entering into a fugue state or trance, behavior that also occurred when she was in therapy with Orne. At times she spanked Linda with brushes or sneakers; other times she threw her against a wall. "Something comes between me and Linda," she confessed to Orne. "I hate her. I want her to go away and she knows it" (209).

Much of what Linda comes to know she credits to psychotherapy, both her own and her mother's. The therapy tapes and transcripts are unsettling to her, but they offer her a second perspective on her mother's illness— and a third perspective, Orne's, as well. Few people have found themselves in the daughter's situation, a witness to a parent's many years in therapy. The audiotapes and transcripts reveal not only her mother's darkest feelings and thoughts but also the ways in which she parented her children. The tapes thus have educational and psychological value for Linda, allowing her to understand her tortured past.

As Anne Sexton's illness worsened, she expected Linda to be her caregiver and confidante, roles no young daughter should be forced to play to a parent. As a child, she felt compelled to be her mother's watchdog, protecting a woman who could no longer protect herself. She fulfilled these roles as best as she could, but eventually she realized she could not be what her mother needed. There were times when she felt she had no choice but to accede to her mother's demands, as when the poet asked Linda to be her literary executor. She describes her silent thoughts when she heard her mother making the request as a fait accompli: *"don't make me responsible for your life,* my reluctance said. *I don't want to be the gatekeeper anymore"* (*Searching for Mercy Street*, 182). She reluctantly agreed to her mother's request, and after her mother's death, she read boxes of the unpublished writings, including therapy notebooks. She was tempted to burn many of these writings, but she couldn't. Her mother's suicide left her feeling overwhelmed by guilt. One of Linda's most brutally honest admissions is that she had grown weary of her "mother's drama" and wished for the "forbidden," namely, that "she would bring the curtain down and let us rest at last" (185). Anne Sexton's suicide provided her daughter with only temporary relief, however. Agreeing to be her mother's literary executor placed new demands upon her. In life, Anne Sexton invaded her daughter's privacy; in death, she continued the invasion.

Devastated by her mother's suicide, which she had been expecting for years, Linda immersed herself in work as a way to mourn the loss. "The only activities that made me feel better were working on my senior thesis on Virginia Woolf—plunging myself into a world of words—and taking care of Mother's work" (201). Slowly her feelings toward her mother began to change. As a result of her mother's therapy notebooks, her own therapy, and becoming a mother herself, she found her anger giving way to understanding and forgiveness.

Loyalty to One's Art or Family?

The publication of Sexton's biography horrified her relatives, as Linda admits ruefully. Appalled by the invasion of their privacy, they challenged the accuracy of the poet's characterization of the family, including statements that her parents drank heavily and that her father had sexually molested her as a child. Sexton's poems are filled with painful accounts of childhood experiences, including enemas and genital examinations performed by her mother. Growing up in a family filled with terrible secrets, Linda recalls her fear that betraying these secrets would result in a terrible fate: *"If you tell they will not love you anymore. If you tell they will send you away again"* (21). The daughter was thus confronted with the same agonizing dilemma her mother had experienced decades earlier: is a writer's primary loyalty to one's art or to one's family?

Anne Sexton never doubted her primary allegiance was to her poetry. When her third and final psychiatrist, "Dr. Constance Chase" (a pseudonym), accused her of breaching Linda's privacy by writing about her in "Mother and Daughter," a poem that describes Linda as looking forward to her mother's death, Sexton—sounding like F. Scott Fitzgerald and Philip Roth—angrily insisted that no one could dictate to her the people she could or could not write about. "I strongly resent the fact that you feel I am using Linda. . . . You so winningly said, 'People come first' meaning before the writing. You forced me to say the truth. The writing comes first. . . . This is my way of mastering experience" (*Searching for Mercy Street*, 146; ellipses in original). Years later, Linda remains conflicted about these invasive poems.

One of Linda Gray Sexton's challenges in writing *Searching for Mercy Street* is to be as truthful as her mother was without invading others' privacy, as her mother did. She makes no effort to spare her own feelings. The last years of her mother's life were characterized by intensifying physical and psychological crises that left the daughter feeling angry and drained. "I hated her selfishness and her sickness, and I could no longer tell where one stopped and the other began" (172). The hope that her mother would die eventually came true, leaving the daughter guilt-ridden and bereft. "How I regret my own hardness" (185). Grieving her mother's death was a process that took years. In many ways the mourning is ongoing. She struggled to define her own identity when her mother was alive, and she found herself wearing her mother's clothes and jewelry after her death. Editing her mother's unpublished books was another way to remain close to her—a healthy identification. She also became angry, depressed, alcoholic, and suicidal.

Linda could not have chosen a better literary critic than Diane Middlebrook, whose biography is remarkably nuanced. Nevertheless, Linda admits that Middlebrook found it impossible to tell Anne Sexton's story without hurting the living members of the family, including Linda's father, Kayo. "No family member will ever like this book," Linda wrote to Middlebrook. "You must not care about that any longer: it is an impossible task. We are all hurt by it." Linda's conclusion is that the "only way to transcend the hurt is to tell it all, and to tell it honestly" (276).

Linda never minimizes the widespread disapproval of her decision to allow Middlebrook to use Anne Sexton's audiotapes and therapy notebooks. She summarizes one of the major arguments against making the public disclosure of her mother's therapy. "Some critics would later contend that a loving daughter would have protected her mother even from her own obsessive need to display herself through her poetry" (231). After listening carefully to the tapes, she concluded that they "revealed important and formerly unknowable aspects of the manner in which she created, even as they exposed certain unattractive aspects of her life" (231). In the end, Linda believed that her mother would have wanted the tapes to be made public. As a youth Linda wanted to become a therapist but then decided, against her mother's advice, to become a writer. Both believed in the power of language to convey truth and change lives, and both committed themselves to this belief regardless of the consequences. "If we must hear the truth," Linda writes to Middlebrook, "let us hear it all. Let us be able to say at the end: this is the price and reward of madness; this is the price and reward of genius" (276).

An Accident of Hope

Dawn Skorczewski, the author of *An Accident of Hope,* has listened to all the therapy tapes. She's in a unique position to comment on the Anne Sexton controversy. A professor of English at Brandeis University, Skorczewski was the 2009 recipient of the CORST Essay Prize for Psychoanalysis and Culture from the American Psychoanalytic Association and the 2007 recipient of the Gondor Award for Contributions to Psychoanalytic Education. She brings exemplary insight and empathy into her discussions of Sexton's therapy.

Skorczewski understands the importance of confidentiality in psychotherapy; she also understands the value of studying Sexton's therapy tapes. She never doubts that Sexton would have wanted a biographer to use the therapy tapes, nor does she doubt that Orne made the right decision in turning them over to Middlebrook. She agrees with Middlebrook that the tapes are indeed a goldmine. The tapes offer insights to those who may be struggling with mood disorders, as Sexton did, as well as to those who seek to heal themselves through writing, again like Sexton. The tapes give mental health professionals an opportunity to evaluate Orne's long-term therapeutic relationship with Sexton. Finally, the tapes show the ways in which poetry and therapy influence each other and how both reflect and shape culture.

Skorczewski remains convinced that Sexton's relationship with Orne was valuable for many reasons. "In eight years of talking together two and three times a week, Orne had helped Sexton to discover a self that she could actually like" (*An Accident of Hope,* 187). Focusing on the last six months of the audiotaped sessions, she reveals how Orne helped Sexton forge her poetic identity.

Listening to the tapes nearly half a century after they were recorded allows Skorczewski to place Orne's long and complex treatment of Sexton in a historical and cultural perspective. She is both sympathetic to and critical of his responses to Sexton. Orne was not trained as a psychoanalyst, but he accepted many of the Freudian assumptions of the age, including belief in a one-person model of treatment in which the therapist remained "objective" and "detached." He responded to Sexton's most anguished self-disclosures with either silence or a noncommittal "mmm." He encouraged Sexton's creativity, but he praised her role as a mother more than that of an emerging major American poet. He was unable to appreciate the extent to which she had been sexually traumatized as a child: posttraumatic stress disorder, which did not begin to appear in the psychiatric literature until the late 1970s, was not yet a clinical diagnosis. Skorczewski suggests plausibly that Sexton might have done better with a therapist who took a relational, intersubjective approach, one who would give her the empathic mirroring she needed. Sexton would have valued a therapist who was more attuned to feminist issues. "If we hear Orne differently today, it is in part because educated readers have gained a different appreciation of a woman's worth than was common in Orne's time" (119–120).

An Accident of Hope casts further light on the Anne Sexton controversy through Skorczewski's surprising discovery that Dr. Orne never precisely answered Sexton's question, which she raised during a session on October 24, 1964, when Orne had returned briefly to Boston, about why he wanted to keep the tapes. Sexton implicitly points out to Orne that having served their stated therapeutic purpose, there seemed to be no further need for him to keep them. She even raised the question as to why the two of them did not simply record each session on the same tape, thus erasing the previous recording after it had served its stated purpose. Six years later, M. M. Stern actually proposed such a protocol for using audiotapes for therapeutic purposes. Orne's stumbling reply includes this stunning passage: *"I certainly don't intend to uh you know give them to anyone else or that sort of thing. Uh. I don't know. Uh. I guess it's just part of the same tendency that I tend to collect all kinds of uh data which may be much more than needed to be saved"* (215; emphasis added). At this point Sexton's tentative but assertive stance melted, and, perhaps afraid to anger someone to whom she was profoundly attached and was in the process of losing, she said sympathetically, "It's a creative thing," to which he responded, "Well, I don't know what it is.

I kind of have the feeling sometimes maybe that 10 years hence that I'll wish that I could go back and really know what happened then."

Anne Hayman, the British psychoanalyst who refused to testify about a patient, would probably assert that Sexton's permission for Orne to keep the audiotapes of her therapy was based on her positive transference rather than a thoughtful consideration of what was in her own best interest. It is difficult to escape the impression that his remark about not giving the tapes to anyone else was an implicit promise that was intended to allay her concerns about his future use of the tapes for nontherapeutic purposes. It's also difficult to escape the impression that the statement he made to Alessandra Stanley in the *New York Times* and that we quoted earlier—"I was often more concerned about her privacy than she was"—was not the complete truth.

The tone of Sexton's voice throughout the October 24, 1964, therapy session, according to Skorczewski, was "politely curious but adamant" in her determination to understand Orne's reasons for keeping the tapes. By contrast, his response to her pointed question was "stuttered," suggesting that he was unable or unwilling to answer the question. Their roles seem reversed here: the patient affirms and mirrors the therapist, helping convince him he has made the right decision to hold onto the tapes. Their conversation is ironic on many levels, as Skorczewski acknowledges. Sexton appeared more curious about Orne's motives for keeping the tapes than when she saw him regularly in therapy. Did she want to keep the tapes herself now that their therapy was over? Did she long for Orne to tell her that he wanted to keep the tapes because she was a special patient and that, although he was losing her as a regular patient, he still wanted to cling to some part of her? Was Sexton projecting herself a quarter of a century into the future, when a biographer might be interested in listening to the tapes? Skorczewski raises all these possibilities. She cannot help gasping when she realizes that Sexton committed suicide almost ten years to the day from the date of Orne's prophetic remark. "Of course this is an uncanny coincidence," Skorczewski remarks, "and yet it leads to important questions about how Sexton's therapy influenced her most important decisions" (216).

Orne's failure to articulate to Sexton his reasons for keeping the tapes may have been because he didn't understand the reasons himself. He never raises this question in his foreword to Middlebrook's biography, where he justifies his decision to give her the tapes, but he does provide a partial explanation in "The Sexton Tapes," where he reveals that he offered to return the tapes to Sexton in 1964, when he left Boston. "She asked that I keep them to use as I saw fit, though she retained a few for herself." Would Orne have made the same decision had he heard himself promise Sexton that he would never give the tapes to anyone? Perhaps. As he notes in "The Sexton Tapes," "After the death of a patient who advocated disclosure and gave consent to it, the consent of the family, in my opinion, is foremost."

Transference and Countertransference

"Anne Sexton was a very difficult patient to treat," Orne remarked to Samuel M. Hughes. "She was very seductive. But you know, if you can't deal with that, you should not be a psychiatrist" (23). Her early poems for Dr. Orne were a figurative seduction of or flirtation with her reserved therapist. Many of these poems were either odes to Orne or self-revealing accounts of what might be called an exhibitionistic nature. As we suggested earlier, Sexton's poems were a key enactment of her growing attachment to Orne and thus part of her transference relationship to him. The "creation" of this famous poet could then be said to be the result of what in effect was a "co-creation": Sexton's obedient deflection of sexual and more primitive feelings for Orne into her poetry, and the tightly wrapped psychiatrist's encouragement of her writing as a way for her to find a sense of self-worth, a way to make a man "feel powerful" without the need literally to engage in prostitution. One can only imagine the extent of the pair's surprise, if not shock, when Sexton's success made her a celebrity. Orne seemed to believe that in a certain way Sexton the poet was partly his creation. Sexton shared that belief. She viewed her psychotherapy and its unexpected outcome as the psychiatric equivalent of (in a word she herself chose) a "Pygmalion" story (Skorczewski, 219). In the final of all the known tapes, dated February 20, 1965, she admits to Orne that the majority of her early work was written for him, a statement she had expressed more fully in a November 21, 1963, session: "I remember the first time I met you, you giving me the Rorschach test, I liked you. . . . Funny that you don't care about me as a poet. But you created me as a poet. You cared about me, then you, I, *we created the poet*" (Skorczewski, 19; emphasis added).

Both Sexton and Orne viewed her identity as "Anne Sexton, the poet" as a kind of offspring stemming from their relationship. For a beginning psychotherapist, such a fantasy is understandable, but behind this relatively benign belief lie the seeds of a serious countertransference problem. Fifteen years after Sexton's suicide, Orne remarked, in a 1991 interview justifying his involvement with the Middlebrook biography, "I guess that, since I still think *Anne's somewhat part of me,* I'd like her to be understood" (Hughes, 39; emphasis added).

Orne's departure from the Boston area for a research position in Philadelphia, after eight years of psychotherapy with Sexton, brought their frequent sessions to an end. Nevertheless, neither of them was able to let the other go. Orne referred her to the psychoanalytically trained Dr. Frederick Duhl, but despite the fact that she began seeing the new therapist, Orne continued to see her during his monthly return visits to Boston. As Orne recounted to Middlebrook, "'I saw Sexton almost every month, though I was not her primary therapist. I maintained a relationship *more like that of a parent or friend,*' he commented later. 'Thus, when she was in trouble, real trouble, she'd call on me'" (314; emphasis added). This bizarre split treatment went on evidently for years. Orne failed to report Duhl to the licensing authority and instead met with them both to advise ending the "treatment," including their sexual encounters, for which, incidentally, Sexton was paying.

The relationship between Sexton and Duhl ended when the psychiatrist's wife found out about the affair and he decided to remain in that marriage. Ultimately, Sexton was placed with yet a third psychiatrist, the psychoanalyst Constance Chase, who pressured her to cut off her relationship with Orne. Supposedly Sexton tried to do this, although in his foreword to Middlebrook's biography, Orne suggests that they had still been planning to meet at the time of Sexton's suicide. Sexton's condition deteriorated, and eventually the third therapist terminated the treatment without referring her to someone else. Nine months later, Sexton committed suicide.

Sexton's loss was devastating to Orne. One senses from the few statements he made publicly on the subject that he was enraged with those he felt had betrayed both him and Sexton. The betrayal resulted in the loss of a friend, a star patient, and perhaps what he may have viewed as his most impressive therapeutic achievement. One wonders, however, if he wasn't also haunted by self-doubt as to whether he had been qualified to treat Sexton in the first place, whether he had betrayed her by moving away, whether he had done her a massive disservice by referring her to Duhl, and even whether she had betrayed him by her sexual involvement with Duhl. He was still troubled years later, when he was under fire for having released the tapes. "There's nothing that makes a therapist more pleased than to be able to help someone. That's what I find most sad. And I guess I am still angry at what happened to her, because I wish she were alive today. She had a lot to give. And she was a suicide that was not necessary" (Hughes, 23). Orne closed his foreword to Middlebrook's biography with these bitter words, aimed apparently at the psychiatrist who in his view severed his relationship with Sexton, and to convey his conviction that his relationship with her continued to be a source of "vital support."

There is no doubt that Orne expected to see Sexton again.

> Although I felt obligated not to interfere with the guidelines that had been established for Anne's treatment [by Dr. Chase?] in the last year of her life Anne called to say that she would be in Philadelphia to give a reading at the public library and that she hoped she could see me. I expected to see her, but she never made it. Sadly, if in therapy Anne had been encouraged to hold on to the vital supports that had helped her build the innovative career that meant so much to her and others, it is my view that Anne Sexton would be alive today. (xviii)

From this account of Orne's involvement with Sexton we may infer additional motives in his decision to release the tapes of her therapy, an act whose wisdom he later called into question. "If I had to do it over again, I don't know what I would do" (Hughes, 39). At that time he had convinced himself that he was doing what Sexton wanted, even though, as we noted, he had denied, during one of their taped meetings in 1964, that he intended to give the tapes to anyone when she pressed him on his reason for keeping them. Was Orne guilty, angry with himself for the outcome of this tragic case? Was he now seeking vindication through disclosing his work with her to public scrutiny?

Martin Orne paid a heavy price for his decision to release the tapes. He faced an ethics complaint within the American Psychiatric Association and was found to have committed an ethics violation. Appealing that decision, which almost three years later resulted in a reversal of the original finding of an ethics violation, cost Orne close to $100,000 in legal costs. He suffered in other ways, as Max Rosenbaum reports:

> When Orne wrote to this writer he stated that his health had been impaired, his son had been roughed up at day camp and scientists at his research laboratory were told that it was assumed that his laboratory would close. Many people told him that he should not appeal the original charges against him. While patients should at all times be protected from exploitation, he asked and stated: The accused physician should be entitled to the same rights to due process that an ordinary citizen has.

He described his experience as harrowing. His advice to colleagues is: Do not suffer alone, do freely discuss the matter with as many colleagues as possible, and do not allow procedures to intimidate you into not getting the help you need. (167)

Ending the Story but Not the Controversy

Linda Gray Sexton was forty when she published *Searching for Mercy Street*, and it appeared that she had come to terms with her mother's suicide. Five years later, when she turned forty-five, the age when her mother ended her life, the daughter found herself staring into the same abyss.

The story of Linda's near-fatal identification with her mother, including the three times Linda nearly succeeded in killing herself, appears in *Half in Love (Surviving the Legacy of Suicide)*. The title comes from John Keats's poem "Ode to a Nightingale" in which he writes about being "half in love with easeful Death." As she observes in the preface to *Half in Love*, writing *Searching for Mercy Street* helped her come to terms with her mother's life but not with her death. "I needed to confront my own struggle with depression, bipolar illness, and our family's history of successful suicides. This struggle reflects the emotional, and perhaps biological, legacy that was passed on from my mother to me" (xii). Linda's second memoir casts further light on the Anne Sexton controversy by showing how the daughter's understanding and forgiveness of her mother turned out to be far more complicated than she imagined in *Searching for Mercy Street*. *Half in Love* shows how Linda's life paralleled her mother's in expected and unexpected ways. Both tried to self-medicate with alcohol, which only exacerbated their problems. Both were horrified by the breakup of their marriages and estrangement from their children, resulting in a catastrophic loss of their support system. Both broke promises to their children that they would never harm themselves.

Suicide is often an intergenerational problem, revealing the sins of the fathers—or, in Linda's case, the sins of the mothers. Central to her assumptive world was the belief that she was fated to end her life as her mother had, in thrall to suicidal depression. In her research Linda discovers that the "biological concordance between bipolar patients (or those with other affective disorders) and parents (or siblings) was eighty percent" (*Half in Love*, 102). Even more frightening is the discovery that 90 percent of those who commit suicide suffer from a mental illness such as depression or bipolar disorder. In her mother's family alone there were three suicides within two generations, evoking a terrible sense of history repeating itself. "Once, I'd vowed that my children would never feel the same pain as I had when I was little. Now, I was horrified to see how easily I had hurt them" (214). She doesn't mention the son of Sylvia Plath and Ted Hughes, Nicholas Hughes, who committed suicide in 2009 at the age of forty-seven, but Linda Sexton's self-destructive tendencies were part of her maternal legacy.

Half in Love chronicles Linda's descent into hell and slow ascent. She was so severely depressed and in an alcoholic haze that for many years she was unable to read or write. "I hadn't known that depression could gut you and leave you there like a flounder, useless, your one eye staring up at the ceiling" (227). She quotes J. M. Coetzee's statement in his novel *Elizabeth Costello* that some experiences are too dangerous to write about—even more dangerous for the writer than the reader. Linda reveals near the end of *Half in Love* that her psychiatrist, Barbara Ballinger, gave her a new diagnosis, borderline personality disorder, which is characterized by an "enormous fear of abandonment, to the point of incapacitation, as well as including, nearly always, self-mutilation" (251). The diagnosis helps Linda understand her pattern of idealizing and then demonizing those people she perceives as abandoning or failing her. With the help of her psychiatrist and the right medication, Linda writes her highly self-revelatory story, one she hopes will help those who suffer from mental illness and their families.

Linda acknowledges in *Half in Love* that she had a "destructive symbiosis" with her mother (16). If so, Linda was perhaps not the best person, as her mother's literary executor, to determine whether her mother's psychiatric records and tapes should be revealed to the public. It is probable that in many similar cases the executor of a patient's estate could be a close relative with a similar problematic relationship with a deceased patient, a relationship fraught with intense feelings: guilt, anger, hatred, rage, and despair. These dark emotions might make it difficult for a literary executor so burdened to make a wise decision regarding a privacy issue.

Apart from Linda Gray Sexton, none of the other major participants of the Anne Sexton controversy are still alive. Martin Orne died in 2000 at the age of seventy-two, having served as professor at the University of Pennsylvania School of Medicine for thirty-two years before becoming emeritus professor in 1996. Frederick Duhl died in 2010 at the age of eighty-one. A January 4, 2011, obituary in the *San Antonio Express-News* referred to him as one of the pioneers in the family therapy movement. His wife, Verne Lee Cooper, was quoted as saying that he "never forgave himself" for his affair with Sexton. "He said he had betrayed his own honor."

Diane Wood Middlebrook died in 2007 at the age of sixty-eight, after having taught at Stanford University for thirty-six years. Her award-winning biography of Sexton was followed in 1998 by *Suits Me: The Double Life of Billy Tipton*, a story about a female jazz musician who, disguised as a man, married five times. Middlebrook's acclaimed biography *Her Husband: Ted Hughes & Sylvia Plath, a Marriage*, appeared in 2003. Middlebrook's talent as a biographer was cited in a December 15, 2007, Stanford University obituary, which quoted her statement that "the dead cannot be shamed." Taken out of context, the statement is harsh and misleading, for Middlebrook brings profound intelligence and compassion to all her biographical subjects. The obituary summarized her credo as a biographer: "The more that each of us knows about each of the other human beings in the world, the better off [we] are. . . . It's true that it is very painful to be exposed to people's curiosity. But it's painful in a way that can only lead to self-knowledge." Middlebrook would have been sympathetic to the observation made by the Roman playwright Terence more than two thousand years ago. *Homo sum, humani nihil a me alienum puto:* "I am a human being. I consider nothing that is human alien to me."

The Anne Sexton controversy has different meanings to different readers. Sexton would appreciate Robert Penn Warren's famous ambiguity in his novel *All the King's Men*: "The end of man is knowledge" (9). Is knowledge the end, or goal, of human life? Or do certain dark truths lead to the end, or death, of life? Many of the haunting questions surrounding the Anne Sexton controversy, like Warren's enigmatic statement, will never be resolved, but there is value in raising them, for they dramatize, in the poet's life and death, the clash between the right to privacy, on the one hand, and the search for truth, on the other.

Take-home lessons:

1. Disclosure of confidential tape recordings of a patient's psychotherapy sessions, with the permission of the patient's executor, is probably not illegal and is probably not considered unethical by the psychiatric profession. Nonetheless, such an undertaking is bound to be controversial.

2. An important principle of privacy protection is that information held about individuals should be used only for the purpose for which it was gathered in the first place.

8. The Tarasoff Case: Must the Protective Privilege End Where the Public Peril Begins?

What to look for: Is a psychotherapist potentially liable for injury caused by a patient to a third party?

"Am I my brother's keeper?" This question from the book of Genesis has traditionally been understood by some as meaning that all of us living in a civilized society have a responsibility to other people, even what might be called a "moral duty" to them. For instance, few people would say that someone seeing a person drowning or bleeding on a sidewalk is free of any moral obligation to that person and should simply walk by. In fact, some states encourage such intervention by strangers through so-called Good Samaritan laws, named after the parable told by Jesus and mentioned in the Gospel of Luke in the New Testament, which shield such strangers from liability if their well-intended intervention itself causes injury.

Alternatively, some believe that the meaning of Cain's question is whether one is morally obligated to "restrain" another person's actions. Are we obligated to keep another from straying away, as a shepherd guides sheep?

Irrespective of the precise meaning of the biblical question, when it comes to the more pointed question of whether a person has a "legal duty" in such a situation, that is, whether the passerby *could be sued for not taking some action* to help, the requirements of the law generally take a different direction. Traditionally, under the law, we do not become liable to the other person—we cannot be sued—for failing to act to help or protect that person.

There are exceptions, however. If a person owns a dangerous dog that tries to attack another person, then the dog owner may be sued for injuries if he or she failed to act to protect the victim, such as by stopping the dog, keeping the dog on a leash, or keeping the dog fenced in. On the other hand, if someone, a *mere passerby*, sees a person being attacked by a dangerous dog that does *not* belong to the observer, then the observer cannot ordinarily be sued for failing to stop the attack or even for failing *to try* to stop it.

Although these brief examples do little to address the legal issues involved in the harrowing story we tell here, they point to related confidentiality questions that now bedevil psychotherapists. If a therapist treats a potentially dangerous patient, is the therapist morally obligated? That is, does the therapist have a "moral duty" to try to prevent that patient from carrying out a grave threat to a specifically identified person? Almost everyone would answer that question in the affirmative. A second and different question, however, is whether the same psychotherapist has a "legal duty" to step in or risk being sued by the victim. The answer to the second question has profound implications that could create a legal minefield for psychotherapists, who are professionally and ethically obligated to keep patient communications confidential.

The Story in Brief

Almost every psychotherapist has heard the name "Tarasoff" and may even have used or heard the expression "Tarasoff warning." Many, though, have no idea who Tarasoff actually was or of the tragic human story that gave the name Tarasoff its iconic significance.

The name refers to Tatiana ("Tanya") Tarasoff, a nineteen-year-old undergraduate at the University of California at Berkeley in the late 1960s, who found herself pursued by Prosenjit Poddar, a twenty-four-year-old naval architecture master's student at Berkeley. Brilliant but unstable, and alarmingly unfamiliar with American life, which was so different from his Indian culture, the thin, introverted Hindu fell deeply and irrationally in love with Tanya. In June 1969, depressed that Tanya was spending the summer in Brazil, Prosenjit reluctantly began seeing a Berkeley psychologist, Dr. Lawrence Moore, who sought to help him let go of his obsessive love for her.

At first Moore seemed to succeed, but three weeks before Tanya returned home at the end of the summer, the psychologist became alarmed at Prosenjit's rapid mental deterioration and threats of violence against her. Moore took the controversial step of breaching the confidentiality of his relationship with his patient by notifying the Berkeley campus police. Prosenjit was picked up for questioning and released a few hours later. Moore's colleagues, including the director of the Berkeley Counseling Center, were sharply critical of his decision to breach confidentiality, and he began feel-ing like an outcast.

Subsequent events, however, confirmed the psychotherapist's fears about his patient's potential for violence. On October 27 Prosenjit went to Tan-ya's home to speak with her, but she wasn't there, and her mother insisted that he leave. He returned later that day, armed with a pellet gun and a thirteen-inch butcher's knife. Alone in the house, Tanya refused to speak with him, and when he persisted, she screamed. He shot her, and when she ran from the house, he caught her and repeatedly and fatally stabbed her. He then telephoned the police and quietly surrendered upon their arrival.

Prosenjit's legal defense was that he was innocent by reason of insanity. Several mental health professionals testified that he committed the act un-der "diminished capacity." The jury rejected the insanity defense and in 1970 convicted him of second-degree murder. The judge sentenced him to five years to life in the California Medical Facility in Vacaville, near San Francisco. In 1974 the California Supreme Court overturned Prosen-jit's conviction on the grounds that the trial judge had failed to instruct the jury on the meaning of diminished capacity, which would have required the defendant to be tried not for first-degree murder but for a lesser crime, manslaughter. Under a compromise agreement, rather than being retried, Prosenjit was allowed to return to India, thus ending his legal involvement with the Tarasoff case. The legal ramifications of the case, however, were far from over.

In 1970 Tanya's parents filed a wrongful death suit against the Regents of the University of California and the City of Berkeley, claiming they had failed to notify them of Prosenjit's threats against their daughter. The first question that had to be answered was whether Dr. Moore and his employ-ers had any responsibility for Prosenjit's murderous act. If the psychothera-pist in the case had no "legal duty" beyond his "moral duty" to do something to interfere with the patient's likely danger to Tanya, then the parents' case would have ended there. This question had to be answered *before* a trial could take place to determine what the psychotherapist actually did and whether his actions were sufficient.

The question had never been addressed by a court before and for good reason. Until the time of Tanya's murder, everyone seemed to believe that a psychotherapist's legal and ethical obligation to keep everything the pa-tient said confidential meant that a dilemma like the one faced by Dr. Moore was settled by the psychotherapist's judgment call. Psychotherapists were obliged to figure out on a case-by-case basis a moral balance between their confidentiality and therapeutic obligations to a patient, on the one hand, and the need to protect their fellow citizens from that patient, on the other hand. There was no thought that the psychotherapist could be sued for what his or her patient did to another person no matter where the therapist came out in the moral calculus. That situation changed drastically as a result of the Tarasoff murder and its legal aftermath. Though the case struck the psycho-therapy community like a thunderbolt, it did not fulfill the catastrophic predictions made at the time. Nevertheless, the psychotherapeutic com-munity has never been quite the same.

In 1974 the California Supreme Court ruled, in what has come to be called *Tarasoff I*, that a psychotherapist who believes a patient is likely to harm a specific person has a *legal duty to warn the potential victim*. Having failed to issue such a warning, the therapist could be sued for the physical harm that his or her patient had done. In 1976 the California Supreme Court, for reasons it never stated, revisited the case, vacated (nullified) its earlier decision, and instead decided that the psychotherapist has a *legal duty to do something to protect, but not necessarily to warn, an intended victim of a dangerous patient*. This legal duty might involve informing the police of the threat or hospitalizing the patient. The court's second bite at the apple is now re-ferred to as *Tarasoff II* or simply Tarasoff. Again, the term "legal duty" means here that a therapist who fails to follow Tarasoff is vulnerable to a lawsuit in the name of the victim. It does not mean that it is "illegal" for the thera-pist to fail to take some protective step.

Fatal Attraction

Few people outside of the legal and mental health communities are aware of the personal dimensions of the Tarasoff story, which, in retrospect, re-sembles a fateful Greek drama about the madness of love leading to obses-sion, horror, and murder. Tarasoff remains a cautionary tale for all therapists who find themselves pondering, as a result of the case, the unpredictable consequences of when, if at all, to breach confidentiality.

Deborah Blum's 1986 book *Bad Karma: A True Story of Obsession and Murder* gives us some of the details of the story not found elsewhere. Tanya Tarasoff's parents grew up in the Manchurian city of Harbin, which was founded by Russia in 1898 with the creation of the Trans-Manchurian Rail-way. Married shortly before the end of World War II, they emigrated first to Brazil, where their first child, Tanya, was born in 1949 and their son,

Alex, in 1950, and then migrated again from Brazil to Berkeley, where their younger daughter, Helen, was born in 1963. Tanya's downtrodden mother, Lidia, barely spoke English. Tanya's abusive father, Vitally, was a car mechanic who had been arrested repeatedly for felony assaults, disorderly conduct, and driving while intoxicated. A student at Merritt Junior College, in Oakland, California, Tanya was anxious to fall in love with a young man who would rescue her from a dreary home life filled with violence and bitter family arguments. She yearned for the excitement of nearby UC Berkeley, which in that era, the late 1960s, was the epicenter of the anti-Vietnam protests and sexual revolution that were transforming every aspect of American society at that time.

Prosenjit Poddar was born in a small, isolated rural Indian village two hundred miles north of Kolkata. He was the oldest of six sons, raised in a thatched hut without running water, electricity, or toilet. He was a Dalit, a member of India's lowest caste. Though Gandhi ended discrimination against the Dalits in 1949, they were still looked down upon by Indian society and condemned to lives of ignorance, social ostracism, and desperate poverty.

According to Blum, Prosenjit's father was one of the first Dalits in his district of East Bengal to learn how to read. With the support of his father, Prosenjit attended high school, where he displayed an unusual aptitude for technology and mathematics. He scored in the 99.9th percentile on the entrance exam for India's most prestigious engineering college, the Indian Institute of Technology. Prosenjit was single-minded in his pursuit of academic excellence and graduated at the top of his class at the institute. With the help of his parents, he was able to raise the money necessary for his study in the United States. Imbued with his parents' conservative values, he impressed his Berkeley professors with his hard work and engineering skills. He hoped to gain his graduate degree and then return to India and marry a woman of his parents' choice as tradition would dictate. He arrived at Berkeley in the fall of 1967, one of the very few Dalits to study at an American university up to that time.

Prosenjit was an outstanding student of science, mathematics, and engineering, but he did not know how to read American culture, particularly American women. His difficulties were compounded by the vast cultural differences between the two countries. Blum points out that he was an "innocent" when it came to women. "Not only did he come from a culture that enforced strict separation between the sexes, but he had grown up in an all-boy household" (44). Ninety percent of all marriages in India were still arranged, a custom he never dreamed of violating while still in his native country. He assumed, like most of his countrymen, that love would arise from marriage. He valued, above all, chastity in a woman; sex was reserved for marriage. His mother had warned him, when he left India, that he would be dead to her if he had a relationship with an American woman. Given this stern admonition, he might have suspected that mixing with Tanya, whom he met at a folk dance held at Berkeley's International House, would destroy his close relationship with his parents, and perhaps himself as well.

Tanya, like many young Americanized women of her age, had different assumptions about relationships. For her, romantic love *preceded* marriage. She could not wait to lose her virginity; marriage was not a precondition for sex. Tanya and Prosenjit thus had strikingly different interpretations of their relationship, as the court account documents:

> They saw each other weekly throughout the fall, and on New Year's Eve she kissed defendant. He interpreted the act to be a recognition of the existence of a serious relationship. This view was not shared by Tanya who, upon learning of his feelings, told him that she was involved with other men and otherwise indicated that she was not interested in entering into an intimate relationship with him. (*People v. Poddar*, 1974)

However, Prosenjit continued to believe that Tanya intended to marry him. He wrote to his mother that he planned to marry an American girl, and, as Blum poignantly describes, with his very limited funds he purchased a traditional wedding sari for her to wear. In addition, he used his knowledge of electronics to plant microphones around his room so that he could record his conversations with Tanya. Later he would play them back, over and over, in a vain attempt to figure out why she didn't love him. He even edited together parts of different tapes to create a tape with her voice saying "I love you," which he then listened to innumerable times.

Neither Prosenjit nor Tanya realized that behavior deemed acceptable, even desirable, in one culture may be unacceptable, even reprehensible, in another. One of the mocking ironies of Prosenjit's life is that even as the first Dalit from India to have been given a grant to study in the United States, he subscribed to the widespread belief that chaste women were, in a different way, untouchable or sexually pure. In Indian culture, "innocent" girls are idealized, like goddesses; impure girls are reviled, like demons. The madonna/whore dichotomy, which Freud believed was an essential component of the Oedipus complex as well as an example of the process of psychic splitting, seemed conspicuously evident in Prosenjit's Indian culture. Blum states that he appeared mesmerized when he looked into Tanya's haunting sea-green eyes, her "cat" eyes; she represented to him the incarnation of the demonic goddess Kali, who seduced men and then led them to their death.

Tanya inhabited a radically different world, with a radically different mythology. Seeking escape from her oppressive family, she had no interest in retaining her virginity. Unlike Prosenjit, who had internalized his country's cultural and religious values with a vengeance, Tanya could not wait to free herself from her parents' stifling working-class values. She was more realistic than Prosenjit, more hardened, more cynical. If love did not work out with one man, she was ready to move on to another. Prosenjit could not move on.

Tanya and Prosenjit were in love with the idea of love. She imagined herself in love with men who were interested only in having sex with her. Disillusioned when these brief relationships inevitably ended, she never lost her sense of reality or her sense of herself. She fatally misread Prosenjit's feelings toward her, but she was never "crazy" in a clinical sense. He, in turn, misinterpreted her affections, believing that she desired to marry him and spend the rest of her life in India, a devoted wife and mother, deferring to his every wish, as the culture demands. His idealization of her soon turned into virulent devaluation, and he concluded she was a "whore" who was intent on humiliating him and destroying his life. Tanya's rejection of him had the effect of destroying his entire inner world, leading to his loss of reality and the dissolution of his identity.

Blum was a sophomore at Berkeley when Tanya was murdered, and when she began researching the story, she came to realize that she could "identify with the emotions of not only the victim, but also of her murderer" (Blum, author's note). She portrays Tanya as a victim though not completely blameless. She seemed to have learned from her conflicted relationship with her father that loving and being loved by a man required a dangerous dance of death, one that involved an exciting chase with an uncertain outcome—both for the chaser and the chased. The cat-and-mouse game was also part of Tanya's relationship with Prosenjit. Not that the young woman was consciously aware of this pattern. She remained, in many ways, shy, naïve, inexperienced, insecure. Yet she felt more self-confident in the presence of the even more vulnerable and needy Indian, flattered that he was enamored of her. Annoyed by his clinging dependence on her, she basked in her power over him.

On the surface, Vitally Tarasoff and Prosenjit Poddar could not be more different, but there were many similarities between them. Tanya's father played on her insecurities, telling her that she was not intelligent enough to succeed in college. She was appalled when Prosenjit remarked, upon hearing that she wanted to transfer to Berkeley, that the academic curriculum might be too difficult for her and that women did not need as much education as men. Both men were possessive and controlling, and both vented their anger toward Tanya through violence. Both men resorted to surveillance to keep her under control. Vitally Tarasoff sifted through the contents of her wastepaper basket looking for proof that she was involved with a man. Prosenjit went even further in his surveillance, secretly recording their conversations with a sophisticated bugging system. His most egregious plan for surveillance, he confessed to Dr. Moore, was paying a friend to secure her sexual favors and then bursting into his room before they had sex to rescue her.

Where does surveillance end and paranoia begin? Vitally Tarasoff's efforts to keep his daughter under surveillance were destructive but probably not pathological. Alcoholism impaired his judgment, but he was still able to function, if only barely. By contrast, Prosenjit became increasingly paranoid. His bizarre efforts to control, spy on, punish, and rescue Tanya demonstrated his growing alienation from reality.

"A Sacred Oath to Guard Your Secrets to the Grave"

Prosenjit told his friend Jal Mehta he had lost his self-control and that he would electrocute himself if he could no longer be with Tanya. It took weeks of pleading and cajoling before Jal was able to convince him to visit the Berkeley Counseling Center. Like most traditional Indians, Prosenjit did not understand the ways in which speaking to a trained professional could bring welcome insight and relief to someone suffering from psychological distress. He also worried about disclosing information to someone who might betray his trust. Jal reassured his friend that psychotherapists take "a sacred oath to guard your secrets to the grave" (Blum, 199).

In retrospect, Jal's statement is fraught with irony. Before the California Supreme Court's historic rulings, it was rare for a psychotherapist to break confidentiality to report a patient to the police. Psychiatrists would hospitalize patients they believed were a threat to themselves or others. But the ability to hospitalize patients without their permission or a legal proceeding led some people to raise civil rights issues. These critics viewed the Tarasoff decisions in a more positive light—as a less intrusive way of dealing with potentially violent patients. Ironically, breaking the confidentiality of the therapist–patient relationship created other problems, including the loss of what might be called the patient's privacy rights. This is what makes the Tarasoff case so compelling.

Jal accompanied Prosenjit to the university counseling center, where the distraught young man remained nearly mute during his interview with Dr. Stuart Gold in June 1969. One of three staff psychiatrists in charge of inpatient services, Dr. Gold allowed Jal to speak for his friend. The psychiatrist was uncomfortable with the situation, partly because of Prosenjit's background. Blum implies that like many of the other therapists, Gold believed that foreign students, especially those from the Middle East and Asia, did not respond well to psychotherapy. After Jal left, Prosenjit declared that his friend was trying to "steal" Tanya from him, adding that everyone was "laughing" at him, from which Dr. Gold concluded that the young Indian was suffering from paranoid schizophrenia. Writing out a prescription for Thorazine and Compazine, two antipsychotic drugs widely used at the time, Gold then referred him to Dr. Moore.

Blum acknowledges in the author's note to *Bad Karma* receiving the help of Dr. Moore with her book, though she doesn't elaborate on the details of his help. Did she interview him? Did he offer confidential details of his treatment with Prosenjit? We don't know. She does reveal much about the psychologist's personal life. During his postdoctorate fellowship two years earlier, when he worked long hours away from home, his Swedish wife became severely depressed, and she fed a potentially lethal number of sleeping pills to her two-year-old daughter and then swallowed the remainder of the pills herself. Mother and daughter narrowly survived. The parents then waged an ugly court battle over custody of the child. These events may have sensitized Moore, in Blum's view, to Prosenjit's threats to murder Tanya.

Prosenjit appeared to make therapeutic progress during Tanya's three-month absence. He was able to redirect his energy to work, and he was taken off medication. In every session Dr. Moore emphasized that Prosenjit's life would improve once he stopped thinking about her. Unbeknownst to the psychologist, however, Prosenjit continued to listen to his secret tapes of Tanya, thus fueling his obsession. He also maintained his friendship with Tanya's brother, Alex, a link to Tanya. Worse, Prosenjit accompanied Alex to a gun shop and began to talk about owning a gun himself. Dr. Moore was horrified when he learned about this detail from Jal, whose help he had enlisted. "Buying a gun is no idle gesture," the psychologist told Prosenjit's friend. "It means he intends to act out his fantasies" (242).

The Psychologist's Dilemma

Dr. Moore believed that an act of violence might be averted if he contacted the police, but he was well aware that almost certainly the therapy would come to an end. If, however, the psychologist remained silent, building on his positive relationship with the young man, a woman's life might be imperiled. Moore's dilemma, Blum suggests, "was one that went to the very heart of his profession" (243).

Maintaining confidentiality is the cornerstone of the patient-therapist relationship. Without the trust that arises from such confidentiality, therapy cannot succeed. No one can predict with certainty whether a patient will carry out a death threat. The best predictor of violence is a history of violence, but there was none in Prosenjit's case. As Alan Stone, who is both a psychoanalyst and a Harvard law professor, points out, many if not most patients struggle with destructive or self-destructive impulses ("The Tarasoff Decisions," note 52, 369); verbalizing one's violence is often the best way to defuse it. Stone's observation reflects Moore's own point of view. "It had been Moore's experience that most patients who came in for help were struggling with destructive impulses, and often made threats, even expressed the wish to see someone dead. The venting of anger was not only routine to the therapeutic process, it was also promising. If a patient had an outlet for his rage, the chances were greater that he wouldn't have to act it out" (244).

Nevertheless, a patient's threat to buy a gun must be taken seriously. The danger is said to be even greater if the patient suffers from paranoid schizophrenia. Prosenjit's delusional world was apparent to his friends and psychologist, and his bizarre actions, along with his inability to focus on his academic studies, were glaringly out of character. He was a desperate, confused, and lonely man who seemed to be crying out for help. A few weeks earlier, a Berkeley student had committed suicide. The suicide, along with the angry clashes between students and police in "People's Park," may have served as an additional trigger to Prosenjit's own violence.

Unable to convince Prosenjit to commit himself voluntarily to a hospital for observation, Moore turned to Dr. Gold for guidance. Fearful of incurring the disapproval of the chief of psychiatry, who was out of town, Gold tried to dissuade his colleague from breaching confidentiality by contacting the police. Gold cautioned his junior colleague not to act precipitously. Disregarding the advice, Moore telephoned the campus police, and two officers visited his office. He expressed his concerns and then wrote an official letter requesting Prosenjit's involuntary hospitalization, noting that he had seen the patient seven times. Prosenjit was picked up by the police for questioning, only to be released a few hours later. "You seem as normal as any of us," an officer tells him. "Like a rational young man" (Blum, 258).

In a lengthy discussion of the Tarasoff case in his 1998 book *Psychotherapy and Confidentiality*, Ralph Slovenko mentions that Dr. Moore notified the campus police along with Tanya's brother. "The police spoke to Prosenjit in Alex's presence," Slovenko writes. "Alex did not take the threats seriously and did not report them to his family" (280). These details do not appear in Blum's account.

Dr. Powelson, the chief of psychiatry, was predictably enraged when he returned and discovered Moore's actions. One of the ironies of Powelson's situation is that even as he chastised Moore for betraying confidentiality, Powelson himself had ordered the therapists under his supervision to report their patients' drug and marijuana use to the police, in violation of the university policy on confidentiality. In this case, he requested that the police return the correspondence about Prosenjit and took what seems the unusual step of ordering that correspondence and all clinic records related to Prosenjit destroyed. "He also ordered that no further action be taken to detain or commit Poddar" (Stone, "The Tarasoff Decisions," 360).

The Tarasoff story dramatizes the familial, clinical, and legal ambiguities of preserving and violating confidentiality. It may be that no one could have stopped Prosenjit from killing Tanya. Bad luck worked against both of them. Prosenjit wrote anguished letters to his parents in faraway India begging for their help, but they responded with cold reprimands. When Tanya heard that he had tried to give a friend one hundred dollars to proposition her, she became frightened and called the International House to speak with the foreign student advisor. She was told that unless she felt it was a "matter of life or death," the earliest appointment she could receive was the following week. Hesitating to sound alarmist, she made the appointment for October 28, which turned out to be one day after her death.

Prosenjit seemed to be in a frenzy when he confronted Tanya at her home, first shooting her with a pellet gun until it clicked empty, then stabbing her repeatedly with a butcher's knife, and finally chasing her out of the house as she bled to death. He made no attempt to escape. He called the police from her home and, when they arrived, placed his hands behind his back to be handcuffed. Tanya was pronounced dead on arrival at a local hospital. A coroner determined that she died of hemorrhaging as a result of eight major stab wounds to her chest, abdomen, and back. She was found to be six weeks pregnant at the time of her death.

Before the murder, no one at the Berkeley Counseling Center had supported Moore's decision to contact the police, and even he had begun to doubt himself. The murder left the Berkeley Counseling Center deeply shaken and in a state of crisis. Accused of attempting to destroy evidence in a murder case, Powelson fired nine therapists, including Moore. Powelson resigned in disgrace in 1972. Blum never tells us how Moore felt about the murder, but she does note that he felt "vindicated" by the evidence that came to light during Prosenjit's trial and Powelson's subsequent resignation (311).

Dr. Moore never wrote about his role in Tarasoff, and he went into private practice after he was fired at Berkeley. He became active in the San Francisco Academy of Hypnosis and, later in his career, specialized in the psychological treatment of pain, burns, and chronic hand injuries. He was also a member of MENSA. He died on April 14, 2013, at the age of seventy-eight. An obituary in the *San Francisco Chronicle* describes him as a "brilliant, unique, and larger-than-life man, irreverent and wickedly funny, witty, opinionated, caring, and kind-hearted. He was an irrepressible spirit, a delightful, unusual, and unforgettable character with strong values. At the same time, he was a very sensitive man with a great capacity for empathy and insight into others."

Fading from Sight

Researching the story, Blum traveled twice to India, presumably to interview Prosenjit. She never mentions speaking with him, though in her author's note she acknowledges the help she received from his friends and family. We know little about Prosenjit's post-Tarasoff life apart from the fact that he returned to Kolkata in 1974, married a lawyer selected by his father and, two years later, received a scholarship to study naval architecture in Hannover, Germany, where, according to a friend, he is "leading a normal life with his wife and daughter" (Blum, 310). Yet if he has disappeared into obscurity, the Tarasoff story continues to haunt psychotherapists.

Early Reactions

Immediately after the first *Tarasoff* decision was announced, many mental health professionals expressed "outrage" about the court's action. It is worth bearing in mind that in the late 1960s and 1970s the so-called antipsychiatry movement was burgeoning, with criticism of psychiatry's role in society coming from a number of directions both inside and outside the profession. Much of this criticism centered around the legal authority traditionally granted to psychiatrists to have patients allegedly suffering from mental disorders but not necessarily seen as dangerous locked away in mental hospitals (even without receiving any treatment). Such cases are now seen, understandably, as a violation of the ill person's civil rights. One year after the first *Tarasoff* decision in California, the U.S. Supreme Court ruled in *O'Connor v. Donaldson* (422 U.S. 563 [1975]) that patients could not be held against their will in mental hospitals except in cases of danger to themselves or others. The growth of so-called forensic psychiatry was in full swing. Many psychiatrists were convinced that lawyers were looking for more and more ways to sue physicians in general and psychiatrists in particular.

Vanessa Merton observed, in a long article published in the *Emory Law Journal* in 1982, that the formerly relaxed relationship between lawyers and physicians was becoming a "guerilla war" (263). The enormous cry of outrage from the psychiatric profession, Merton suggests, compelled the California court to revisit its 1974 decision and ultimately vacate that decision, replacing it with *Tarasoff II*. Far from satisfying the critics of the original decision, *Tarasoff II*, which still upheld the right of the victim's family to sue the psychologist and the clinic, led to further acrimony.

In an article published in the *American Journal of Psychiatry* in 1977, Howard Gurevitz, a California psychiatrist who played a central role in the preparation of a brief in the case, offered his summary of the events leading to *Tarasoff II*. In a tone alternating between anger and apology, he implied that the American Psychiatric Association was blindsided by the second hearing of the case by the appellate court. Gurevitz gave several reasons for the mental health profession's ineffective response, beginning with the fact that the American Psychiatric Association had only two weeks' notice to assemble an amicus brief. No amicus brief had been filed before the *Tarasoff I* hearing except by the Trial Lawyers Association in favor of the complainant. Nor was any national psychoanalytic organization willing to join the American Psychiatric Association in its brief. Additionally, the profession's response between the *Tarasoff I* and *Tarasoff II*'s decisions had been "isolated and sporadic." Gurevitz further complained, as did others, that the court had disagreed with the brief's contention that psychiatrists are incapable of making predictions of violence despite the fact that psychiatrists still held the power to "lock up" patients on the grounds that they might be violent.

Merton predicted that the Tarasoff doctrine would "exacerbate psychiatric role conflict and compromise loyalty to patients while achieving little in the way of compensatory objectives" (341). Using a metaphor from *Hamlet*, she observed wryly,

> It has been asserted, reasonably enough, that the Tarasoff court hoist psychiatrists with their own petard. Having failed to correct the perception of judges, correctional authorities, legislators, and the public that they are incapable of detecting "dangerousness" and having acquired in large part the power to determine society's response to the "dangerous individual," psychiatrists are now confronted with the logical implications of that power. (296–297)

Tellingly, Merton drew noteworthy parallels between confidentiality in law and confidentiality in psychotherapy. Her statement about the Buried Bodies case, which we discussed in Chapter 2, is significant enough to be quoted in its entirety:

> Outsiders are scarcely sympathetic when lawyers place the preservation of client confidences above the arguably more compelling interest of preventing harm to identifiable individuals. We need only recall the notorious "where the bodies are buried" case. To many lawyers, the refusal of Robert Garrow's attorneys to divulge his secrets, despite the plea by the father of one of Garrow's victims to know whether his daughter was alive or dead, was the only ethical choice. To most onlookers, the attorneys' behavior appeared not merely unheroic but downright inhuman. The eventual dismissal of the indictment against them and the bar association's vindication of their position could not have altogether assuaged the revilement they suffered. (280)

Both Merton and Alan Stone remarked that, on statistical grounds, mental health professionals' predictions of violence are fraught with difficulty. Because violent behavior such as murder is relatively rare, making predictions, even by someone with a hypothetically high degree of accuracy, results in an unacceptably large number of false positives. As an example, Stone indicated that if one in one thousand suspected individuals in fact kills someone, a psychiatrist with an unimaginable 95 percent accuracy in predicting violence would, out of a population of one hundred thousand people, correctly predict violence in ninety-five of the hundred individuals who were actually destined to commit a violent act. On the other hand, the psychiatrist would incorrectly predict violence in 4,995 others. A more realistic predictive skill level, if such prediction is possible at all, might be 70 percent accuracy. The psychiatrist would correctly predict violent behavior in seventy people but would incorrectly predict violence in 29,970 others out of the hundred thousand.

Thomas Szasz, a psychiatrist who was well known as a major figure criticizing psychiatrists' use of the power granted them to incarcerate patients, put it this way:

> The Tarasoff ruling is an example of the proverbial chickens coming home to roost. . . . The claim that psychiatrists possess special knowledge and skills concerning the diagnosis and prediction of dangerousness continues to undergird the psychiatrist's power to commit. The Tarasoff decision extends this classic rationale and duty from the mental hospital to any professional relationship between therapist and patient. (43)

Mandatory reporting laws sometimes generate a flood of reports only a few of which, if any, may be taken seriously. An example is the NY SAFE Act, a sweeping gun control act that requires doctors, nurses, and mental health professionals to report patients who are "likely to engage in conduct that will cause serious harm to self or others." According to Curtis Skinner in an article published in the *New York World* on June 3, 2013, in the three months since the law went into effect, six thousand reports were filed but only eleven involved people who have or applied for gun permits. County directors of community services, who under the law must approve or dispute evaluations made by mental health providers, have complained that the deluge of SAFE Act reports forces administrators to take valuable time away from more important responsibilities "for a minimal return." Another criticism, expressed by a senior staff attorney at the New York Civil Liberties Union, is that the NY SAFE Act "requires psychiatrists, psychologists and other mental health professionals to report a patient they deemed dangerous, even if that mental health clinician believes that the patient would respond well to treatment and has no reason to believe the patient owns a gun."

Dire Predictions for Psychotherapy

Judge Clark, who wrote a dissenting opinion in the *Tarasoff II* decision, expressed his fear that the duty to warn would have dire consequences for psychotherapists' ability to help potentially violent patients.

> Given the importance of confidentiality to the practice of psychiatry, it becomes clear the duty to warn imposed by the majority will cripple the use and effectiveness of psychiatry. Many people, potentially violent—yet susceptible to treatment—will be deterred from seeking it; those seeking it will be inhibited from making revelations necessary to effective treatment; and, forcing the psychiatrist to violate the patient's trust will destroy the interpersonal relationship by which treatment is effected. (*Tarasoff*, 1976, Clark, J. dissent)

In 1976 Alan Stone, then president-elect of the American Psychiatric Association, echoed Judge Clark's concern. Stone predicted disaster if the court imposed a duty to protect; such an imposition, in Stone's words, "which may take the form of a duty to warn threatened third parties, will imperil the therapeutic alliance and destroy the patient's expectation of confidentiality, thereby thwarting effective treatment and ultimately reducing public safety" (Alan Stone, "The Tarasoff Decisions"). Eight years later Stone changed his view and concluded that the duty to warn is "not as unmitigated a disaster for the enterprise of psychotherapy as it once seemed to critics like myself" (Alan Stone, *Law, Psychiatry, and Morality,* 181).

Nonetheless the effects of the *Tarasoff* decision have been enormous. One expert commentator, Douglas Mossman, wrote, "no court ruling has had a broader or more enduring impact on day-to-day mental health practice. . . . Thirty years after its promulgation, Tarasoff remains, to mental health professionals, the most influential ruling in mental disability law" (523, 526). Another expert, Thomas G. Gutheil, wrote, "The Tarasoff case (1974, 1976) represents one of the most significant developments in medico-legal jurisprudence of the past century. . . . The original case of *Tarasoff v. Regents of the University of California* in 1974 burst like a bomb over the clinical scene" (345). Gutheil added:

> In the Author's consultative experience, this novel breach in the age-old mandate to maintain confidentiality appears to have shaken clinicians' conviction of the "rightness" of confidentiality itself. Together with mandated reporting requirements, "snitch laws" *that require reporting of impaired physicians and Tarasoff-like rulings in other states, treating clinicians*

have lost their moral and clinical compass as to whether confidentiality should be broken. (346; emphasis added)

Legal Implications of Tarasoff

The Tarasoff case is routinely taught in medical school and law school, and it continues to receive enormous scholarly attention. "The Tarasoff case sparked a firestorm of controversy among psychotherapists, lawyers, academics, and judges regarding the status of the therapist–patient privilege," Brian D. Ginsberg notes. "Since the ruling was handed down, the literature has burgeoned with medical and psychological commentary, case law analysis, extensions to other disclosure scenarios, analogies to the lawyer-client privilege, and even application to the privacy of the mental health of the President of the United States" (33).

The case "fascinates," writes Slovenko, a law professor with psychiatric training, "and it will continue to fascinate for years to come" (*Psychotherapy and Confidentiality*, 275). In two often quoted sentences, the California Supreme Court concluded in *Tarasoff II* that "the public policy favoring protection of the confidential character of patient-psychotherapist communication must yield to the extent to which disclosure is essential to avert danger to others. The protective privilege ends where the public peril begins."

Paul B. Herbert, one of the most vociferous critics of Tarasoff, pointed out two "explained ironies" of the court ruling: "declaring a professional relationship special expressly to undermine precisely that which makes it special, trust and confidentiality" and "finding that an outpatient psychotherapist is in a position to control a putatively dangerous person but that police officers who have detained the person for questioning are not" (419). Herbert regarded the four-to-three *Tarasoff II* ruling as a "stretch" that went far beyond the facts of the case itself. His main objection was that it was "quintessential judicial legislation—law-making, grounded explicitly on policy-weighing" (420).

The Tarasoff case was settled out of court for a figure rumored to be fifty thousand dollars, "a nominal sum in a wrongful death action," in Slovenko's view.

Insurance companies, we know, do not relish being placed at the peril of the jury. In this case the cost of litigating probably would have exceeded the settlement. It is a pity, in a way, that the case was not tried as it is more likely than not it would have been decided in favor of the defendant and thus would not have caused such a stir in the psychiatric community. In this case, settling was penny wise and pound foolish. (*Psychotherapy and Confidentiality*, 288)

The American Psychiatric Association filed an amicus brief during the Tarasoff case arguing against the imposition of a legal duty for the therapist to warn or protect a potential victim.

In a strictly legal sense, the bark of the Tarasoff ruling has been worse than its bite. The number of Tarasoff suits remains "surprisingly small," James C. Beck pointed out in 1990, "and only a few psychiatrists have been found liable for breach of the Tarasoff duty" (9). Three factors help determine "foreseeability": a history of violence, a threat to an identifiable victim, and a plausible motive. "If two or more of these are present, the courts usually find that violence was foreseeable. If only one is present, the courts sometimes do and sometimes do not find foreseeable violence. If none is present, courts rarely find that violence was foreseeable" (Beck, 10). According to Robert I. Simon, the "dangerousness of psychiatric patients is equivalent to that of persons in the general population" (24). Depending upon the degree of violence in the general population, this observation may be reassuring or alarming. Simon admits that nothing "unnerves psychiatrists and disrupts their free-floating attention more than patients who make threats of violence toward others" (23). Simon's conclusions, made more than twenty years ago, remain generally true, but later studies have added qualifications. Reviewing the results from the National Epidemiologic Survey on Alcohol and Related Conditions, published in a 2009 issue of *Archives of General Psychiatry*, Steven Dubovsky commented that "the data suggest that mental illness alone does not usually lead to violent behavior unless the patient also has a substance use disorder or history of violence."

California and other states have attempted to limit the scope of the Tarasoff duty, as George C. Harris observes:

Through a 1985 addition to its Civil Code, California now provides that a therapist cannot be liable for "failing to warn of and protect from a patient's threatened violent behavior or failing to predict and warn of and protect from a patient's violent behavior except where the patient has communicated to the psychotherapist a serious threat of physical violence against a reasonably identifiable victim or victims." The statute goes on to create a safe harbor for the therapist whose patient has made such a threat, providing that, "[i]f there is a duty to warn and protect . . . the duty shall be discharged by the psychotherapist making reasonable efforts to communicate the threat to the victim or victims and to a law enforcement agency." Most states that have adopted the *Tarasoff* duty have, like California, limited the therapist's duty to instances in which the patient has identified a specific victim. (48)

Clinicians still debate the Tarasoff case. Some psychologists continue to believe that Prosenjit might have stayed in therapy and worked through his violent feelings had Dr. Moore not breached confidentiality. Slovenko raises this question without attempting to resolve it.

> If Poddar's therapist had not reported him, would he have remained in psychotherapy and the life of Tatiana saved? In nearly all of the reported cases, where harm has occurred, psychiatric care had been terminated. More often than not, the fault lay in the termination of treatment. In litigation, however, given the harm done, the therapist may be hard put to justify outpatient psychotherapy over legal commitment. (*Psychotherapy and Confidentiality*, 306)

To date, most mental health professionals have been able to live with Tarasoff. Slovenko probably speaks for the majority of therapists and patients when he argues that the "general public or prospective patients as well as patients in therapy do not lose faith in the psychiatrist as a keeper of secrets when, in cases of emergency, he acts contrary to strict and absolute confidentiality" (*Psychotherapy and Confidentiality*, 272).

However, no matter how clear it must be to most people that a psychotherapist, as much as any other citizen, has some "moral duty" to do whatever is reasonable under the circumstances to prevent imminent physical harm to a fellow human, the "legal duty" imposed on psychotherapists by the *Tarasoff* decision and its so-called progeny raises a troubling question. Ordinary citizens who hear a person issue a credible threat to injure an identifiable person would not be at risk of a lawsuit if they failed to do anything about the threat. Nor would psychotherapists be at risk of such a suit if they heard such a threat from someone with whom they had no treatment relationship. Does it make sense, then, Herbert asks, that psychotherapists, in the context of a treatment relationship, *the very relationship that depends most heavily on confidence and trust*, should become potentially liable if they fail to act to protect an identifiable victim?

Writing a quarter of a century after the Tarasoff decision, Herbert pointed out that twenty-seven states impose an actual duty to warn; nine others, plus the District of Columbia, depart from Tarasoff and allow only to grant permission to warn; one state, Virginia, rejects Tarasoff; and the remaining thirteen states have no definite Tarasoff law (417). There is thus no national Tarasoff law or policy. In an article coauthored with Kathryn A. Young, Herbert commented that the variety of duty-to-warn laws, with no agreement among the states, is virtually unprecedented with respect to legal doctrine. "Confusion is an inevitable product, and confusing law is inefficient at best, and often harmful. Further, when the law is so diverse, one must wonder whether a meaningful consensus obtains about basic premises or policy objectives, let alone whether any coherent empiricism underlies the law in question" (280). The National Council of State Legislatures has created a map and table showing the great variation in "duty-to-protect" or "duty-to-warn" laws among the states and the way in which the "duty" issue becomes entwined with laws that, without necessarily mentioning a "duty," grant permission to make disclosures in certain circumstances.

Few Breaches of a Tarasoff Duty

A 2010 review by Matthew F. Soulier and his colleagues of twenty-one years of legal cases against psychotherapists that centered on their "duty to protect" showed that cases in which a therapist was actually found liable for the actions of a patient are surprisingly uncommon in recent decades. The review concluded in part,

> The primary finding of this review is that defendants are now rarely held to be negligent on grounds of failing to warn or protect. In reviewing 21 years of legal history, we found just four cases in which psychotherapists were found liable for breach of a *Tarasoff* duty. . . . This is a welcome change in how the law views the legal duty of practicing clinicians. In the early days after *Tarasoff*, courts found negligence in cases with fact patterns similar to those that now are being decided for defendants [i.e., in favor of the therapists] [reference omitted]. This evolution is the result of legislative actions, increased judicial sympathy toward the clinician confronted with threats of violence, and social climate change. (Soulier et al., 469)

However, psychotherapists need to keep in mind that although court findings against them in such cases are rare, the actual number of filed cases was close to one hundred. Win or lose, each case required a court decision with all the legal commotion and disruption of the therapist's life that one can imagine. So while losing such a case must be wrenching for a psycho-therapist, going through the process of "winning" such a case can be damaging in itself. Such cases would have been hard to imagine before Tarasoff.

Another interesting finding in the Soulier review is that psychotherapists have a better chance of prevailing in such a case if the state has a law mandating (twenty-eight states) or permitting (ten states) such disclosures than they would in a state with no such mandate or permission (thirteen states). This may be because the circumstances in which a disclosure is mandated or permitted in states with such laws are relatively specific, whereas in states that have no such laws at all courts or juries apply standards on a case-by-case basis that can vary considerably.

Furthermore, as Griffin Sims Edwards argues in a paper presented at the 2010 meeting of the American Law and Economics Association, there is evidence that rather than preventing violent acts, the duty created by the Tarasoff decision actually tends to increase the homicide rate! Edwards calculates that the mandatory duty to warn increases the homicide rate by about 8.9 percent, or .76 people per hundred thousand. This may occur because the duty to protect dissuades potentially violent persons from seeking help with their impulses. Another explanation is that the duty to protect may discourage therapists from accepting such patients because of the potential legal complications that could occur if such a patient gives voice to a credible threat during the psychotherapy. Edwards's research, suggestive as it is, would require a much larger number of subjects to be statistically significant. Agreeing with Edwards but in the absence of any further authoritative empirical studies, the philosophy professor Kenneth Kipnis speculates that Tarasoff will have a counterproductive effect on society: "paradoxically, ethical and legal duties to report make it less likely that endangered parties will be protected."

When the U.S. Supreme Court established the federal psychotherapist-patient privilege in *Jaffee v. Redmond* in 1996, it noted that despite the absolute privilege it had fashioned, there may be limits on the confidentiality of psychotherapist-patient communications. The court added in a footnote: "Although it would be premature to speculate about most future developments in the federal psychotherapist privilege, we do not doubt that there are situations in which the privilege must give way, for example, if a serious threat of harm to the patient or to others can be averted only by means of a disclosure by the therapist" (*Jaffee v. Redmond*, note 19). In an earlier footnote in *Jaffee* the court had written, citing as authority the ethics code of three major *nonpsychoanalytic* mental health organizations: "At the outset of their relationship, the ethical therapist must disclose to the patient 'the relevant limits on confidentiality'" (ibid., note 12).

Issuing a Miranda Warning?

Some experts have inferred from these footnotes that a psychotherapist must issue to patients at the beginning of psychotherapy what is ironically or cynically referred to as a "Miranda warning": "Anything you say can and will be used against you in a court of law." Actual Miranda warnings are routine warnings that police give to suspects when they are arrested. The automatic warning arose from a 1966 U.S. Supreme Court ruling in *Miranda v. Arizona* on the right of suspects to avoid self-incrimination under the Fifth Amendment. The use of the term "Miranda warning" in connection with psychotherapists' duty to warn appears to have originated in a 1974 law review article by John G. Fleming and Bruce Maximov written while the *Tarasoff I* court was deliberating. According to Alan Stone, the article had a significant influence on the court's view of the case ("The Tarasoff Decisions").

The requirement that a psychotherapist issue a "Miranda warning" to patients, however, represents a grave contradiction of the instructions given to psychoanalytic patients regarding the "fundamental rule" of psychoanalysis. As we pointed out in our introduction, Freud insisted that patients speak openly and freely about their free associations: "Say everything which comes to your mind, holding nothing back." Freud also warned analysts not to make deals with a patient at the beginning of treatment that certain subjects need not be mentioned. Whether the 1976 *Tarasoff II* decision, coupled with the Supreme Court's understanding of psychotherapy in *Jaffee*, creates an untenable situation for psychoanalytic psychotherapists and their patients remains the subject of an ongoing debate.

The Tarasoff story raises disturbing questions not only about the psychotherapist's dual responsibility to patient and society but also about the ambiguities of confidentiality and privacy. Prosenjit Poddar's tape recording of Tanya Tarasoff was clearly unacceptable, but what are the limits of the state's control of psychotherapy? When does the duty to warn and protect become counterproductive? One of the lost ironies of the story is that Dr. Moore did more than most therapists would have done to avert a tragedy, but a murder nevertheless occurred. Could *anything* have been done to prevent Tanya's death? Would Prosenjit have been able to work through his violent intentions toward Tanya had he remained in therapy with

Dr. Moore? What would have happened if the Berkeley police recognized that the tormented young man was a danger to Tanya? What if she had told the foreign student advisor at the Berkeley International House that her problem was indeed a matter of life or death? These are only some of the story's haunting questions. In the final analysis, Tarasoff dramatizes the role of bad luck—or perhaps bad karma. No one can say what could or should have been done, if anything, to change the ending of this fateful story.

Take-home lesson: Psychotherapists in California, under a limited set of circumstances, can be held liable for a patient's act of violence against a third person when the psychotherapist could have intervened in some way to protect the intended victim. There is no "national policy" regarding this issue. Many other states have adopted positions similar to California's. Nevertheless, the variation among the states as to whether psychotherapists may breach confidentiality to make such an intervention, and whether they are liable to a third-party victim for failing to do so, is surprisingly wide.

9.   *Jaffee v. Redmond*: The Supreme Court Speaks

What to look for: Is confidential communication between a psychotherapist and a patient privileged information in the federal courts of the United States?

To understand the significance of *Jaffee v. Redmond*, one must recognize that the federal courts of the United States operate under a different set of rules from those that apply in the individual states, each of which sets its own rules for evidence presented in a court of law. While "evidence rules" may seem like an arcane subject, the evidence rules in our courts have a profound effect on the way the rest of our society views the relative confidentiality of different kinds of information. Evidence rules are largely about what can and cannot be introduced as "evidence" in a court proceeding. This includes a ban on what is called "privileged" information, if the person "holding the privilege" objects. For example, the fact that lawyer-client communications are privileged in the federal courts (as well as in all state courts) is constantly on the minds of lawyers and their clients all over the country.

Setting the Stage

In the states, the rules of evidence are established through laws passed by the state legislatures, but in the federal courts, the courts themselves, under a law passed by Congress, establish the federal evidence rules. Federal courts are extremely reluctant to establish additional privilege rules, so a decision in the Supreme Court adding a particular kind of relationship to those rare few that are privileged has enormous significance. There is no privilege in federal courts for physician-patient communications, accountant-client communications, parent-child communications, sibling communications, or journalist-source communications.

Cases from the lowest-level federal courts in the states of Illinois, Indiana, and Wisconsin, when they are appealed, are heard in the Seventh Circuit Federal Court of Appeals, housed in the Dirkson Federal Building in Chicago.

In 1991, the tragic event described here took place in the Village of Hoffman Estates, a mostly white middle-class suburb about thirty miles from Chicago, when a police car was dispatched to a disturbance at a largely African American housing development. The following account is quoted directly from the subsequent legal case written by the judges of the Seventh Circuit Court (*Jaffee v. Redmond*, 94–1151 [7th Cir. 1994]). Mary Lu Redmond, a twenty-nine-year-old female police officer with five years experience on the police force, was on duty at the time.

The Shooting

On June 27, 1991, Officer Redmond, who was alone on patrol duty in that area on the day shift, responded to a dispatcher's report of a fight in progress at the Grand Canyon Estates apartment complex in the Village of Hoffman Estates, a suburb of Chicago, Illinois. Redmond was the first police officer to arrive at the scene. Officer Redmond testified that, as she pulled into the apartment complex parking lot, she saw two African-American women running toward her car, waving their arms above their heads, one of whom stated that there had been a stabbing inside the building. Redmond relayed this information to her dispatcher and requested assistance and an ambulance.

As Redmond approached the apartment building, five men ran out the front door yelling and screaming. One of the men was waving a pipe above his head. At this time, Officer Redmond testified that she ordered the man carrying the pipe to drop the pipe and also ordered everybody to the ground. After repeating the command "Drop the pipe" several times to no avail, Officer Redmond was forced to draw her service revolver. Almost immediately thereafter, two more men—a Caucasian man followed by an African-American man in hot pursuit—came running out of the door of the building. Officer Redmond testified that the African-American man, later identified as Ricky Allen, Sr., was armed with a butcher knife, was chasing and gaining on the Caucasian man, and was "directly behind" and poised to stab him when she fired the fatal shot. Officer Redmond testified that she commanded Allen to drop the butcher knife several times before firing:

"I ordered the black male subject with the knife to drop the knife several times. I told him to drop the knife and get on the ground. . . . I was yelling at him to drop the knife and get on the ground. . . . [H]e did not drop the knife and he did not get on the ground. . . . [I yelled] at least three times. I just kept yelling the minute I saw him."

Officer Redmond explained the moment before the shooting as follows:

"As [Allen] was gaining speed on the first subject until they were directly—he was directly in front of him, like the first subject's back, and then the second subject, as he was gaining on him the second subject, the male black subject with the knife took the knife back, raised it above his head and I waited, and as he started to come down with the knife and made the downward motion, I fired one shot at him."

Redmond testified that she "didn't even have time to square up," when she fired her weapon "[b]ecause the second subject was about to kill the first subject with the knife." She noted that only three or four seconds elapsed from the time Allen emerged from the apartment building door until the time she fired. Four of Allen's brothers and sisters, all of whom witnessed the shooting, testified that Allen was unarmed when Officer Redmond fired her weapon.

Officer Redmond testified that after she fired the single shot at Ricky Allen, Sr., he fell to the ground and she ran toward him with her gun at her side. She observed the butcher knife lying on the grass approximately two or three feet from his body. She repeated her request for backup support and an ambulance on her portable radio, as "people came pouring out of the buildings." Redmond stated that several people within the crowd "started to charge" at her, as they were "yelling," "screaming," "swearing" and "quite hysterical." She raised her gun when one person from the crowd came within arm's length, and ordered everyone "to get back, get down, get beyond the sidewalk, get on the ground." In her testimony, Officer Redmond made it clear that no one from the crowd attempted to come to Allen's aid, and that the knife was not moved from the place it landed when Allen fell to the ground until it was retrieved by one of the investigating officers.

Allen's siblings remembered the events just after the shooting differently. Connie Allen, his sister, testified that she attempted to approach her brother's body when Officer Redmond, with her gun raised, ordered her to get back and also stated that Redmond ordered her sister Sharon to step back at gunpoint. Connie Allen did not observe a knife lying on the grass near her brother's body until after the ambulance had taken the body away. When asked at trial why she had not reported what she had seen to any of the police investigating the shooting, Connie stated that she had not felt like talking to anyone immediately after the shooting.

Officer Joe Graham arrived at the Grand Canyon Estates apartment complex shortly after the shooting. When he arrived, he saw a large crowd of people—approximately twenty-five to thirty African-Americans and five to ten Caucasians—gathered on the grass, and a number of people "rushing out of the apartment buildings—to see what was going on." Officer Graham observed Officer Redmond standing on the lawn, behind Allen's body, with her gun drawn and aimed at the crowd. Officer Graham testified during his deposition that Officer Redmond appeared "somewhat bewildered" when he first arrived at the scene, and later at trial he explained that he meant she was "visibly shaken or upset or disoriented." Officer Graham testified that the members of the crowd were "fluctuating back and forth . . . in a very chaotic movement," yelling that Redmond "had shot Mr. Allen and that she didn't have to shoot him in the head" and that "they were going to sue the white bitch for shooting Mr. Allen." Graham testified that when he knelt down to check for Allen's pulse, he saw a butcher knife lying on the grass approximately an arm's length away from the body.

Officer Redmond's Counseling

After the shooting, Mary Lu Redmond sought counseling from Karen Beyer, a licensed clinical social worker certified by the state of Illinois as an employee assistance counselor and employed by the Village. Officer Redmond met with Beyer for the first time three or four days after the shooting incident and continued counseling for approximately two or three sessions per week through at least January of 1992, six months after the shooting.

During pre-trial discovery, the plaintiffs learned that Officer Redmond had participated in a number of counseling sessions with Beyer, the licensed clinical social worker. At Officer Redmond's deposition, the plaintiffs inquired regarding the substance of her communications with Beyer. Officer Redmond refused to respond to this line of questioning, contending that her communications with a licensed clinical social worker were privileged. The plaintiffs subsequently subpoenaed Beyer to testify at a deposition and to produce her credentials as a counseling professional as well as all her "notes, records, [and] reports pertaining to Mary Lu Redmond." The defendants, Redmond and the Village, moved to quash the subpoena, maintaining that all of Officer Redmond's communications occurring within the context of the counseling relationship, as well as Karen Beyer's notes and reports pertaining to those communications, were privileged. The trial court denied the defendants' motion to quash, based on the judge's belief that the psycho-therapist/patient privilege recognized in other circuits does not extend to a licensed clinical social worker, and ordered Karen Beyer to testify as to "the disclosures made to her by Ms. Redmond of the incidents of the day that relate to [the shooting]." We wish to point out that the Illinois statute specifically grants the psychotherapist/patient privilege to social workers. . . . The court ordered Officer Redmond to appear for a third deposition session to answer questions concerning her communications with Karen Beyer. Officer Redmond appeared for the third deposition session, and again the answers she gave regarding her counseling sessions with Karen Beyer were evasive and incomplete, obviously an attempt to protect her privileged communications.

When Karen Beyer appeared for her deposition, she limited her answers only to those facts concerning disclosures made by Officer Redmond about the circumstances leading up to the shooting incident on June 27, 1991. Beyer also refused to produce any notes or reports from Redmond's counseling sessions. The plaintiffs filed another motion to compel Karen Beyer to answer certain questions to which objections had been made and to produce all of her notes and reports on Mary Lu Redmond. After Karen Beyer's second deposition session, the plaintiffs filed another motion to compel further responses, and the trial judge responded with an order permitting unrestricted and unlimited inquiry into statements made by Officer Redmond to Beyer during their counseling sessions. At her final deposition session, and again during trial, Officer Redmond responded "I don't recall" to the majority of questions dealing with the substance of her counseling sessions with her therapist and licensed clinical social worker, Ms. Beyer. Ms. Beyer likewise refused to divulge her communications other than the officer's factual description of the events leading up to the shooting. Karen Beyer did produce three pages of redacted notes.

On April 6, 1993, the district court ordered that Officer Redmond would be barred from testifying at trial as to her version of the shooting incident "because plaintiffs' attorneys have been blocked from effective cross-examination." Just prior to trial, on December 10, 1993, the trial judge reconsidered and vacated this ruling, but he made it clear that the jury would be instructed that it could draw an adverse inference from the defendants' failure to produce Karen Beyer's notes and in fact gave such an instruction. At trial, the district judge instructed the jury that the defendants had no legal justification to refuse to produce Karen Beyer's notes of her counseling sessions with Officer Redmond. Over the defendants' objection, the district judge instructed the jury that it was *"entitled to presume that the contents of the notes would be unfavorable to Mary Lu Redmond and the Village of Hoffman Estates"* (emphasis added).

The trial judge also instructed the jury on the factors it could take into account in determining whether Officer Redmond's use of deadly force was proper. The court gave the plaintiffs' proffered Jury Instruction No. 5, over the defendants' objection, and it rejected the defendants' Proposed Jury Instruction No. 7. The defendants' Proposed Jury Instruction No. 7 included two points omitted from the plaintiffs' Jury Instruction No. 5: first, that the jury should not consider a police officer's subjective intentions or motivations in using deadly force; and second, that the jury should make allowance for the fact that police officers frequently "have to make split second judgments under tense, uncertain, and rapidly evolving circumstances."

The jury returned a verdict in favor of the plaintiffs based on these jury instructions and awarded $45,000 for the federal constitutional claim and $500,000 for the state wrongful death claim. The defendants appeal.

The Federal Courts

The U.S. federal courts are organized by districts, each of which covers a particular geographical area, and each district has a trial court. The ninety-four such districts are organized into twelve "circuits," each of which generally includes all the districts from several states, and each of these "circuits" has a court of appeals that takes appeals from any of the district courts in its circuit. When an appeals court hands down a ruling on a particular issue, the decision may set a precedent that is binding on all the federal district courts in its circuit.

For example, the lawsuit against Officer Redmond and the Village of Hoffman Estates was tried in the Federal District Court for Northern Illinois, which covers the nineteen counties in the northern part of that state.

As we noted above, appeals from cases in that court are heard in the U.S. Court of Appeals for the Seventh Circuit, which hears appeals from all the federal district trial courts in the "Seventh Circuit" (all the federal districts in Illinois, Indiana, and Wisconsin). A decision made in the appeals court based on this case would be binding on all the other federal courts in those three states but not in any of the other federal circuits. Only a decision made in the U.S. Supreme Court is binding on all the federal circuits and therefore on all ninety-four federal trial courts in the United States.

Although the Seventh Circuit Court of Appeals had never ruled on the question of whether a psychotherapy patient in the territory covered by the Seventh Circuit has the privilege of preventing her therapist from testifying, several other appeals courts sitting atop the district courts in other parts of the country had issued rulings on this question, a question that had by the early 1990s been under discussion for over forty years. Those courts had not all reached the same conclusion despite the fact that a committee appointed by the Supreme Court had recommended such a privilege in the 1970s.

Previous Decisions in the Federal Appellate Courts

The earliest case cited by the court, from the Fifth Circuit, which in 1976 included Florida, Georgia, Alabama, Mississippi, Louisiana, and Texas, was the Florida case of a man named William Joseph Meagher who actually appeared never to have been in psychotherapy. According to the court record, he'd been involved for three years in a program in which he communicated by mail with Dr. Samuel Yochelson,

> a psychiatrist employed by the National Institute of Mental Health, an
> agency associated with the Department of Health, Education and Welfare
> in Washington, D.C. The defendant had been a voluntary member of
> Dr. Yochelson's program in the research of criminal behavior between
> December, 1971 and July 1971 [*sic*] and between May, 1973 and September
> 1973; the robbery of the Jacksonville bank occurred in October of 1973.
> Throughout the period beginning in July of 1971, the defendant main-
> tained a fairly regular correspondence with Dr. Yochelson. Dr. Yochelson
> testified that, in his professional opinion and as a result of his long
> personal contact with defendant, he did not believe the defendant to have
> been insane at the time of the bank robbery. In addition, the prosecutor
> introduced into evidence the correspondence between the defendant and

Dr. Yochelson to support its contention that Meagher was sane when he robbed the Jacksonville bank. (*United States v. Meagher*)

The court ruled that the correspondence could be admitted as evidence because there is no physician-patient privilege in the federal courts. The court was noncommittal on the issue of a psychotherapist-patient privilege and mentioned it only in passing. Curiously, this case is cited as an example of an appellate decision denying the existence of a psychotherapist-patient privilege.

In a second case, dating from 1983 in Michigan, a psychiatrist by the name of Jorge S. Zuniga, who apparently was under investigation for an insurance claim irregularity, was ordered by a grand jury to produce billing records for a large number of patients. In that decision, the Appellate Court for the Sixth Federal Circuit (Kentucky, Michigan, Ohio, and Tennessee) did a careful analysis of the issue of a psychotherapy-patient privilege and concluded that, in theory, such a privilege actually exists in the federal courts. Nevertheless, the court refused to allow Dr. Zuniga to claim the privilege for billing records that were considered not to fall within the confidential communications protected by such a privilege (*In re Zuniga*).

In 1986, in Florida, Ray L. Carona, a man who allegedly had a history of cocaine abuse, was convicted of violating the law when he evidently lied about his drug history when purchasing firearms. He had been treated as a private patient by Dr. Roberto Ruiz during weekly sessions for about six months. He told Dr. Ruiz during those sessions that at times he was a user of cocaine, spending up to ten thousand dollars a month for the drug. Over Carona's objection, the trial court decided to allow Dr. Ruiz to testify, based on the judge's contention that there is no psychotherapist-patient privilege in the Eleventh Circuit (*United States v. Carona*).

It turns out that the Eleventh Circuit was formed in 1981, when the federal districts in the states of Alabama, Florida, and Georgia were separated from the Fifth Circuit to create a new Eleventh Circuit. Evidently, all the precedents established before that date in the old and larger Fifth Circuit were still binding on the new Eleventh Circuit. This would include, of course, the decision in the Meagher case, the first case described above. The appellate court in this case ruled that there was no physician-patient privilege in the Eleventh Circuit and pointed out that there is no common law history for a psychotherapist-patient privilege. The court made clear that it was relying on the earlier decision in Meagher. This negative view of the privilege in the Eleventh Circuit has been listed (and, in effect, "double counted") in subsequent court opinions as indicating an additional federal circuit denying the privilege.

In a 1989 California case, Jane Doe, an unnamed woman, claimed that her infant died of sudden infant death syndrome, but Ms. Doe was suspected of murder during a grand jury investigation. Dr. Roe, a psychiatrist who had treated her, declined to release her records to the grand jury as part of the investigation unless ordered by the court to do so. The officials of the two hospitals where she had been treated similarly refused to release her records. Jane Doe appealed to the Sixth Circuit Court of Appeals when the trial court declined to support her demand that her records be kept confidential. The Appellate Court refused to create a psychotherapist-patient privilege in its circuit but declined to comment on the possible merits of such a privilege. The court did note that such a privilege should be created, if deemed desirable, by an act of Congress. This point completely ignored the earlier statement from Congress that new privileges should be created in the federal courts in the "light of reason and experience" and, in particular, that the failure of Congress to enact a psychotherapist-patient privilege was not to be taken as a negative view of the possible value of such a privilege (*Doe v. United States*).

The Second Circuit Court of Appeals took up the issue in 1992 in a case, *In re Doe*, of a man named Steven Diamond who was on trial for extortion. A witness against Diamond, "John Doe," had a history of psychiatric treatment. In an effort to discredit Doe's testimony, Diamond's attorneys demanded that Doe provide psychiatric records and allow his psychotherapists to be interviewed, albeit *in camera* (that is, in the privacy of the judge's chambers and under a protective order prohibiting public disclosure except where disclosure at the trial was ordered by the judge). In its decision on the appeal, the court decided that a psychotherapist-patient privilege does exist in the federal courts but went on to add:

> However, we also recognize, as appellant concedes, that the privilege is highly qualified and requires a case-by-case assessment of whether the evidentiary need for the psychiatric history of a witness outweighs the privacy interests of that witness. . . . Indeed, the privilege amounts only to a requirement that a court give consideration to a witness's privacy interests as an important factor to be weighed in the balance in considering the admissibility of psychiatric histories or diagnoses. (*In re Doe*)

It is extremely important to note here that the appellate court is describing a *qualified privilege*, which means that in each case in which the issue comes up, the judge can decide whether the need for information by the justice system outweighs the privacy interest of the psychotherapy patient, employing what is referred to as a "balancing test." In support of this line of reasoning, the court cited the 1983 *Zuniga* decision in Michigan. Thus, in the only two appellate court decisions supporting the existence of a privilege at all, both had advocated for a qualified privilege, a privilege that is rightly viewed as "weak" and hardly comparable to the kind of privilege that the federal courts recognized at that time only for a small, select group of special relationships: the lawyer-client, spousal, and priest/clergy-penitent relationships. Because of the great importance our society grants to law, marriage, and religion, the federal courts recognize what is called an *absolute privilege* in these relationships, meaning that a trial judge must recognize the privilege and cannot use any kind of balancing test to determine if the privilege applies. Those who find themselves in federal court have a virtually unassailable right to prevent their lawyer, spouse, or clergyman from testifying against them in court or from even placing in evidence any information regarding confidential communications between them.

Finally, in 1994 the Tenth Circuit Court of Appeals took up the question of the privilege in a case in which Wilkie Bill Burtrum was on trial for criminal sexual child abuse committed on Indian Territory (and hence within the jurisdiction of a federal court). After the alleged incidents had taken place, Burtrum had sought psychotherapy and marital counseling with Joe Miller. During sessions with Miller, Burtrum admitted to the alleged incident of child abuse and was diagnosed as a pedophile. He sought to keep Miller from testifying against him in the trial. In its decision in an appeal by Burtrum of the lower court's ruling against Burtrum's privilege claim, the appeals court took the position that even if a privilege exists, it would not apply in cases of criminal child abuse. The position went no further than that. However, the court reaffirmed that federal courts are reluctant to recognize new privileges (*United States v. Burtrum*).

In looking back over these cases, one sees that it is impossible to conclude that the federal appellate courts have supported a psychotherapist-patient privilege. As noted above, even in the two instances where some support can be found, only the most qualified privilege was envisioned. In neither of those specific cases was the information sought protected by the privilege despite the two courts' assertions that the privilege exists. This antecedent history makes what actually happened when *Jaffee* was taken up by the Supreme Court all the more remarkable.

Psychotherapist-Patient Privilege in the States

Except for constitutional issues, state courts are not under the jurisdiction of federal courts. Rather, state courts operate under that state's own separate set of evidence rules. Privilege rules in the state courts are generally created by the legislatures, not by the common law process the U.S. Congress established for privilege rules in the federal courts. By the time the *Jaffee* case arose, the legislatures in each of the fifty states had created a psychotherapist-patient privilege, but like the states themselves, these rules were all over the map. The rules have many inconsistencies from one state to the next, and the rules even vary within a state as to which psychotherapy patients are protected by a privilege and which are not.

For instance, there is no separate and specific psychiatrist-patient privilege at all in New York State. Confidential communications between a psych*iatrist* and a patient are protected only by the exception-riddled New York physician-patient privilege. By contrast, confidential communications between a psych*ologist* and a patient are protected by a privilege that has the identical force of the absolute New York lawyer-client privilege! Patients of social workers in New York State are protected by a privilege that lies in its effectiveness somewhere between that of the psychiatrist-patient and psychologist-patient relationship. Nevertheless, despite all the inconsistencies and variations, it is possible to create a list of all fifty states showing that some kind of psychotherapist-patient privilege exists in each of them.

The Seventh Circuit Court Rules in Jaffee v. Redmond

Following the jury's verdict in *Jaffee,* the Village of Hoffman Estates and Officer Redmond appealed the decision to the Seventh Circuit Court of Appeals, the appellate court for that area and one of the federal appellate courts that had never ruled on the existence of a psychotherapist-patient privilege in the federal courts. While a psychotherapist-patient privilege had been recognized or denied in some of the federal circuits as noted above, there had been no such privilege ruling either way in the Seventh Circuit, so, at the least, this was bound to become a precedent-setting case in the Seventh Circuit. It became, however, much more than that.

The appellate court had been asked to decide the question of whether Officer Redmond actually did have the right (that is, a *privilege*) to prevent the trial court from compelling Karen Beyer, her therapist, from testifying in the trial, which would include handing over Beyer's notes of the contents of the fifty to seventy psychotherapy sessions she had had with Redmond since the shooting. *Except where the courts have recognized a privilege applying to the communications, no person has a legal right to prevent a court from compelling another person to testify.*

As it happens, the state of Illinois has a very strong privilege protecting psychotherapist-patient communications, including communications to social workers functioning as psychotherapists, from being disclosed in court if the patient objects. However, because this case had been brought as both a federal civil rights case and a state wrongful death case, the "rules" for mixed cases dictated that the case had to be tried in a federal court in Chicago and not in an Illinois state court, so the federal rules applied instead of the state rules.

The federal trial judge, Milton I. Shadur, had taken a hard line on the question of whether Beyer would be required to testify. He stated that there is no privilege for psychotherapist-patient communications and that even if there is such a privilege based on rulings in other circuit courts, that privilege applied only to communications to licensed physicians and licensed psychologists, not to social workers. The judge said, in his instructions to the jury:

> During the course of this lawsuit the Court ordered the Village of Hoffman Estates to turn over all of Ms. Beyer's notes to plaintiff's attorneys. The Village was provided with numerous opportunities to obey the Court's order and refused to do so. During the course of this lawsuit Mary Lu Redmond also testified that she would not authorize or direct Ms. Beyer to turn over those notes to plaintiff's attorneys.
>
> During Ms. Beyer's testimony she referred to herself as a "therapist," although she is not a psychiatrist or psychologist—she is a social worker. This Court has ruled that there is no legal justification in this lawsuit, based as it is on a federal constitutional claim, to refuse to produce Ms. Beyer's notes of her conversations with Mary Lu Redmond, and that such refusal was unjustified.
>
> *Under these circumstances, you are entitled to presume that the contents of the notes would be unfavorable to Mary Lu Redmond and the Village of Hoffman Estates.* (*Jaffee v. Redmond,* 1995, note 9; emphasis added)

The appellate court, in its opinion on the appeal, disagreed strongly with the trial judge. The court decided that the psychotherapist-patient privilege should be recognized in the Seventh Circuit. In addition, citing as a precedent the Illinois law on privileged communication in the psychotherapist-patient relationship, a law that extends such protection to the patients of social workers and physicians, the court ruled that the privilege in federal courts should also apply to patients of social workers. However, the privilege recognized by the court, as had been decreed in the 1983 and 1992 cases in the Sixth and Second Circuits, respectively, was highly "qualified," meaning that it was a "weak" privilege, allowing a trial judge to order a psychotherapist to testify if the judge believed such testimony would serve the "interests of justice." In the words of the court, "However, we also note that the privilege we recognize in a case of this nature requires an assessment of whether, in the interests of justice, the evidentiary need for the disclosure of the contents of a patient's counseling sessions outweighs that patient's privacy interests."

So at this point in April 1995, when the Seventh Circuit Appellate Court decided the case, it appeared that even if a psychotherapist-patient privilege was destined to be recognized some day across all the federal courts, that privilege would be highly qualified and would be much more like the highly questionable executive privilege held by the president (a qualified privilege) than like the extremely protective absolute privilege held by clients of attorneys in which no balancing by a trial judge can even be considered.

The Supreme Court Accepts Jaffee

Unlike other federal courts, the U.S. Supreme Court is not obligated to accept every case that is filed with it. As we observed in Chapter 6, the Supreme Court accepts only a tiny percentage of the cases brought to it each year. The court received petitions for review (that is, appeals) for approximately seven thousand cases in 1995 but accepted only about one hundred of those cases for decision. The Jaffee case was one of those hundred or so, probably in part because the main issue in the case, the existence of a federal psychotherapist privilege, was "ripe" for a decision that would resolve once and for all the differences among the federal appellate courts.

The Supreme Court, in actuality, had to decide on two questions in Jaffee. First and most important, is confidential communication between a psychotherapist and a patient privileged in all the federal courts? Second, should such a privilege, if it exists in the federal courts, apply to psychotherapy patients of social workers as well as to those of physicians and psychologists? The Supreme Court's decision to accept the case was of great significance to mental health professionals as well as to the legal profession. The court's agreement to take up the issue was widely reported in the press.

The Amici Curiae Briefs

Following the announcement that the Supreme Court had accepted the case, fourteen amici curiae briefs, representing many of the country's leading mental health organizations and other interested parties, such as Joseph Lifschutz (see Chapter 3) and the U.S. Solicitor General, were filed, all urging the court to recognize a psychotherapist-patient privilege in the federal courts. Significantly, no amicus briefs were filed in support of the petitioner, Carrie Jaffee, Ricky Allen's mother, who wanted the court to see the records. The lack of briefs in support of the petitioner was characterized by Justice Antonin Scalia as unusual "spin" in his dissenting opinion. Of the fourteen briefs, we will focus on two and reference a third.

The American Psychoanalytic Association (APsaA) joined three other leading professional mental health organizations whose members practice psychoanalysis and/or psychoanalytic psychotherapy in filing an amici brief on January 2, 1996. Our interest in the APsaA brief lies, first, in its compelling argument for a new psychotherapy privilege in the federal courts and, second, in its suggestion of the psychological conflicts Officer Redmond may have experienced as a result of the shooting of Ricky Allen.

The APsaA brief opens with the central question confronting the Supreme Court: "Whether federal courts in a civilian action should treat as privileged confidential communications made by a patient to a licensed clinical social worker in the course of psychotherapeutic treatment." (All quotations come from the HTML version, which Paul Mosher placed on the *Jaffee-Redmond* website, at http://Jaffee-Redmond.org/briefs/ApsaABrief. htm.) The APsaA brief states that the petitioner is mistaken for believing that there is no need for psychotherapist-patient confidentiality and that the disclosure of these communications results in "therapeutic benefits." The brief then offers three reasons for the protection of privacy in psychotherapy: (1) "Confidentiality enables individuals to obtain needed psychotherapy," (2) "Confidentiality is essential to the establishment of the 'therapeutic alliance,'" and (3) "Without confidentiality psychotherapy would lose much of its efficacy."

To support the statement that confidentiality allows people to enter psychotherapy, the APsaA brief emphasizes the social stigma still attached to therapy. The document cites a 1985 study in which "55 percent of supervisors surveyed indicated that they would have a negative attitude towards an employee who had seen a psychiatrist." (Social stigma, as we shall see in Chapter 11, prevented New York State Chief Judge Sol Wachtler from seeing a psychotherapist.) The APsaA brief suggests, additionally, that social stigma forces some patients to avoid using insurance, for which they have paid, to cover psychotherapy sessions because of the fear that an employer will find out about their need for treatment.

To illustrate that confidentiality is essential to the creation of the "therapeutic alliance," the APsaA brief points out that confidentiality is a "cornerstone of therapy," the way in which a therapist demonstrates trustworthiness and concern with the patient's well-being. "The possibility that a therapist might reveal in a court of law a patient's most troubling inner secrets would stand as a permanent obstacle to development of the necessary degree of patient trust in the therapist, and would pose a significant, and for many patients an insurmountable, barrier to effective treatment." Karen Beyer confirmed the truth of this statement when she observed, in her talk published in *The American Psychoanalyst*, that the subpoena was the "beginning of the end" of her therapeutic relationship with Redmond, "because of the extreme pressure brought upon both of us to reveal the content of the records."

To show that psychotherapy would lose much of its effectiveness without confidentiality, the APsaA brief calls attention to the petitioner's misunderstanding of psychotherapy: Carrie Jaffee's concession that confidentiality is essential to psychoanalysis but not to psychotherapy. On the contrary, the APsaA brief argues, both classical psychoanalysis and psychoanalytic psychotherapy help people achieve insight to understand conscious and unconscious motivation. The APsaA brief cites a study concluding that psychoanalytic psychotherapy "is the most common form of treatment among mental health professionals and is a highly valued form of treatment." The APsaA brief then reminds the court that Karen Beyer testified that her principal training was in "psychodynamic" psychotherapy. Such therapy, the APsaA brief explains, quoting T. Paolino in *Psychoanalytic Psychotherapy*, operates from the premise that "psychopathology is a consequence of repression and a lack and/or avoidance of self-knowledge about how one's mind works. . . . Accordingly, all psychodynamic therapists attempt to help their patients recover and reintegrate those parts of the patient that were lost to the unconscious and attempt to achieve some degree of insight."

The APsaA brief refers only a few times to Mary Lu Redmond, not always by name. The brief mentions, as an example of a person who seeks psychotherapy for help dealing with problems deemed "too shameful or troubling" to be shared with relatives, a "law enforcement officer who experienced strong fears in the aftermath of a violent incident and did not want to disclose this to colleagues." Noting the increasing number of people who have sought the documented benefits of psychotherapy, the brief then refers to Mary Lu Redmond by name. "In particular, psychoanalytic therapy can help police officers such as Officer Redmond to face and learn to accept unavoidable violent encounters rather than dealing with them through emotional numbing."

What would happen if there were no psychotherapist-patient privilege? The APsaA brief answers this question in a footnote. In the absence of a privilege, "a prudent attorney might well advise law enforcement officer clients not to seek psychotherapeutic treatment where a wrongful death charge or other legal claim is pending or threatened." Law enforcement officers would then confront, the APsaA brief continues, a "Hobson's choice"—that is, an apparently free choice where there is no real alternative—"between seeking treatment which will increase their legal exposure, or foregoing the benefits of psychotherapy."

One of the central arguments of the APsaA brief is that the "benefits to litigation" from the absence of a psychotherapist-patient privilege are "minuscule." A patient's disclosures to a therapist are "inherently unreliable as evidence since psychoanalytic therapy focuses not on reconstructing objective facts, but on furthering understanding of the patient's internal reality through fantasies, dreams and imaginings." Here the APsaA brief might have cited Donald P. Spence's 1984 book *Narrative Truth and Historical Truth* to suggest that psychoanalysis should be viewed more as a construction than as a reconstruction of the truth. Even with a psychotherapist-patient privilege, the truth-seeking process of litigation can still take place through deposition and trial testimony. Thus, Officer Redmond testified during the trial about her recollections of the shooting, and the petitioner had an opportunity to cross-examine her.

What would happen if the truth-seeking process of litigation devolves into the cross-examination of a patient's therapist? The amici brief filed by the National Association of Social Workers and related professional organizations envisions this disturbing possibility.

> If a party may open up the effect of psychotherapy in order to impeach a witness's testimony, may not the opposing party offer expert testimony to contradict the theory of undue influence suggested by the questions?

Such a scenario would divert juries and judges from their essential tasks in any civil action of determining what happened and whether it was justified, to determining the effects of psychotherapy on a witness's subsequent recall of events—an inquiry which is at best fraught with difficulty and risk of prejudice.

The question of whether the privilege, should it exist, would also extend to criminal cases was not addressed. However, that issue had been alluded to in a 1983 case in California in which a judge was horrified by the possibility that a psychotherapist could be compelled to testify against a defendant in a criminal trial: "there is obviously something revolting about the spectacle of a psychotherapist testifying to a patient's confidences in which the patient is the defendant" (*People v. Stritzinger*).

The amici brief of the American Psychiatric Association (APA) and American Academy of Psychiatry and the Law is noteworthy because it raises a crucial objection to the Seventh Circuit Appellate Court's decision, an objection with which the Supreme Court ultimately agreed. The petitioner had suggested a case-by-case "balancing" test to be left to the discretion of the district court. Disagreeing, the APA brief observed that

such an approach would seriously damage the very values protected by the privilege; exceptions must remain well-defined and truly exceptional, or else the assurance of protected confidentiality, on which the patient's interests depend, would be lost, and adverse litigants could exploit the uncertainty of standards to make patients pay a prohibitive price for protecting their privacy by invoking the privilege.

The APA brief then pointed out that the Supreme Court had made the same argument with respect to the lawyer-client privilege in the case of *Upjohn Co. v. United States*. The Supreme Court had said in that decision that an "uncertain privilege, or one which purports to be certain but results in widely varying applications by the courts, is little better than no privilege at all." Quoting the Supreme Court's argument in *Upjohn* turned out to be brilliant strategy: the Supreme Court used the same argument in *Jaffee*.

The amici curiae briefs make a convincing argument for the creation of a psychotherapist-patient privilege, but we may point out one sad irony. In extolling the many benefits of psychotherapy to law enforcement officers such as Mary Lu Redmond, the APsaA brief asserts that through treatment, "officers such as Officer Redmond can learn to face and accept the strong feelings raised by unavoidable stressful and violent encounters rather than blocking these feelings out. . . . This will, in turn, help such officers to remain empathetic and effective in their interactions with the community." This is not what happened to Mary Lu Redmond, however. Redmond went on leave from active duty after the shooting, never went back on active duty, and a few years later resigned from the police force and went into real estate.

The oral argument before the court took place on February 26, 1996. An audio recording of the Court proceeding can be heard at the Oyez Project webpage devoted to the case (Oyez Project, *Jaffee v. Redmond*).

The Supreme Court Creates a New Absolute Privilege

The Supreme Court announced its *Jaffee* decision on June 13, 1996. In a powerful eight-to-one vote, unusual in such a divisive case, the court stunned legal scholars by not only deciding the case in favor of the privilege but also going well beyond what could be viewed as the "common law" basis for its action. The majority opinion in the case was written by Justice John Paul Stevens, at that time the court's most liberal justice.

Relying heavily on the existence of some form of psychotherapist patient privilege in all fifty states and coupling the existence of such privileges with federal law giving the court the power to create new privileges based on "reason and experience," the Supreme Court decided that the existing state privileges constituted a history of "experience" in support of the need for and the merits of a privilege. This was the court's view despite the fact that the state psychotherapist–patient privileges had been created by state legislatures, with all the implications of political influence and lobbying that implies, rather than having been created by an unfolding process of common law.

As noted above, the psychotherapist-patient privilege across the states are quite varied, however. Many of these state privilege laws are subject to statutory exceptions, while some apply only to civil but not to criminal cases. Only a few include a "balancing test" that would allow a judge to decide on a case-by-case basis whether the privilege applies in that particular case and whether the privilege established by the appellate court in *Jaffee*, a civil case, would also apply in criminal cases.

Nonetheless, the court reached back into its own history of privilege case decisions to create a new absolute privilege essentially analogous to the lawyer-client privilege, the most revered privilege rule in both state and federal courts. Citing its own opinion on the lawyer-client privilege in *Upjohn Co. v. United States* (1980), the court wrote:

> We part company with the Court of Appeals on a separate point. We reject the balancing component of the privilege implemented by that court and a small number of States. Making the promise of confidentiality contingent upon a trial judge's later evaluation of the relative importance of the patient's interest in privacy and the evidentiary need for disclosure would eviscerate the effectiveness of the privilege. As we explained in *Upjohn*, if the purpose of the privilege is to be served, the participants in the confidential conversation "must be able to predict with some degree of certainty whether particular discussions will be protected. An uncertain privilege, or one which purports to be certain but results in widely varying applications by the courts, is little better than no privilege at all." (*Jaffee*, 1996)

In other words, the court held that a patient speaking with a psychotherapist, no less than a client speaking with a lawyer, must be able to rely on the promise of confidentiality to fulfill the intended purpose of the relationship. The court also decided, this time by a seven-to-two vote, to apply the new privilege to patients of social workers: "We have no hesitation in concluding in this case that the federal privilege should also extend to confidential communications made to licensed social workers in the course of psychotherapy. The reasons for recognizing a privilege for treatment by psychiatrists and psychologists apply with equal force to treatment by a clinical social worker such as Karen Beyer."

From June 13, 1996, onward, the status of the privacy of confidential psychotherapy communications in all federal courts was thus dramatically changed. The court stated, however, that a similar privilege was not needed and did not exist in the federal courts for other kinds of confidential health care communications. It is a long standing view of the Supreme Court that there is no physician-patient privilege in the federal courts, despite the existence of such a privilege in forty-two of the fifty states and despite the fact that the creation of such a federal privilege has been urged by a determined minority of legal scholars. Kenneth S. Brown, a professor of law at the University of North Carolina and a consultant to the Federal Rules of Evidence Advisory Committee, is one of many legal scholars who proposes broadening the federal psychotherapist-patient privilege to include physicians. The title of Brown's 2004 article—"Should It Matter Whether Your

Ego or Your Elbow Hurts?"—suggests why he believes patients of psychotherapists and physicians should be treated equally in the courts.

Justice Scalia Dissents

The only dissenting vote on the question of the recognition of the psychotherapy privilege was Justice Antonin Scalia. He was joined *in part* in his dissent by Chief Justice William Rehnquist, who actually supported the court's creation of the privilege but objected only to the court's extension of the court's new psychotherapy privilege to patients of social workers.

Justice Scalia begins his dissent with a telling sentence: "The Court has discussed at some length the benefit that will be purchased by creation of the evidentiary privilege in this case: the encouragement of psychoanalytic counseling." This is the only place in the majority or minority opinions where the reason for a psychotherapist-patient privilege is linked to psychoanalysis rather than to psychotherapy in general. This is extremely important because the basis for the majority view of a utilitarian or classical rationale for a privilege arguably may fail to apply to other forms of psychotherapy not based on psychoanalytic theory or psychoanalytic principles.

To understand this distinction, it may be helpful to divide all versions of "psychotherapy" into two groups. The first group, which consists of psychotherapy based on psychoanalytic theory or psychoanalytic principles (psychoanalysis and psychoanalytic psychotherapy), assumes the existence of unconscious conflicts, fantasies, and motivations and, to varying degrees, tries to bring these into consciousness through the technical device of free association. The second group, which consists of psychotherapy not based on psychoanalytic theory or psychoanalytic principles (behavior therapy, supportive counseling, cognitive-behavior therapy, and many other forms), does not assume, or at least does not focus on, the existence of an unconscious and, therefore, does not strive to "make the unconscious conscious." The initial rationale for the psychotherapist-patient privilege was devised at a time, in the 1950s, when psychotherapies in the second group mostly did not exist and when it was implicitly assumed that all psychotherapy was "psychoanalytic."

The Supreme Court's willingness to carve out a privilege for "psychotherapy" patients still rests on the ability to distinguish the confidentiality needs of "psychotherapy" from the confidentiality needs of the "rest of health care." Despite the strong interest medical patients have in protecting their privacy, such as their ob-gyn history, mental disorders, genetic disorders, and other disabling conditions, the federal courts do not, and likely will not, extend a privilege to medical patients. If the federal courts were to extend a privilege to medical patients, it would likely be a weak privilege, as is found in forty-two states. The argument for a privilege for psychotherapy patients in the second group isn't much stronger than it is for medical patients.

On the other hand, the rationale for a privilege for psychotherapy patients in the first group (psychoanalysis and psychoanalytic psychotherapy) is strong. The patients in the first group satisfy all four criteria for a privilege based on Dean Wigmore's classic treatise: (1) "The communication must originate in a confidence that it will not be disclosed," (2) "The element of confidentiality must be essential to the full and satisfactory maintenance of the relationship between the parties," (3) "The relationship must be one which in the opinion of the community ought to be sedulously fostered," and (4) "The injury which would inure to the relation by the disclosure of the communication must be greater than the benefit thereby gained for the correct disposal of the litigation." Only if these four conditions are present, Wigmore concluded, should a privilege be recognized (2285).

The Supreme Court cited the second Wigmore criterion because psychotherapy based on psychoanalysis or psychoanalytic psychotherapy depends on the use of free association. Patients must be able to comply with the "fundamental rule" of psychoanalysis, which states that they must be able to speak aloud every thought that comes to mind. Without the assurance of an ironclad privilege, these patients will not feel free to express themselves without censorship, and therefore, the entire treatment cannot take place. As a result, if there is no treatment without a privilege, then there is no information to disclose in the first place, making the evidence lost to litigation minimal (the fourth Wigmore criterion).

To accept the distinction between the two groups of psychotherapy is not to argue that patients in the second group do not need and deserve privileged communication. Rather, it is to argue that the Supreme Court's decision to create a psychotherapy privilege based on the second Wigmore criterion makes it illogical to extend the privilege to other forms of psychotherapy. The federal courts' willingness to sustain the privilege in its "full contours" will depend on the ability to draw a distinction between "psychotherapy" and "medical care." Justice Scalia seemed to be hinting at this crucial issue in the opening sentence when he refers to "psychoanalytic counseling," yet elsewhere he confuses the situation by referring twice to "psychotherapeutic counseling" and another two times to "psychological counseling."

Justice Scalia might have simply concluded, in opposition to the majority opinion, that a new federal privilege protecting the confidentiality of the psychotherapist-patient relationship does not produce the benefits to outweigh the need for probative evidence. Or he could have been content with stating, as he did, that the court has a duty "to proceed cautiously when erecting barriers between us and the truth." Instead, his argument calls into question, without appearing to do so, the societal value of psychotherapy itself.

Justice Scalia claims he has "no quarrel" with the court's majority opinion that the "psychotherapist privilege serves the public interest by facilitating the provision of appropriate treatment for individuals suffering the effects of a mental or emotional problem. The mental health of our citizenry, no less than its physical health, is a public good of transcendent importance." He immediately follows this with the statement that effective psychotherapy "undoubtedly is beneficial to individuals with mental problems, and surely serves some larger social interest in maintaining a mentally stable society." On the surface, the characterization appears positive, yet the reference to "mental problems" implies a hint of judgment. He next questions why psychotherapy has become so important as to require a new privilege that will "justify making our federal courts occasional instruments of injustice." He then widens the question and asks whether those who seek out therapists may be better served by turning to other members of society:

> When is it, one must wonder, that the psychotherapist came to play such an indispensable role in the maintenance of the citizenry's mental health? For most of history, men and women have worked out their difficulties by talking to, *inter alios*, parents, siblings, best friends and bartenders none of whom was awarded a privilege against testifying in court. Ask the average citizen: Would your mental health be more significantly impaired by preventing you from seeing a psychotherapist, or by preventing you from getting advice from your mom? I have little doubt what the answer would be. Yet there is no mother-child privilege.

Justice Scalia evidently cannot imagine the stigmatizing nature of mental illness, the profound shame and fear that prevent those suffering from anxiety and depression, drug and alcohol addiction, post–traumatic stress disorder, sexual abuse, and suicidal thinking from speaking with their parents, siblings, friends, or bartenders. Nor does he consider the possibility that psychotherapists have the professional training and experience others lack to facilitate such conversations. Why see a therapist, he muses, when you can talk to your bartender—presumably as an adjunct to drowning your sorrows in alcohol?

Justice Scalia recognizes that people derive benefit from speaking with their mother despite the lack of privilege for such conversations. The privilege is not necessary in Scalia's mind because people who need psychological support of a nonpsychoanalytic nature and based on advice are not prevented from giving another person, such as a therapist, the needed information without an ironclad promise of privacy. He seems here unable to distinguish a "therapeutic conversation" with a therapist from an ordinary conversation with an acquaintance.

Never referring to Mary Lu Redmond by name, Justice Scalia offers a worst-case scenario of the situation, a police officer who seeks counseling because she "shot without reason, and wounded an innocent man." He cannot understand why she should "be enabled both not to admit it in criminal court (as a good citizen should), and to get the benefits of psychotherapy by admitting it to a therapist who cannot tell anyone else. And even less reason why she should be enabled to deny her guilt in the criminal trial or in a civil trial for negligence while yet obtaining the benefits of psychotherapy by confessing guilt to a social worker who cannot testify." Here he seems to be unable to distinguish a "therapeutic conversation" with a trained therapist from a "confessional."

Justice Scalia refuses to imagine other situations, however, in which the police officer's statements (not confessions) to a therapist do not imply shooting a person "without reason." Cannot a police officer be exonerated from any wrongdoing in a shooting, as both the Hoffman Estates Police Department and the state of Illinois concluded in Redmond's case, yet, at the same time, be so traumatized by the event that she seeks the help of a psychotherapist? In this scenario, the officer's statements to a counselor are completely irrelevant to a trial.

Even when Justice Scalia admits that all fifty states recognize a psychotherapist-patient privilege for psychiatrists and clinical psychologists, he excludes patients of licensed social workers from this privilege because their training is different from that of other mental health professionals. He ignores here the research reported in the National Association of Social Workers' amici brief indicating, as stated in footnote 13, that studies on the "therapeutic progress of patients treated by psychiatrists, clinical psychologists and clinical social workers have shown consistently that the three professions deliver essentially equivalent outcomes." The same footnote reports the results of a large survey published in the November 1995 issue of *Consumer Reports* concluding, in the brief's words, that patients reported "similar progress whether they saw a social worker, psychologist or psychiatrist."

Justice Scalia asserts that social workers' training is different from and inferior to that of others who have a testimonial privilege. "With due respect, it does not seem to me that any of this training is comparable in its rigor (or indeed in the precision of its subject) to the training of the other experts (lawyers) to whom this Court has accorded a privilege, or even of the experts (psychiatrists and psychologists) to whom the Advisory Committee and this Court proposed extension of a privilege in 1972." He ignores the fact that the professional training of mental health workers involves its own rigor. Those who treat mental illness, whether they are physicians, psychologists, or social workers, often confront challenges that are greater than those who practice law. And he ignores the fact that confidentiality is no less essential to mental health professionals than to lawyers.

Justice Scalia ends his argument by noting in passing he is not surprised that, unlike the fourteen amici briefs supporting the respondents, the Village of Hoffman Estates and Mary Lu Redmond, not a single amicus brief was filed in support of the petitioner, Carrie Jaffee. His explanation? "There is no self-interested organization out there devoted to pursuit of the truth in the federal courts." Surely it's odd to view professional mental health organizations as a powerful private lobby. Members of the National Association of Social Workers, for example, often serve the least economically advantaged people in society. Therapists who treat the mentally ill rarely have had the power or prestige to create the kind of public injustice claimed by Justice Scalia.

How are we to interpret Justice Scalia's dissent in *Jaffee*? Ralph A. Rossum notes, in his largely sympathetic study of Scalia's jurisprudence, that the associate justice has "confessed that he writes with the verve and panache he does in part to ensure that his comments are included in constitutional casebooks where they will influence the next generation of lawyers and legal scholars" (205). It's true that Scalia is a consummate prose stylist, a man who revels in intellectual combat and who takes delight in lampooning and skewering his opponents. A legal provocateur with a no-compromise approach to the truth, he writes with Swiftian indignation to expose the follies of others, including what he believes to be utopian thinking. Whether he drives home his points with thrusting rapier wit or folksy, homespun humor, he is never boring. He is, by far, the most quoted of the nine Supreme Court justices and one of the most sought-after public speakers in the country. One's appreciation of his stylistic brilliance, however, does not make the content of his arguments more palatable. Reading his often provocative, if not jarring legal opinions, one is struck by his misunderstanding of and hostility to psychotherapy, something on which Rossum does not comment.

Nonetheless, few scholars who have commented on Scalia's dissent in *Jaffee*, including Kevin A. Chan, who is both a psychiatrist and a lawyer, have remarked on the justice's startling mistrust of psychology in this and in his other legal opinions. Scalia appears to go out of his way to express this mistrust, regardless whether he is writing for the majority or minority. In his dissent in *Lee v. Weisman* (1992), he argued that the Establishment Clause in the First Amendment to the Constitution protects state establishments of religion from federal interference. Insisting that the Supreme Court should allow prayer at public school graduation exercises, he invoked a long tradition of invocations and benedictions established by the framers of the Constitution. This tradition, Scalia added, is the opposite of what he calls a "psycho-coercion test, which suffers the double disability of having no roots whatever in our people's historic practice, and being infinitely expandable as the reasons for psychotherapy itself" (Rossum, 134). Psychotherapy, he implies here, is without legitimate history and capable of meaning both everything and nothing. Rejecting "psycho-coercion" tests, Scalia reminds his colleagues on the bench that they have "made a career of reading the disciples of Blackstone rather than Freud" (Rossum, 134). It's easier to understand Scalia's praise for William Blackstone, an influential eighteenth-century English judge and legal scholar whose philosophy was infused with Judeo-Christian principles, than his condemnation of Freud. Inveighing against the court's majority opinion that requiring junior high school students to pray at a graduation exercise was a form of psychological coercion, Scalia derides all things "psychological":

> I find it a sufficient embarrassment that our Establishment Clause jurisprudence regarding holiday displays has come to "require scrutiny more commonly associated with interior decorators than with the judiciary." But interior decorating is a rock-hard science compared to psychology practiced by amateurs. A few citations of "research in psychology" that

have no particular bearing upon the precise issue here cannot disguise the fact that the Court has gone beyond the realm where judges know what they are doing. (134)

Whether one agrees or disagrees with Scalia's view of psychology, one must concede the power of his prose. The literary critic Stanley Fish, himself a gadfly and provocateur, characterizes the second sentence in the above passage as a "rock thrown at Scalia's fellow justices in the majority; it is a projectile that picks up speed with every word . . . and the sentence hurtles toward what is both its semantic and real-life destination: the 'amateurs' who are sitting next to Scalia as he spits it out." Fish likens the sentence to target shooting, praising Scalia's ability "to load, aim, and get off a shot before his victims knew what was happening" (6–7).

Justice Scalia's withering objections to "psychology" extend beyond amateur and professional psychologists. His passion for originalism usually leads him to oppose expanding the rights of criminal defendants—he is a fierce opponent of "Miranda rights"—yet he has championed the rights of criminal defendants, particularly the power to confront their accusers, even if that constitutional right requires children who have been psychologically traumatized by sexual abuse to face the possibility of retraumatization when compelled to testify at a trial. In *Coy v. Iowa* (1988), he wrote the majority opinion in striking down a state decision to allow child victims of sexual abuse to testify in court behind a large screen placed between them and their alleged assailants. The justice opined that this was a violation of a defendant's right to face-to-face confrontation. The most he was willing to concede was that it is a "truism that constitutional protections have costs" (Rossum, 183). Two years later, in his dissent in *Maryland v. Craig*, he objected to the court's decision to allow the state to use one-way closed-circuit television for jurors to hear the testimony of child witnesses who were alleged to be the victims of abuse.

> The Court characterizes the State's interest which "outweigh[s]" the explicit text of the Constitution as an "interest in the physical and psychological well-being of child abuse victims," an "interest in protecting" such victims "from the emotional trauma of testifying." That is not so. A child who meets the Maryland statute's requirement of suffering such "serious emotional distress" from confrontation that he "cannot reasonably communicate" would seem entirely safe. Why would a prosecutor want to call a witness who cannot reasonably communicate? And if he did, it would be the State's own fault. (Rossum, 183)

It is never good when Justice Scalia invokes Freud, who comes to represent everything detestable in contemporary society. In his 2002 article "God's Justice and Ours," a defense of the constitutionality of the death penalty, Scalia offers a contrast between Christian and secular morality, leaving little doubt which he prefers. "The doctrine of free will—the ability of man to resist temptations to evil, which God will not permit beyond man's capacity to resist—is central to the Christian doctrine of salvation and damnation, heaven and hell. The post-Freudian secularist, on the other hand, is more inclined to think that people are what their history and circumstances have made them, and there is little sense in assigning blame." The second sentence would surely astonish the creator of psychoanalysis.

However, it is not difficult to see that a psychological theory that can "excuse" misbehavior (that is, "sin") based on the assumption that some behavior is determined by events beyond a person's control is incompatible with Scalia's theological worldview. As described by Scalia's biographers and by Scalia himself, he comes from and is to this day mightily influenced by an extremely conservative Catholic upbringing. According to his biographer Joan Biskupic,

> Scalia was a product of his immigrant background, his traditional upbringing and his devout Catholicism. President Reagan and his team probably did not consider Scalia's Catholicism when they chose him, except as it might have suggested his views against abortion. Yet it was a very traditional Catholic Scalia who joined the Court in 1986, a jurist who only a few months before his nomination had said in a speech that legal views are "inevitably affected by moral and theological perceptions." More than two decades later he would still suggest that such influences, while not determinative in rulings, were inevitable in one's overall thinking. He recalled the lasting lesson from his years with the Jesuits at Georgetown University: "[Do] not . . . separate your religious life from your intellectual life. They're not separate." (210)

In explaining why he and his wife had nine children, Scalia responded with the following answer to a question during a network television interview: "We didn't set out to have nine children. . . . We're just old-fashioned Catholics, playing what used to be known as 'Vatican roulette'" (Biskupic, 187). On another occasion he expressed pride in the fact that all nine of his children attended regular Sunday Mass (188), and when the liturgy in the church changed in a more liberal direction in the 1970s, he sought out a church, the Cathedral of St. Matthew the Apostle in downtown Washington, for himself and his family that adhered to the more traditional Latin Mass despite the fact that the church was more than an hour's drive from his home in suburban Virginia (185).

In a 2005 interview in the *New Yorker*, published around the time Scalia was being considered to replace the retiring William Rehnquist as chief justice of the Supreme Court, Margaret Talbot summarizes the associate justice's belief that a "Living Constitution" approach is an "expression of judicial arrogance that all too often leads to the 'discovery' of bogus new rights—such as the 'right to privacy' that undergirds two decisions that Scalia loathes, *Roe v. Wade* (1973) and *Lawrence v. Texas* (2003), which declared unconstitutional a law forbidding homosexual sodomy" (42). Talbot might have added *Jaffee v. Redmond* to these loathsome decisions. A contrarian critical of the court of public opinion, Scalia would never allow his judicial thinking to be influenced by "ethicoscientific concerns" (Biskupic, 133), the derisive term he used in *Stanford v. Kentucky*, a 1979 case that allowed states to execute sixteen-year-old murderers. Seemingly, no amount of medical, scientific, or, least of all, psychological evidence could convince Justice Scalia of the importance of the psychotherapist privilege.

In the final analysis, Justice Scalia will never be convinced that people with mental health problems would be better advised to seek advice from psychotherapists rather than from their mothers. And perhaps he is right, as Emanuel Muravchik noted wryly in a letter published in the *New York Times* on June 15, 1996. "Apparently he has never heard the old story of the mother who boasted about the devotion of her son: 'Not only did he buy me a condo, a Cadillac and a mink coat, but he also pays a psychiatrist $250 for a visit every week and all he talks about is me.'"

Karen Beyer

We don't regard Justice Scalia as the antagonist in *Jaffee*, but we do regard Karen Beyer as the story's protagonist. She seemed an unlikely person to play a decisive, even heroic role in the effort to create a psychotherapist-patient privilege in the federal courts. Before *Jaffee*, she had never had a privacy conflict as a clinical social worker, nor had she heard of Joseph Lifschutz or Anne Hayman. She viewed herself in the 1990s primarily as an administrator, not a therapist. She was not a person who enjoyed being in the limelight or staking out controversial opinions. Modest and self-effacing, she did not see herself as different from others in her field. Yet when a federal judge ordered her to turn over her psychotherapy notes on Mary Lu

Redmond, Beyer chose to defend a principle even if it meant accepting the harsh consequences of breaking the law.

Most of our information about Beyer comes from our March 30, 2013, telephone interview with her. The rest comes from her article "First Person: *Jaffee v. Redmond* Therapist Speaks" published in *The American Psychoanalyst* in September 2000. We did not ask her any questions about her treatment of Mary Lu Redmond. Nor did she volunteer any information about the police officer apart from what was reported in the media. Beyer admitted she felt "uncomfortable" about our interview partly because she is a private person and partly because of her commitment to confidentiality. We knew we were in a delicate situation where we could not discuss many of the details of the privacy story, which we all agreed must remain private.

Nevertheless, Beyer was willing to allow us a glimpse into the human story behind her role in the landmark *Jaffee* case: her views on confidentiality, her conflicts over deciding whether to comply with or defy a federal judge's insistence to testify and turn over her psychotherapy notes, her efforts to seek advice from lawyers and mental health professionals, and her growing apprehension when the Supreme Court unexpectedly accepted the case.

Leaving the World a Better Place

Some children know what they want to do in later life, but not many five-year-old girls hope to become social workers, particularly if they are the first in their families to go to college. But this was Karen Beyer's early dream. As a teenager she thought to herself, "The important thing in life is that when you leave this world, people will say, 'The world is a better place because that person lived.'"

An only child, Beyer credits her mother, who was a "real caretaker," for awakening her interest in people and her desire to help others. She and her mother attended a Unitarian church that contributed to her keen awareness of right and wrong. "It's in my nature to want to be an ethical person," she told us, unassumingly.

Beyer became a clinical social worker in the late 1960s, but slowly she became more interested in administrative and supervisory work. As the director of the Department of Health and Human Services for the Village of Hoffman Estates, she was responsible for the provision of immunization and health screening clinics and group therapy for community residents; she was also responsible for Employee Assistance Program (EAP) services for Village employees. She knew, when police officer Mary Lu Redmond sought help from the village's EAP, that this was going to be a high-profile case, one that was likely to attract intense media attention because of the explosive state of racial relations in Chicago and its surrounding suburbs. Like others, she sensed that race was bound to be a polarizing issue in a case in which a white police officer fatally shot a black man. She could imagine the horror of everyone present at the shooting, including Ricky Allen's family and Mary Lu Redmond. "It was awful for everyone involved, including the family members who saw him shot." For this reason, Beyer decided not to assign Redmond to one of her master's-level staff, who were likely to feel uncomfortable with intense media coverage. But she, too, felt uncomfortable with the potentially incendiary case, which she herself took on.

Accepting the Limits of Confidentiality

We were surprised to discover that Karen Beyer has never been a confidentiality absolutist either before or after *Jaffee*. She knew about the Tarasoff case from her professional training, and she made sure that students under her clinical supervision understood the limitations of confidentiality. As a social worker trained in family and marital therapy, she was interested mainly in here-and-now problems, using cognitive, behavioral, and systems-based therapy rather than psychoanalytic therapy. She regards a therapist's ability to warn an intended victim more as a freedom than as a constraint. In her view, breaching confidentiality to report child abuse is reasonable, and in the past she actually had breached confidentiality to report child abuse. "Children can't defend themselves." Beyer could imagine the possibility that a patient's threat of violence might be an unconscious cry for help, though she generally did not make such inferences about unconscious motivation.

Beyer had explained to Officer Redmond at the beginning of therapy, as she had to other clients, the limits of confidentiality, informing her, "There could be circumstances where I would have to reveal information that was legally required," such as reporting certain criminal activity. Beyer believes it's possible that some people did not talk to her about certain subjects because of this warning, but she doesn't think this occurred often. Because there was no hint of any criminal activity on the part of Officer Redmond—both the Hoffman Estates Police Department and the state of Illinois found that she had acted appropriately during the shooting—Beyer never believed it would be acceptable to breach confidentiality by testifying about the content of her counseling sessions or giving over her psychotherapy notes to the court. She maintained from the beginning that her psychotherapy notes revealed only her client's private struggle to come to terms with the tragic event.

"I Was Very, Very Worried"

Beyer used the word "traumatized" to describe Mary Lu Redmond's reaction to the shooting. Redmond would have been devastated if the social worker's psychotherapy notes were made public. "It was an awful situation for my client because she was a really private person who did not want any of her counseling experience to be introduced into the court. . . . I knew if I handed over the records, then everything she might have said to me about her personal life . . . would be fair game. I couldn't do that to a client." All of the lawyers, including Sandra Nye, who was hired by the Village of Hoffman Estates to provide legal counsel to Beyer, told her that the safest thing for her to do would be to turn over the therapy notes. The clinical psychologist who worked for Hoffman Estates was supportive but refused to offer advice. Suddenly Beyer found herself alone, with only her own conscience to guide her.

According to the Seventh Circuit Court, "Officer Redmond met with Beyer for the first time three or four days after the shooting incident and continued counseling for approximately two or three sessions per week through at least January of 1992, six months after the shooting." Beyer observed in *The American Psychoanalyst* that she worked with Redmond for nearly two years, almost eighteen months longer than court documents indicate. We asked Beyer about a possible discrepancy in these accounts of her work with Officer Redmond. She replied, "I'm not sure that there is a discrepancy if you interpret the court's statement as being I saw her for a minimum time ('through at least January 1992'). That leaves open the possibility that I saw her for a longer time, which is true." She had earlier reported that her therapeutic relationship ended when she received a subpoena to turn over the records of their work together, at which point she helped her client find another therapist whose identity remained unknown to court officials and the plaintiff's lawyer.

Beyer did not hesitate to describe her involvement with the case as an "ordeal," especially since she regarded herself, in general, as a worrier. "It was a very anxious time because I was dealing with things that were foreign to me." Her guiding assumption was that she would do everything to protect her client's best interests, even if it meant sacrificing her own well-being. "I was referred to the U.S. Attorney's office by the federal judge, and that office was instructed to consider charges against me, charges that could have led to my incarceration," Beyer wrote in *The American Psychoanalyst*. "The U.S. Attorney failed to pursue the charges, but I was not informed of this, and so I believed for some time that I might go to jail."

The timing of the case was strangely serendipitous. Had the case occurred a few years earlier, when her daughter was in high school, Beyer, a single mother from the time her daughter was four, would have felt torn between conflicting family and professional obligations, mainly because she did not have family living near her who, should she be jailed, could take care of her daughter. Now that her daughter was in college and on her own, however, Beyer no longer felt anxious about that part of her life. She was prepared to make whatever personal sacrifice was necessary to help her client, including going to jail. She imagined she might be sent to a low-security "Club Fed" where her friends would be able to send her chocolate. "I'm not going to go to jail for long," she said, with a mixture of pragmatism and gallows humor.

Beyer knew that the trial was bound to be controversial. "At the federal level everybody was treating this issue like a hot potato," she told us. "Congress and the courts didn't want to deal with it, and I had been told [by the Hoffman Estates lawyers] that I should just not make an issue over it." But, of course, she did make an issue over it, albeit reluctantly. "I was very, very worried about the outcome once the case started moving through the courts to the Supreme Court." Her greatest fear was that she would be contributing to a "bad law," one that had the potential to damage not only future patients and clients but also the nature of psychotherapy itself. "I felt like I was making decisions on my own that people who knew a lot more than I did felt were risky."

One of the people who was most supportive of Beyer was Michael O'Malley, an adept politician who at that time was the mayor of Hoffman Estates. Their offices were adjacent, and after 5:00 PM, he would turn on raucous Irish music that she enjoyed hearing. She describes O'Malley as a "larger-than-life" person who took a personal interest in both her and Mary Lu Redmond. Beyer had many talks with him in his office, after hours, about the case. "He viewed Redmond almost like a daughter. He was so protective of her, and he did everything he could to make sure she was going to be okay. He cared about me too."

Beyer reported in *The American Psychoanalyst* that she was "stunned" when the Supreme Court ruled in favor of the primacy of patient privacy. "When the Supreme Court decision was announced, licensed therapists who had been following the case were relieved. Probably none were more relieved than I. 'Forever grateful' best describes my feelings toward the American Psychoanalytic Association for submitting its powerful amicus brief explaining the importance of privilege and confidentiality."

Life for Beyer After Jaffee

Beyer had a small private practice at the same time she worked for Hoffman Estates, but even before the events that culminated in the *Jaffee* decision, she was gradually leaving the psychotherapist role for administrative work. With the financial help of Hoffman Estates, she earned a master's in public administration and resigned from her job in 1993. She is currently executive director of the Ecker Center for Mental Health, a nonprofit organization near Chicago.

Beyer is proud of her role in one of the landmark cases in American psychotherapy. In 2000 she received the Special Presidential Award from the American Psychoanalytic Association for the courageous stand she took in defense of patient privacy. Fame has been a "mixed bag," she told us. She has met wonderful people and visited interesting places as a result of being invited to speak at professional organizations, but the "downside" is that she doesn't regard herself as an expert on confidentiality and therefore has not always been able to answer questions that have been posed to her.

Karen Beyer sees herself as an ordinary therapist who was caught up in an extraordinary historical event. Yet it's obvious that few clinicians would have been willing to make such a personal sacrifice to defend a principle. The creation of a new kind of privileged information called "psychotherapy notes," a category written into the federal privacy rules four years later by the Department of Health and Human Services, may well be a result of her principled stand. She speaks about being touched by history, but we should add that she has also touched and changed history, in the process fulfilling her childhood dream of leaving the world a better place.

Sandra Nye

Sandra Nye is no less grateful for the opportunity to have played a role in the *Jaffee* decision. Whereas Beyer acknowledged her mother's profound impact on her life, Nye singled out her father for his formative influence on her choice of profession. A lawyer and judge with an interest in the religious laws of the Talmud, her father loved to "pontificate," she said, wryly, during our April 27, 2013, telephone interview. He encouraged her to develop her own strong views of the law, which were not always the same as his.

Sandra Nye began her career as a lawyer, graduating DePaul University Law School in 1962. Handling high-conflict divorce and child custody cases, she felt frustrated after eleven years. "The law did the best it could do, but there had to be a better way to handle these cases. I needed to find more that I could do for the people I serve." She decided to leave her law practice, where she worked with her then husband, a law professor, and enrolled in a master's program in social work. She immediately became fascinated with psychotherapy. She was then recruited to serve as director of law and social work at the Institute for Juvenile Research, which is currently housed in the Department of Psychiatry of the College of Medicine of the University of Illinois at Chicago. Convinced that the law and social work could join together for the benefit of children and families, she returned to her law practice after two years, though she still maintained a small private practice of psychotherapy patients.

A commitment to privacy and confidentiality in psychotherapy led Nye to work with the distinguished psychoanalyst Jerome S. Beigler (1916–2009), who taught at the University of Chicago and the Chicago Institute for Psychoanalysis for five decades. As the obituary in the *Chicago Tribune* indicated, he became an expert on patient-doctor confidentiality, "feeling like many of his colleagues that people didn't fully open up in analysis if they thought their feelings might someday be opened to the prying eyes of insurance companies or other interested parties" (November 13, 2009). Dr. Beigler—to whose memory we have dedicated our book—was the founder and the first chair of the Confidentiality Committees of both the American Psychoanalytic Association and the American Psychiatric Association. (Paul Mosher succeeded him as the chair of the former and served as a member and consultant to the latter.) "'He knew all the laws, and many of us would call him whenever we were in a quandry,' said Dr. Arnold Goldberg, a Chicago psychiatrist who knew Dr. Beigler since the 1950s. 'He was not so much opinionated as he was informed.'" Nye worked with Beigler and others to draft the 1994 Illinois Mental Health and Developmental Disabilities Confidentiality Act, which created the psychotherapist-patient privilege in the state. The landmark law, one of the first in the country, was later cited by the Supreme Court in the *Jaffee* decision.

Given Sandra Nye's background in law and in social work, her involvement with the National Association of Social Workers, and her advocacy for confidentiality issues, it was not surprising that Karen Beyer sought her legal counsel. The two women quickly developed a close working relationship based on mutual respect. Nye felt it was not her place to advise her client about what to do in *Jaffee*; instead, she informed Beyer of the legal implications of complying with or resisting the court's orders. "Karen is such a brave woman that she decided that this was going to be her commitment and that whatever happened, happened." Once Beyer made the decision not to testify about her treatment of Mary Lu Redmond or to turn over her psychotherapy notes, Nye supported her decision. "I was 100 percent on her side."

Nye was quick to point out to us that while *Jaffee* was an ordeal to her, as it was to Beyer, they were in different situations, which perfectly illustrates the difference between a relationship in which the communication was privileged and a relationship in which the communication was, at that time, *not* privileged. As Beyer's lawyer, Nye knew that she would not be forced to testify as to what her client disclosed to her. Nye identified with Beyer's situation, but she was not herself faced with the possibility of imprisonment and the loss of her license to practice. She sympathized with Beyer but felt it was important to maintain her professional perspective; otherwise, she told us, she would have lost her effectiveness as a lawyer. She likened her situation to that of a physician, who treats patients without suffering from their illnesses.

Like Beyer, Nye believes that there are certain situations when a psychotherapist is required, legally and morally, to breach confidentiality. She also believes, again like Beyer, that psychotherapists have an obligation to break confidentiality if they believe a patient intends to do imminent harm to another person. "The safety of others is important." Nye does not agree with Joseph Lipschutz that the psychotherapy privilege belongs to the therapist as well as to the patient. If Redmond had allowed Beyer to testify, Nye would have advised her client to do so. "I don't think Karen would have the right to withhold that information." But that was not the situation in this case. As a lawyer, Nye does not work with clients whom she believes are guilty of violent crimes.

Nye worked with Beyer only during the first phase of the trial, when it was heard in a federal district court in Chicago. She revealed that Judge

Shadur told her, privately, that social workers lacked rigorous training and education (the same attitude expressed by Justice Scalia in his dissent). "That's why, I think, he ruled the way he did. That's when I dug my heels in," all the more determined to help her client. She was "thrilled" but not surprised when the Supreme Court accepted the case; she was fairly certain that it would create a new federal psychotherapist privilege. "We felt we were right and that the Supreme Court would go along with us."

What would Nye have told Beyer if the highest court in the country ruled against a psychotherapy privilege for social workers? "I would have warned her that if the Supreme Court ordered her to testify and give her psychotherapy notes, she had two choices: either obey the law of the land or drop out, because she would have been thrown out of the profession, lost her license, and Lord knows what else." Fortunately, this frightening scenario did not occur, mainly because, in Nye's opinion, of a "wise Supreme Court." How does she feel about Antonin Scalia's dissent in *Jaffee*? "I don't pay much attention to him. He's so self-righteous and negative."

Nye was reluctant to speculate whether the federal psychotherapist-patient privilege created by *Jaffee* will be threatened by the increasingly blurred distinction between mental health care delivered through psychotherapy and mental health care delivered through psychopharmacology. But she believes that just as the lawyer-client privilege is discussed constantly in law school and law practices, so should the psychotherapist-patient privilege be discussed often and understood fully in clinical programs and clinical practices. She remains a privacy advocate in law and psychotherapy.

Nye had difficulty recalling some of the circumstances surrounding *Jaffee*, which had occurred more than two decades earlier. "That was a long time ago, and I'm old," she exclaimed. Feisty and tenacious, she still practices law after half a century. She represents the Illinois Department of Professional Regulations in defense matters, and she is also a professor of jurisprudence in psychiatry at the University of Illinois' Abraham Lincoln School of Medicine. She remains keenly aware of the momentous significance of *Jaffee*. "I've always felt proud of the case and its result." She regards the case as a precursor to HIPAA and other right-to-privacy laws. How has *Jaffee* affected her life? She paused when we asked her the question and then told us that her daughter, who is a gynecologist, and her son, a lawyer ("my partner and my boss"), are both devoted to privacy and confidentiality issues. *Jaffee* has thus become part of her family's life, the legacy she has bequeathed to her children.

After Jaffee

Subsequent to the court's decision in *Jaffee*, efforts within the psychotherapy professions to educate practitioners about this new status of their work
have been modest at best. Many psychotherapists are still unaware that their
communications with patients are protected by this extremely powerful
legal development. Lawyers, having the heritage of a two-hundred-year
history of the lawyer-client privilege, regularly inform their clients and the
public of the privilege and usually add a statement about the privileged nature of every communication they issue, such as on letters, faxes, and e-mails.
Psychotherapists rarely do this.

*Take-home lesson: Confidential information deriving from a treatment conversation
between a licensed psychotherapist and a patient is now protected from compelled disclosure in all federal courts of the United States by a very strong "absolute" privilege
rule similar to the well-known and highly regarded attorney-client privilege. Many
psychotherapists have failed to grasp the importance of this powerful protection for
their patients.*

10.   *The People v. Robert Bierenbaum*:
       "Long-Ago Warnings Cannot Justify
       Abrogating the Privilege Covering Still
       Confidential Communications"

What to look for: If a psychotherapist issues a Tarasoff warning, with the patient's permission, does that constitute a waiver of the patient's privilege to prevent the psychotherapist from testifying in court about the patient?

It was a perfect story for the New York tabloids: the mysterious disappearance of a young physician's attractive wife, Gail Katz Bierenbaum, on July 7, 1985. The only suspect was the husband himself, Robert Bierenbaum. Despite evidence suggesting the young physician's explosive temper and a history of violence, an official police investigation yielded no solid clues, and the case was closed. In 1989 a human torso, with the neck, arms, and legs cut off, was found floating off Staten Island. The remains were initially identified as Gail's, and a death certificate was issued, making her officially dead. The Manhattan district attorney's office was not informed about the torso until years later, however, and the case remained closed. Years passed. Robert Bierenbaum moved to another state, remarried, had a child, and earned the respect of his community. After more than a decade, the case was reopened, and the physician, now an established practicing plastic surgeon, was indicted. Now the story appeared in the *New York Times* as well as the *New Yorker* and *Vanity Fair*.

What makes the murder mystery more fascinating are the privacy issues surrounding the case. Before Gail's disappearance, several therapists had treated or spoken to Robert or Gail. Recognizing the husband's potential for violence, at least three of the therapists warned Gail that she was in grave danger. In the months following her disappearance, the police sought to subpoena the therapists, but they refused, citing confidentiality. The judge agreed with the therapists' position, particularly since there was no evidence of an actual crime. The same confidentiality issues reappeared nearly fifteen years later, when the then cold case was reopened. The therapists felt torn between maintaining confidentiality, on the one hand, and serving justice, on the other. The Bierenbaum case was so important to the mental health community that the New York State Psychiatric Association and the American Psychoanalytic Association filed an amici curiae brief urging the judge to respect the therapist-patient privilege. The judge was confronted with a critical decision that some believed might affect the outcome of the case and the future of the psychotherapist-patient privilege.

The Bierenbaum story was featured on Dominick Dunne's television program *Power, Privilege, and Justice*, which chronicled notorious murder cases involving the rich and famous. The Bierenbaum story is the basis of Kieran Crowley's bestselling 2001 book *The Surgeon's Wife*, part of St. Martin's "True Crime Library." The book serves as one of the sources of our narrative. Although Crowley, an investigative reporter for the *New York Post*, highlights the lurid details of the story and uses clichéd language and a whodunit mode of narration, his account appears generally to be accurate: its details are consistent with those in the published court papers and trial transcripts. Another source of our information is Alayne Katz, Gail's sister. We sent Alayne an early draft of our chapter, and she was kind enough to comment on it. She later granted us an interview. A lawyer who specializes in protecting the rights of women as well as abused and neglected children, she offered us her personal and legal impressions of the case.

A Marriage of Opposites

Gail Katz and Robert Bierenbaum were married on August 22, 1982. From the beginning, the union was an attraction of opposites. Born in Brooklyn, New York, in 1956, Gail Katz came from middle-class parents who later moved to Long Island, where they raised her and their two younger children, Alayne and Steven. Gail briefly attended the State University of New York at Albany, dropped out of college during her sophomore year in 1975, and began experimenting with alcohol and drugs, unsure of her future. Emotionally unstable, she became depressed and suicidal in 1979, overdosing on a combination of pills and alcohol; when that failed, she slashed four of her veins. A friend found her close to death, and she was rushed to a nearby hospital. After her recovery, she received psychiatric treatment for three months at the Payne Whitney Clinic in New York. Her experience solidified her long-standing interest in psychology, and after she graduated from Hunter College, she enrolled in the doctoral clinical psychology program at Long Island University.

Alayne Katz pointed out to us that the above summary, which comes from Crowley's book, fails to do her sister justice. Without romanticizing Gail, Alayne portrayed her sister much more sympathetically. A brilliant student who graduated high school a year early and entered college when she was only seventeen, Gail knew from childhood that she wanted to become a clinical psychologist. Like many people of her generation, she was a "hippie type" in high school but not a "druggie." She didn't drop out of college, as Crowley claims; rather, after attending SUNY-Albany for two years, she decided our Albany weather was too cold and gray to remain here. (By sheer coincidence, both authors of this book live in the Albany area.) "Who could blame her for not wanting to stay in Albany?" Gail decided she would move to Colorado, live there for a year to establish residency, and then attend Colorado State University.

But before actually taking the step of moving to Colorado, she fell in love with a rock musician and stayed in Albany to be with him. There she worked in a bookstore during the day and tended bar at night, ate macrobiotic food, sewed her own clothes, choreographed a rock musical, and worked with developmentally challenged adults for free. This was a good time in her life: she felt loved and appreciated, "happy as a clam." She then moved to New York City to finish her BA at New York University, which had admitted her into its dance therapy program. An injury forced her to drop out. Her life "floundered," and she began to "look for love in the wrong places." As a child she had suffered from depression and low self-esteem, but now her life was spinning out of control. She made a suicide attempt on the night she knew her sister would be driving from SUNY–Stony Brook to visit her. "She knew I would be there to save her." Recovering, Gail then founded a small business, finished her undergraduate degree at Hunter College (earning perfect grades), and was admitted to a PhD program in psychology. She received a prestigious internship at Beth Israel Hospital; her supervisor, Leah McCulloch, who later taught at Harvard, testified at the trial that Gail was one of her most brilliant assistants.

Before attending Hunter College, Gail met Robert Bierenbaum. His world was strikingly different from hers. Born in Newark, New Jersey, in 1955, the son of a cardiologist, he had a privileged childhood. In 1972 he was admitted to the special Albany Medical College/Rensselaer Polytechnic Institute "Physician Scientist Program" that enabled qualified individuals to complete the BS and MD degrees in seven calendar years. The program was limited to individuals who "display the motivation, maturity, and intellectual capacity to pursue an accelerated course of study."

Within five years he graduated at the age of twenty-two. Everyone who met him recognized his brilliance. He spoke many languages (eleven, by his own count), played the classical guitar, was an expert skier and gourmet cook, and had a pilot's license. Gail was awed by Bob's accomplishments and, in love with the idea of love, pictured herself as his perfect wife. One of the most romantic and dazzling experiences of her life was when he flew her over the New York City bridges on Valentine's Day.

"On paper Bob was perfect—until you met him," Alayne told us. Neither she nor her friends could understand what Gail saw in him or why she stayed with him. He was socially inept, a pathological liar, a rigid conservative, and controlling. As an example of his controlling behavior, once, when Gail went to turn off a light switch, he placed his hand on top of hers to show her how it was done. Worst of all was his violent temper. Before he met Gail, he had strangled his previous girlfriend's cat to death in her car, claiming it was endangering him while he was driving. A few months before his marriage to Gail, he tried to drown her own cat in the toilet bowl. When Gail told Alayne that he had tried to kill her cat and wanted to give the pet to a shelter, her sister said sarcastically, "Let's keep the cat and get rid of Bob." But Gail remained oblivious to being in danger and defended him. "Bob's just jealous of the cat. When he comes home and sees that I've given away the cat, everything will be just fine." But nothing changed after she brought the cat to a shelter. He was so hypercritical that he devalued her even when they were engaged. "I'm not marrying you, you don't have your BA, you have a Long Island accent, and you're too short."

Alayne said that her sister was "drop-dead beautiful" and loved to wear sexy clothes but that she began dressing conservatively after she became involved with Bob. Alayne could hardly recognize the person Gail was struggling to become, all to please her boyfriend. Alayne recalls a dinner with her boyfriend, Gail, and Bob during which Bob, using chopsticks, fed the two women the entire night, as if he owned both of them.

Despite her growing alarm and two broken engagements, Gail went through with the wedding, hoping for the best. "My sister was the queen of 'I'm going to change him.'" But the violence continued after they married. As their arguments worsened, Bierenbaum became increasingly abusive, alternating between suicidal and homicidal fury. On one occasion he picked up a knife and threatened to use it against himself; on another occasion, he threatened to jump out of the car while Gail was driving. The most serious incident occurred when she was studying for the Graduate

Record Exam. Seeing cigarette butts on the terrace of their twelfth-floor apartment, after he had forbidden her to smoke, he furiously jumped over the railing, holding onto the ledge only with his hands and feet. Then, deciding he would rather kill her than himself, Robert climbed back into the apartment and tried to strangle her, causing her to lose consciousness. Later he admitted that he had choked his previous fiancée during an argument, which caused her to end the relationship.

The strangling incident became a turning point in the Bierenbaums' marriage. Gail moved out of their Manhattan apartment; called her cousin, a Legal Aid lawyer, for help; relocated to her grandfather's home in Brooklyn; and filled out a police report. If a husband tried to strangle his wife today the way Robert Bierenbaum sought to strangle his wife then, Alayne told us, the police would view it as attempted murder, and he would be arrested. But there was less societal concern about domestic violence two decades ago, and the police did little more than recommend she go to family court. Gail then telephoned Robert's sister, a psychiatrist, who advised the couple to seek professional help. Shelley Juran, a psychologist, had one consultation and three telephone sessions with Robert. During one of those phone sessions, she became so alarmed about Gail's safety that, with Bob's consent, she asked to speak with Gail on the telephone to make sure she was still alive. Robert then met twice with Dr. Stanley Bone, a Manhattan psychiatrist, who requested and received permission from him to call Gail to warn her that she was in danger. After those two sessions with Robert, Dr. Bone referred him to another Manhattan psychiatrist, Dr. Michael Stone, who agreed to see Robert.

"A Psychopath"

At the time Dr. Stone, the author of two books on borderline personality disorder and a third book on schizophrenia, had a developing interest in criminal behavior that would subsequently result in additional books exploring the "anatomy of evil." Dr. Stone agreed to see the Bierenbaums separately, each for ninety minutes, before deciding whether to treat them as a couple. Within a few minutes of beginning his interview with Dr. Stone, Robert recounted, in a cold, emotionally detached voice, examples of his volcanic temper, including choking his previous fiancée and kicking his father in the groin when he tried to stop his son from taking LSD. Dr. Stone quickly diagnosed Robert as highly narcissistic and dangerous—a "psychopath." The psychiatrist believed that Gail's life was in immediate danger and that he might be placing his own life at risk if he agreed to treat Robert. Crowley describes Stone's two preconditions to which Robert had to agree before Stone would treat him, protocols the psychiatrist had never before demanded from a patient:

> "First, I will make two audio tapes of each session," Michael began. "At the end of each session, you will take one of the tapes and I will put the other into an envelope. I will address it to my lawyer and drop it into the mail box down the hall as you leave."
>
> Bob sat and waited for the second condition, something Michael said Bob's parents would have to pay for.
>
> "Lloyd's of London would have to insure my life with a policy that would pay four million dollars—two million dollars for each of my sons—if I am killed." (59)

Neither Crowley nor anyone else who has written about the Bieren-baum case has commented on the bizarre details of these therapeutic preconditions, which seem more appropriate for a murder mystery than actual therapy. One can only imagine Robert's astonishment—*anyone's* astonishment—about these preconditions. Crowley speculates that Dr. Stone had created these impossible preconditions so that he could refuse to treat Bierenbaum without appearing to reject him.

Dr. Stone's meeting with Gail was no less unsettling. She told him, according to Crowley, that her husband was a pathological liar, that she thought his parents were in denial about his violent character, and that she had deliberately provoked him into violence by smoking. Dr. Stone diagnosed her as suffering from borderline personality disorder, characterized by low self-esteem and chronic feelings of emptiness, identity diffusion, impulsivity, unstable relationships that alternate between idealization and devaluation, intense anger, and fears of abandonment. Dr. Stone thought that her self-destructiveness placed her at even greater risk; she exhibited the "Carmen Syndrome," flirting with death by taunting men into violence. Despite the psychiatrist's dire warning, Gail denied she was in grave danger. During their next meeting, Stone made an even more urgent attempt to have her leave her husband, asking her to sign a document stating that she had received a letter, a Tarasoff warning, which he felt professionally obliged to deliver to her:

> I have been advised by Dr. Stone that, for reasons of my own safety,
> I should at this time live apart from my husband, Dr. Robert Bierenbaum,

and until such time as it would appear that the risk of injury to my person has been significantly reduced.

I further understand that, owing to the unpredictable nature of my husband's physical assaults and to the chronic nature of the characterological abnormalities that underlie these assaults no firm date can as yet be fixed as to when it might be safe to resume living together.

If I do not heed this advice, I must accept the consequences, including the possibility of personal injury, or death, at the hands of my husband, and absolve Dr. Stone of responsibility for any such eventuality.

Dr. Stone, meantime, promises to warn me and my parents of when he deems any such risk to be particularly intense and imminent, to the best of his ability—should he undertake to treat my husband, and will also give such warning, if indicated, during this consultation period. (62)

Most likely Dr. Stone knew in advance that Robert Bierenbaum would never agree to his extraordinary preconditions for therapy. Bierenbaum angrily rejected the demands and left, to the psychiatrist's relief. But Dr. Stone was dismayed when Gail Bierenbaum also left, refusing to sign an acknowledgment of having received the letter. Nor would Gail give Dr. Stone the address and telephone number of her parents, so that he could notify them she was in danger. Instead, she took the unsigned copy of the letter as she left. The next day the psychiatrist mailed a copy of the letter to Gail's address by registered, return receipt requested mail, to protect himself, according to Crowley, from a possible medical malpractice case (63). Gail showed the letter to her husband, told him that she refused to sign an acknowledgment of it, and then placed it in her bank safe deposit box in the event that something happened to her. Dr. Stone never saw either member of the couple again in therapy, though he testified that he met Robert's parents to explain why he believed their son was dangerous, a meeting that the parents, Marvin and Nettie Bierenbaum, later denied had ever taken place.

Alayne didn't see Dr. Stone's Tarasoff letter until after her sister's murder. She remembers Gail's anger at Stone and her knee-jerk rejection of his psychiatric diagnosis, referring to it dismissively as a CYA ("Cover Your Ass") letter. "Dr. Stone's a jerk," Gail shouted. "Bob's not dangerous. He simply has anger management issues." Gail told this to Alayne on the telephone, and Alayne remembers thinking, "This is what my sister the budding psychologist thought?" Yet on some level Gail must have feared that the psychiatrist was right—why else would she have placed his warning letter in her safe deposit box? Confused and ambivalent about the marriage, Gail didn't know what to do, and her sister, who was now in law school, didn't know how to help her. Alayne loved Gail deeply and knew about her history of depression and low self-esteem, but Alayne is still to this day amazed at her sister's blindness to the warnings from Dr. Stone and others.

Alayne helped her sister as much as she could, but there were limits to even a sister's help. Besides, Alayne grew understandably exasperated by her sister's problems and wanted to focus on her own life as a law student. "Gail and Bob were masterpieces of psychological drama."

The Bierenbaums' marriage improved for a while. Both found another psychiatrist, unnamed by Crowley, who was willing to treat them. "That psychiatrist did not label Bob a dangerous psychopath or Gail a self-destructive borderline case. He simply labeled them a dysfunctional couple. He diagnosed Bob as suffering from a lesser adjustment disorder 'with mixed disturbance of emotions and conduct.' The shrink concluded that Gail suffered from a 'Partner Relational Problem' that was characterized by poor communication and a pattern of criticism, as well as 'unrealistic expectations'" (65). Gail moved back with her husband in the fall of 1983, but the following spring she met with a divorce lawyer and told her husband that she was moving out. She demanded that Robert pay all her living expenses for the next two years. She also began having extramarital affairs, sometimes inviting a man to their apartment when her husband was working at the hospital. Suspecting her infidelity, Robert became so enraged that he reportedly said to a friend, "I hate her so much, I could kill her" (68). Months earlier, he had told Gail, after they had watched a television movie about the British socialite Claus von Bülow, who was charged with poisoning his heiress wife, that if he killed Gail, he would not make the mistake of leaving a body to connect him with the crime.

Around this time Gail began her doctoral psychology program. Crowley notes tersely that in May 1985 she studied for her psychopathology final exam. "On paper, at least, she mastered the subject—she got an 'A' on the final" (69). The next month she told her mother about her worsening marital problems, including the choking incident eighteen months earlier, the police report, and the letter from Dr. Stone.

One moment Gail loved Bob, the next moment she hated him, Alayne told us. Gail's marital ambivalence was strikingly evident when, while making plans for a surprise thirtieth birthday party for her husband, in July, she was also planning to move out of her apartment on the weekend of her disappearance.

"Surgeon's Wife Vanishes"

Gail's parents, Manny and Sylvia Katz, along with her therapists, suspected immediately that Robert was responsible for her abrupt disappearance, but there was no evidence to indict him for a crime that might not
have occurred. The *New York Post*'s headline, "Surgeon's Wife Vanishes,"
evokes the media's fascination with the case. Robert gave contradictory
statements to the police when he was interviewed, but without a body, forensic evidence, or a confession, Gail was deemed a "missing person," not
a homicide victim. He did not help his cause by refusing to answer detailed questions about his marital problems; instead, he immediately hired
a lawyer who advised him to remain silent. There was a single report of a
sighting of Gail after posters were placed around the Bierenbaums' neigh-
borhood. The witness, Joel Davis, was certain he had seen Gail in a bagel
store, but Davis's description did not fit her. Gail's parents asked Dr. Stone
to send a copy of his Tarasoff warning letter to the police, but he refused,
citing confidentiality. Instead, he sent a copy of the letter to them, knowing they would give it to the police, which they did. The police requested
the psychiatrist's notes on the case, but he declined to release his notes on
the advice of his lawyer.

As Crowley describes in his narrative, during and following the investigation, Robert continued living in New York City, completed a residency
in plastic surgery, and had a number of romantic relationships. He seemed
to feel relief, not anguish, that he was no longer living with his wife. However, two troubling incidents occurred that would later become significant.
He began living with an anesthesiology resident, Dr. Roberta Karnofsky,
who was mystified by Robert's response to a 2 A.M. telephone call nearly a
year after Gail's disappearance. The caller, an officer from the New York
Police Department Missing Persons Squad, informed Robert that a woman
who resembled his wife had been picked up by the Port Authority Police
at the World Trade Center and that he was needed immediately to identify her. Rather than becoming excited and rushing to the precinct, as any
person in his situation would do, Robert seemed annoyed and uninter-
ested, as if he knew the woman could not possibly be his wife. He assured
Karnofsky that she need not vacate the premises. Why would he respond
this way, she wondered, unless he knew Gail was dead? Nor were her suspicions allayed when he returned home a few hours later with the news
that the woman was not his wife. Distressed, Karnofsky confided in a physician friend. The two women searched Robert's apartment and came

across his pilot's logbook. Opening it, they could see that a log entry written on "7/7/85," the date of Gail's disappearance, had clumsily been changed to "8/7/85."

After Karnofsky and Robert had an argument, she defiantly offered her "murder hypothesis" to him. According to Crowley, Karnofsky said, accusingly, "You could have put Gail in one of those big flight bags, as she was so small, and put her on the plane and driven her out to the airport and put her on the plane and then thrown her out of the plane" (145). To her disappointment, he did not react to her provocative words. Fearful for her safety, friends advised Karnofsky to leave Robert, but she stayed with him until she caught him cheating, at which time they separated. He then began openly dating other women.

Despite the many clues suggesting that Gail was murdered, the district attorney's office reluctantly closed the case in April 1987 after a nine-month investigation. There was insufficient evidence for a conviction, and the DA did not want to risk an acquittal, which would have prevented Bierenbaum from being brought to trial again because of the Fifth Amendment's double jeopardy clause, which prohibits a person from being tried twice for the same crime. The story appeared to end at this time—another example of a missing person who would never be found.

The most poignant moment of our interview with Alayne Katz came when she told us how Gail's disappearance and presumed death devastated everyone in the family. Alayne began a letter-writing campaign in which she sent out countless posters with Gail's photo on them, hoping to gain information about her disappearance. Alayne told us that for a while she was consumed by the case, but honesty compelled her to dispel whatever "lovely romantic myth" we might have about her willingness to keep the cold case alive indefinitely. "The truth is that I gave my heart and soul for every minute of my waking life for two years, but I realized I could not continue to live that way. The police closed the case and I said to myself, 'he killed my sister and he's ruined my parents' lives. I've got to close the book on this and live—otherwise he will kill me too.' I actually said to my mother, 'You're no longer permitted to accost me about this. I cannot live like this anymore.'" Yet despite Alayne's efforts to resume her life, there was a part of her emotional life that had ended.

Four years after Gail's disappearance, a disarticulated torso floated ashore near Staten Island, and Alayne was once again dragged back into the case. Medical tests indicated the torso could not have been in the water for more than six months. The radiologist's report was good enough to identify

Gail's body but not strong enough evidence for an indictment and conviction. Gail's broken-hearted parents held a funeral and buried what they thought were their daughter's remains. They refused to invite Robert to the funeral.

A few months later Bierenbaum moved to Las Vegas, where he continued to develop his professional life and began to rebuild his personal life. Several times a year he piloted his own plane to Mexico to perform surgery for free on indigent patients. In 1996 he married a young gynecologist, Dr. Janet Chollet, and within two years they relocated to Minot, North Dakota, where they had a daughter. Robert established a practice in plastic surgery in Minot and became a respected and well-regarded member of the community. According to the *New York Times* of October 30, 2000:

> Then 39, an intense New Yorker with piercing eyes and an in-your-face manner, [Dr. Bierenbaum] seemed misplaced amid the brawny and laconic citizenry. But his exotic skills and civic spirit soon won over the town: he published a bagel recipe in the local newspaper, read Hebrew amid the 20-some congregants of the tiny clapboard synagogue, even flew members of the Rotary Club to Mexico in his twin-engine Comanche to give free medical care to orphans.

Alayne described to us her shock when, twelve years after Gail's disappearance, Andrew ("Andy") Rosenzweig, the chief investigator who had believed from the beginning that Robert was guilty, telephoned her. Believing that the evidence would probably never get better, the Manhattan district attorney's office was considering reopening the inquiry—mainly because Rosenzweig, who planned to retire soon, had always been haunted by the case. "Do you want to do this?" Rosenzweig asked her. She indeed did. When Rosenzweig and his colleagues arrived at her office, they went through her files, including the medical examiner's report identifying the disarticulated torso as Gail's. The detectives were flabbergasted to learn that the Staten Island district attorney had never communicated the identification of Gail's body to the Manhattan district attorney. The case continued to be elusive, however. Improved DNA testing determined, when the torso was later exhumed, that it was not Gail's body. There was, at that point, such little concrete evidence that Alayne Katz did not believe Bierenbaum would be found guilty, but she felt that even if he were not convicted, putting him through the agony of a trial would offer her some consolation for Gail's death. "So what if double jeopardy happens," Alayne thought. "It's now or never—and now is better than never."

But suddenly a break in the case came when investigators heard Roberta Karnofsky's account of the altered flight log and Bierenbaum's strange response to the news that the police thought they had found his wife. On November 30, 1998, Rosenzweig and an associate flew to Minot and informed Robert that he was wanted for questioning. The police were convinced he had murdered Gail, cut up her body, placed it in a bag, and then dropped it somewhere in the ocean during his two-hour plane flight on the day of her disappearance.

Possibly as a result of this initial inquiry, or for other reasons we cannot know, Robert's idyllic life in Minot appeared to be coming apart, as the *New York Times* reported on October 30, 2000.

> Dr. Bierenbaum carried on as a good citizen, writing an article in the *Grand Forks Daily Herald* about how to avoid snowblower injuries. But his life in Minot was already fraying. His wife had decided to leave medicine and begin law school about 200 miles away in Grand Forks, N.D., where she moved, creating a hectic commuter marriage. Upon selling their Minot condominium, Dr. Bierenbaum took to sleeping on a folding cot in the hangar beside his airplane and showering in the hospital where he still treated patients.

The following September a grand jury convened and quickly issued an indictment of second-degree murder, fourteen years after the mysterious disappearance. Would there be enough evidence for a grand jury to indict Bierenbaum? Crowley reports that in a moment of black humor, Bierenbaum's lawyer told him that it was likely he would be indicted and arrested. "That was how grand juries worked—a prosecutor could get a grand jury to indict a ham sandwich" (238). (The expression was coined by New York Chief Judge Sol Wachtler, who used it in a different context.) But was there enough evidence for a conviction? Without a body, witnesses, forensic evidence, or a confession, how would the prosecution convince a jury that Bob had murdered Gail?

A Fascinating Story

The Bierenbaum story generated enormous media attention including, eventually, fourteen stories on different days in the *New York Times*. Robert's refusal to answer any questions about the story only heightened the mystery surrounding the case. In retrospect, the various stories about the case take on added significance because so many of them were one-sided.

In "A Cold Case," appearing in the "Annals of Crime" section of the February 14, 2000, issue of the *New Yorker*, Philip Gourevitch highlights Andy Rosenzweig, who comes across as a larger-than-life investigator.

> Rosenzweig is an understated man, but understated in the implacable manner of Humphrey Bogart, to whom he bears some resemblance: he has the trim proportions; he has the versatile, long, toothy face, at once bemused and brooding, with a smile that bares a hint of a snarl, and a sense of preoccupation in his own private calculus; and his nasal, slightly sibilant speech recalls Bogart's nervous rhythms. (43)

Gourevitch's emphasis in the story is on another criminal, long presumed dead, whom the veteran investigator locates nearly three decades after a brutal double murder, but at the end of the article the writer cites Rosenzweig's role in reopening another cold case, that of Robert Bierenbaum. "For eleven years, the case had been bothering Rosenzweig, and he couldn't say exactly why he had finally reopened it when he did, except that he felt it would be wrong for him to retire with it unresolved" (60). Asked, when he was called to testify at the Bierenbaum trial, why he had decided to pursue a case that nearly everyone had forgotten, Rosenzweig responded: "'What could I tell them?' he said. 'Just that I'm the slowest damn, most tiresomely methodical dot-the-*i*'s-and-cross-the-*t*'s investigator they'll ever meet'" (60).

"The Harriet-the-Spy Club," written by Tad Friend, appeared in the July 31, 2000, issue of the *New Yorker*, two months before the beginning of the trial, and focused on the rumors swirling around Bob. Three of the women whom Friend interviewed, all of whom had dated him in Las Vegas, and who requested their names, professions, and other identifying information be changed in the article, recited stories that soon bore all the hallmarks of gossip. Friend cites the example of "Suzanne's" mother, who was suspicious of Bierenbaum from the start—"not as a murderer, necessarily, but as a lousy potential son-in-law, which, in her eyes, amounted to almost the same thing" (40). Friend refers to another person who exclaimed that there was something about Bob he had never felt about anyone else. "He wouldn't look me in the eye, and then he looked right through me with the intensity and darkness of an obsessional killer" (38). Friend concludes from the statement that a "startling bit of new information about a person, even if it is implausible, prompts a reinterpretation of every puzzling previous interaction. And if the person in question is not someone we know well, the reappraisal is all the more swift and absolute" (38). Friend's article is a cautionary tale about the danger of weaving together disparate stories into a coherent narrative that may sound entirely convincing but nevertheless be totally wrong. Or that may prove prophetically correct, as Friend would perhaps be surprised to discover.

Lisa DePaulo's article "Intimations of Murder," appearing in the September 2000 issue of *Vanity Fair*, offered a probing analysis of the circumstantial evidence surrounding the Bierenbaum case—and, it should be emphasized, *all* the evidence was circumstantial. That's what made the case so intriguing, along with the story's lurid newspaper headlines, endless gossip, and macabre subplots, such as the torso fiasco, the altered flight log, and the mysterious sighting of a disappeared woman. DePaulo captures the conflicting characterizations of Robert Bierenbaum. To some he was a genuine genius devoted to healing his patients and serving his community; to others he was the evil mastermind of a perfect, or almost perfect, crime. One of the women he dated two years after his wife's disappearance, Sandy Schiff, a New York City criminal attorney, characterized him as "very sweet, very gentle, very devoted" (172). A self-described cynic, she couldn't believe that Bierenbaum was capable of harming anyone, much less devising a ghoulish murder. Schiff was with him for three years, until he moved to Las Vegas, and she was horrified by the idea that he was a murder suspect. "I can't imagine him intimidating anybody. We didn't even have an argument in all the years we were together" (172).

Another fascinating aspect of the story is that Gail, according to DePaulo, was "literally *surrounded* by shrinks" (174). Apart from Shelley Juran, Stanley Bone, and Michael Stone, Gail had been seeing a therapist named Sybil Baron up to ten times a month. In addition, reports DePaulo, "her clinical-psychology degree included actual therapy as part of her training" (175). Gail and Robert were also in couples therapy. With so many therapists seeing the couple, how could a murder occur without being detected?

Dr. Stone was the only one of these therapists who was at first willing—indeed, eager—to speak about the Bierenbaum case. Our major interest in DePaulo's article is her extended interview of Michael Stone, who doesn't hesitate to reveal his feelings about Gail and Robert Bierenbaum. DePaulo's account is consistent with Crowley's, who also interviewed Dr. Stone, though *The Surgeon's Wife* was published after the jury had reached its verdict.

DePaulo begins her interview with the psychiatrist's blunt words to Gail Katz Bierenbaum: "So I said, 'Why fuck with suicide when you can marry a killer?'" (164). Expressing astonishment that the soft-spoken, patrician-looking psychiatrist actually said this to a patient, DePaulo quotes his next words: "I said that to her. It's about the nastiest thing I've ever said in 35 years in the business. But I don't waffle." Nor does he waffle when he talks about his satisfaction upon learning that Bob finally has been brought to trial. The case has "always been front and center in my mind like a hole in my brain . . . because I predicted it" (164). Appearing in 1994 as an expert witness in a Las Vegas trial, Dr. Stone admitted looking through the telephone directory to see the "shrinks here" and noticed Bierenbaum's name. Dr. Stone considered hiring a private detective to determine whether the plastic surgeon had remarried but decided against it because it wasn't worth a "bundle of money."

Dr. Stone expressed no hesitation in discussing publicly what is generally considered to be confidential information about a patient. Part of his justification, he asserts, was that Gail's sister had signed an authorized release allowing him to do so. When asked if he had consulted a lawyer to advise him on confidentiality, he replied in the affirmative—Alayne Katz. Crowley's account is similar: "Michael had told a journalist that he'd received free legal advice from Alayne Katz, who told him he had nothing to worry about" (261). DePaulo's version of Alayne Katz's answer to the question is different. "Alayne says she gave Stone her 'opinion' on several occasions. 'I did not advise him'" (164). It would have been a conflict of interest had she offered him legal advice. Dr. Stone also maintained in the interview that he was allowed to discuss his conversation with Bob's parents because it was "group therapy" and thus, *in his view*, outside the therapist–patient relationship.

Surely Dr. Stone must have realized that his willingness, even eagerness, to testify at the Bierenbaum trial would weaken therapist–patient confidentiality. In his own balancing of the need for confidentiality in psychotherapy and the requirements of justice, he concluded that the latter outweighed the former.

Dr. Stone acknowledged during his nearly two-hour interview that he was obsessed by the Bierenbaum case, an observation that seemed to be supported by the 350 true-crime books in his library. "Stone is so fascinated by true-crime stories," DePaulo observes, "that he has devised an elaborate 'coding system' for them, in order to see at a glance how many killers had tattoos, how many were left-handed, how many were doctors" (168).

DePaulo ends her interview with Dr. Stone by calling into question his statement that he had met with Bob's parents. Marvin Bierenbaum insisted he was out of the country on the day the psychiatrist claimed he spoke with them, and Nettie Bierenbaum signed an affidavit stating she had never met or spoken with him. "Well, she's fucking lying" (168), Stone responded, an epithet that DePaulo cleverly uses as bookends in the account of her interview with him.

The "Edge of Possible New Legal Ground"

The presumably random selection of a judge for this case adds another dimension to the story. The presiding judge in the case was Leslie Crocker Snyder, who was viewed as strongly pro-prosecution. Her toughness with criminals was legendary, earning her the nickname "The Ice Princess." She was also known as "Judge 232" because she had once sentenced the leader of a gang to 232 years in prison. A former state prosecutor who became an acting justice of the New York State Supreme Court, she had twenty-four-hour security and seemed undaunted by the death threats that the media did not hesitate to report. "Kill the Judge" screamed the headlines of both the *New York Daily News* and the *New York Post* on August 10, 2000, a reference to a Wall Street swindler who was charged with ordering the assassination of the iron-fisted jurist.

During a pretrial hearing on July 12, 2000, Judge Crocker Snyder was faced with the vexing decision of whether to allow the three mental health professionals who had spoken to or treated Robert Bierenbaum to testify at the trial. Such testimony might well be crucial to the outcome of the trial. On August 2, 2000, Dr. Stone testified willingly that he had met with both Bierenbaums in 1983, had issued a Tarasoff warning letter to Gail, and had met with Robert's parents to warn them about the danger their son posed to his wife. On cross-examination, the defense tried to prove that Dr. Stone had never spoken to Marvin and Nettie Bierenbaum. At the end of the hearing, Judge Crocker Snyder stated that she would reserve judgment about the confidentiality issues until the other therapists testified.

Four days later an article by Katherine E. Finklestein in the *New York Times* emphasized Dr. Stone's key role in the prosecution's case. "He is at the center of a murky and fiercely argued conflict: does the psychiatrist's duty to warn the potential victim of a violent crime nullify that patient's right to confidentiality in a court of law?" Dr. Stone at first expressed no hesitation in answering Finklestein's questions, but then he seemed to have second thoughts. "Dr. Stone said he knew from the first consultation and Dr. Bierenbaum's haunting admissions that he was a dangerous man. 'People who kill large animals . . .' he stopped himself, then continued, 'we know it's a risk factor.'" After summarizing Dr. Stone's involvement with the case and the fact that the "ethics of medicine and the rule of law are worlds apart," Finklestein ends with the psychiatrist's view of the trial. "Dr. Stone remains hopeful that he will be allowed to speak. 'I feel that the information that I hold has great bearing on the case,' he said. 'I have my beliefs; that's apparent already from the [warning] letter I wrote to Gail.'"

The second pretrial hearing was held on September 7, 2000, and it was apparent that Dr. Stone had undergone a change of heart about his willingness to testify. Crowley reports that the judge had received a "weird" letter (261) from Dr. Stone's attorney, Jonathan Svetkey, stating that the psychiatrist who had voluntarily turned over to the court forty pages of his clinical notes on the Bierenbaums no longer wished to cooperate with the prosecution. "I think we are trying to put the genie back into the bottle, so to speak," Svetkey told Judge Crocker Snyder. "I think you would agree that the genie is really out of the bottle," the judge retorted (261).

A moment of comic relief occurred when the judge, ordering Stanley Bone to testify, mistakenly called him "Dr. Stone," to which the witness replied, "Dr. *Bone*." Unlike Dr. Stone at the first pretrial meeting, Dr. Bone was from the start extremely reluctant to testify, citing the confidentiality of the therapist-patient relationship. Dr. Shelley Juran was similarly hesitant to speak at the pretrial hearing, though when directed to testify by the judge she admitted expressing her concern about Gail's safety. After the hearing, she told the *New York Times* on September 8 that she found herself in a difficult position. "'You have your responsibility to your client.' But she added that there were also 'judges and the rule of law; I know we're at the edge of possible new legal ground.'"

The Amici Curiae Brief

When one of the psychotherapists involved in the case received a request for his records and testimony in the lead-up to the trial, he sought to consult with Paul Mosher, who was believed to be especially knowledgeable about such matters among professional colleagues in the American Psychoanalytic Association. Paul, upon hearing about the complicated and potentially momentous nature of the issues raised by the request, immediately suggested that the therapist get in touch with Seth P. Stein, an attorney especially expert in such matters.

The privacy issues surrounding the Bierenbaum case were momentous enough that two of the major mental health organizations then became involved. The New York State Psychiatric Association, representing five thousand psychiatrists throughout the state, and the American Psychoanalytic Association, representing three thousand psychoanalysts nationwide, submitted a joint amici curiae brief on September 6, 2000. The two organizations urged the court to exclude from consideration the medical records of the defendant and statements made by the defendant to his psychiatrists, Michael Stone, Stanley Bone, and Carl Kleban, during the course of his professional relationship with them. (Nowhere does Carl Kleban's name appear in Crowley's book, the judge's rulings on the confidentiality issues of the cases, or any of the newspaper or magazine articles.)

Central to the brief's argument, written by Seth P. Stein, the lead author and executive director of the New York State Psychiatric Association, is the distinction between the concept of privilege and the concept of a physician's duty of confidentiality to a patient.

> A privilege may be asserted to prevent the disclosure of certain communications in a legal proceeding. A party to a legal proceeding has the option to assert or waive the privilege. On the other hand, a physician has a duty of confidentiality to a patient not to reveal any information about the treatment of a patient, including non-privileged information such as fact of treatment on a certain date; privileged information such as the course of treatment of the patient; and even privileged information where the patient waived the privilege. The duty of confidentiality is not affected by whether the physician may be compelled to testify on the witness stand about the treatment of his patient. (Stein et al., 6–7)

The brief made two major points. The first is that the defendant, contrary to what the district attorney claimed, "did not waive his psychotherapist-patient privilege by discussing his marital difficulties with third parties" (4). That is, Bierenbaum disclosed nothing more to his wife and parents than the facts that he was receiving treatment from the therapists and the existence of marital problems. What he revealed, in other words, is not privileged material. "The difference between disclosing general topics (e.g., marital problems) and disclosing details of communication (e.g., adultery, or impotence, etc.) is gargantuan" (7). The second point is that Dr. Stone's warnings to the defendant's wife and parents "did not constitute a waiver or abrogation of all communications between the defendant and his physicians for any and all purposes" (8). The danger that justified Dr. Stone's "limited" 1983 disclosure has ceased, Stein argued, and with no present danger, "the current disclosure of psychiatrist-patient communications serves no beneficial purpose and can only serve to discourage patients from seeking assistance from psychiatrists and communicating their problems" (8). One of the brief's most noteworthy conclusions is that "long-ago warnings cannot justify abrogating the privilege covering still-confidential communications when there is no longer an imminent danger to avert by further disclosure" (15).

The amici curiae brief makes a powerful argument for why the psychiatrists' clinical notes and testimony should be excluded from the Bierenbaum trial. It also presents a compelling case for preserving confidentiality except in narrow and limited circumstances. "The purpose of the physician-patient privilege . . . is to encourage uninhibited communications between physicians and their patients for the purpose of securing appropriate treatment and to shield patients from humiliation, embarrassment and disgrace" (11).

The brief was careful not to weaken the therapist-patient privilege in any way. In its nuanced argument, for example, it does not say that Dr. Stone's actions were justified at the time, only that the danger that justified his "limited" 1983 disclosure has ceased. "If Dr. Stone's 1983 warnings are relevant," the brief points out, using the conditional tense, "it could only be because they somehow justify an abrogation, or override, of the statutory privilege that plainly covers the still-secret communications made within the relationship with defendant" (10). The brief offers an historical overview of therapist-patient confidentiality, citing "Doe v. Roe," the case surrounding *In Search of a Response* (see Chapter 6), to prove that confidentiality is essential to psychiatric treatment. The brief does not comment on Dr. Stone's disclosures to the *New York Times* or his interview with *Vanity Fair* after Robert Bierenbaum's indictment.

Paul Mosher was one of the psychiatric professionals with whom Seth Stein conducted discussions while writing the amici curiae brief for the Bierenbaum case. Paul remembers being conflicted at the time, as were some of his colleagues, by the possibility that protecting the privilege of confidentiality by excluding from testimony Dr. Stone's warning letter and the therapists' statements might result in Bierenbaum's acquittal. As part of the preparation for this book, we asked Seth Stein whether he had shared the same concern. "At the time, I never considered whether upholding the privilege could have resulted in a guilty person escaping justice. First, I didn't have sufficient information to determine whether the remaining evidence would have been strong enough to convict him. Second, I also was very concerned about the potential adverse impact on psychiatry if the prosecution prevailed and my clients were compelled to testify. As an attorney, my duty was to make the best arguments I could on behalf of my clients to uphold the privilege. That's what I was supposed to do." We asked Alayne Katz whether she was angry or upset by the amici curiae brief. "It was painful for me because it would help Bob, but I didn't take it personally. They [the mental health associations] were in favor of a principle, not in favor of Bob." When Paul expressed relief that her family never felt that the mental health community was trying to help Bierenbaum's legal case, she responded, "You can rest assured after all these years that I never felt that way." She might have felt bitter, however, she added, if Robert were acquitted.

Alayne Katz's View of Michael Stone's Disclosures

Alayne Katz admitted feeling a conflict between her personal and professional judgment of Michael Stone. As Gail's sister, she was grateful that, unlike the other therapists who maintained professional silence, Stone was willing to tell anyone who would listen that Bob Bierenbaum was a psychopath who murdered her sister. "Michael Stone is a giant," Alayne effused, with unqualified admiration. He was an indispensable ally, a psychiatrist whose professional judgment validated her own suspicions. "Thank God he's giving interviews," she thought at the time. "It was the best thing in the world for my sister's case that Stone was giving interviews." She still feels the same way. She supported the psychiatrist because she was Gail's bereaved sister, desperate to do anything to bring her sister's murderer to justice. Perhaps anyone in her situation would feel the way she did. She believes he was motivated out of a strong sense of moral conviction, even if that conviction clashed with his professional obligation to respect confidentiality. "It's odd that Dr. Stone said my sister was not self-protective because he was not self-protective himself. He was too interested in what he thought was right and not interested enough in protecting his career." Alayne felt that, without his outspoken support, she would have been a lone voice proclaiming Bierenbaum's guilt. "How credible was I, since I was her sister?"

Alayne realized, however, both then and now, that the psychiatrist was violating confidentiality when he gave interviews to the media. As a lawyer she knew Stone had an obligation either to warn Gail or otherwise to act in some way to protect Gail because he believed she was in imminent danger when he issued the Tarasoff warning letter. Importantly, that obligation ended after Gail's disappearance. "I hate to admit this," she told us, "but you guys got it right," meaning the members of the two mental health organizations that had submitted the amici curiae brief. "He had no business giving the police the letter. There was no danger anymore." She knew then, as she knows now, that what was at stake was not only justice for her sister but also the principle of confidentiality, which is the foundation of both law and psychotherapy.

As a result of her recent research, Alayne was surprised to discover that in New York State the psychologist-patient privilege (but not the psychiatrist-patient privilege) is as strong as the lawyer-client privilege. She was glad to learn this because it protected her therapist colleagues who were being "dragged to court" to testify about their patients. She appreciated learning about the rules of confidentiality in psychotherapy. She expressed incredulity when we told her about the ongoing erosion of confidentiality in psychotherapy, particularly the decision of the American Psychiatric Association to eliminate its Committee on Confidentiality. She continues to believe that the lawyer-client privilege is more important than the therapist-patient privilege, but she readily concedes that she may feel this way because she is a lawyer.

The Judge Upholds Confidentiality

On September 12, 2000, Judge Crocker Snyder issued her long-awaited ruling on the complex confidentiality issues of the Bierenbaum case. Reviewing the reasons for the existence of certain testimonial privileges, she stated that "Our Court of Appeals has long held that the physician-patient privilege is to be given, 'a broad and liberal construction to carry out its policy,' has narrowly construed statutes limiting the privilege and has rejected claims that there is a general public interest exception to the privilege" (Snyder, 2). She then concluded that "the People are precluded from calling as witnesses at the trial of this case any of the three mental health professionals with whom the defendant consulted prior to his wife's death" (5). She thus accepted the central arguments in Seth Stein's amici curiae brief. She conceded, however, that in denying the People's application, "this Court is disturbed that evidence which could be highly probative must be excluded. Unfortunately, the nature of testimonial privileges is to interfere with the truth finding process" (5). She then went on to justify her decision by quoting at length from the U.S. Supreme Court's majority opinion in *Jaffee v. Redmond*, reproducing a passage from that decision in which the court expressed the view that the confidentiality on which the patient may rely is an indispensable prerequisite for the psychotherapeutic process.

Yet if the defense won a major legal victory that seemed to some to ensure the defendant's acquittal in a crime without any direct evidence, the judge allowed the prosecution to call witnesses to present hearsay testimony about what Gail told them regarding the therapists' fears for her safety. Dr. Stone could not testify about the letter he wrote to Gail, but her relatives and friends could testify about that letter and anything else they heard about the therapy. "In some ways," Crowley claims, "the judge's rulings were the worst possible result for Bob's defense" (264). "Nonsense," Alayne told us. She said that if the three therapists who had spoken to or treated Robert Bierenbaum were compelled to testify, their statements would have been far more damaging.

Alayne disagreed with several of Crowley's other assertions. According to Crowley, perhaps the biggest surprise in the trial came when the defense's star witness, Joel Davis, who was convinced that he had seen Gail after the time the prosecution claimed Bierenbaum had murdered her, lost his certainty of conviction, as well as his credibility with the jury, under the prosecution's tough cross-examination. According to Alayne, however, "the only surprise was that the defense would call a witness so easily discredited. He described remembering Gail because she was so well endowed when in reality she was very flat chested."

Crowley and Alayne Katz also disagree over the alternate jurors' perceptions of the outcome of the case. After being dismissed by the judge, one of the alternate jurors told a *New York Times* reporter outside the court that, in Crowley's words, "the jury was divided and that he and other jurors felt the prosecution's case was weak—because there was no body or direct evidence against Bob. He said he felt the defense lawyers might have made a mistake by admitting that Gail was probably dead. The 1983 choking incident, he said, was not damning, because 'everybody has fights in their marriage. Where's the body?' the juror asked. 'How do you even know she's dead?'" (366–367). By contrast, Alayne told us that "one of the alternate jurors left the jury box and to my surprise sat down next to me in the courtroom and assured me that there would be a guilty verdict."

Guilty

On October 24, 2000, the jury came in with its verdict. The *New York Times* reported the next day that Bierenbaum "remained expressionless as the jury forewoman stood and read from the verdict sheet on the sole count of second-degree murder: 'Guilty.'" The opening paragraph of the *New York Post* was less restrained: "A brilliant but evil plastic surgeon was convicted yesterday of strangling his pretty young wife—a brutal murder he'd gotten away with for 15 years." All the major New York City newspapers had closely followed the trial. The headline in the *New York Post*, "Justice for Gail," expressed the prevailing belief that justice was served. A juror told the *New York Times* that the jury was never divided. "People were pretty much in agreement. All the facts, in totality, kind of drew you in the direction [of guilt]." It was the second murder case of the year without a victim's body that the Manhattan district attorney's office had successfully prosecuted. Upon hearing the verdict, Gail's brother and sister sobbed. "There is no such thing as a perfect crime," Steven Katz exclaimed. "I've waited a very, very long time for this day," Alayne Katz added.

During Bierenbaum's sentencing, Alayne delivered an eloquent victim's impact statement, but Judge Snyder denied her request to read to the court Dr. Stone's Tarasoff warning letter to Gail. Alayne told us that although Bierenbaum's lawyers pretended at the sentencing that their client was innocent, there never was any doubt in Judge Snyder's mind about the verdict. "We all know he's guilty," Judge Snyder stated openly during the sentencing hearing. Referring to all the participants' awareness of what the therapists would have said and what the letter would have shown had they been admitted into evidence, the judge continued, "We know what [evidence] didn't come in." Alayne has the highest praise for Judge Snyder. "When I sat in that witness chair, she made me feel very comfortable, like I was sitting in her living room."

On November 29, 2000, Bob was sentenced to twenty years to life, making him eligible for parole no earlier than 2020. He has never confessed to the crime or publicly spoken about it. Admitting she didn't know what the fair sentence was, Judge Crocker Snyder gave him a sentence that fell halfway between the minimum and maximum sentencing guidelines. The judge declared, according to the *New York Times* the next day, that given the "hours and hours" that he spent cutting up his wife's body before he dumped it into the ocean, "the portrait that emerges of this defendant is one that needs a psychiatrist" to sort through. Crowley notes in the epilogue to *The Surgeon's Wife* that for "personal reasons, I hoped Bob was innocent—because it would have made a nice change for me and it would have made a better book, if I could have proved he did not kill Gail" (397). He ends by saying, "I hope to hell Bob Bierenbaum *is* guilty of the crime for which he is serving twenty years to life behind bars—it's the only way I can sleep tonight" (400).

Crowley's admission that he hoped for a not-guilty verdict is one of the reasons Alayne Katz believes that his account of the story is biased. She refused to cooperate with him when he sought her help. "I had a simple request, that the publisher make a contribution to a domestic violence organization, any organization. They said no. I said no." We should point out that she never asked us for a contribution. The reason, Alayne told us, is that cooperating with us on a scholarly book focusing on far-reaching legal and psychotherapy principles is much different from cooperating with a novelist writing an exposé about her sister's death.

"A Murderologist"

The Bierenbaum case has had a paradoxical impact on Michael Stone, making him reluctant to write about the story but perhaps heightening his determination to write about evil and psychopathy. The case made him into a "murderologist," his neologism to describe his scholarly interest in the subject. He testified in the 1987 civil trial in which Dr. Jeffrey MacDonald, convicted of three murders, sued Joe McGinnis, author of the 1983 biography *Fatal Vision*, for tricking MacDonald into believing the journalist thought he was innocent of wrongdoing. Dr. Stone said that he could tell, solely from reading the murderer's transcripts of tape recordings to McGinnis, that MacDonald was a "psychopath."

Crowley, who warmly acknowledges Stone's help in writing *The Surgeon's Wife*, discloses that the psychiatrist feared Robert would file a lawsuit against him if acquitted. According to Crowley, during the Bierenbaum trial, the "Ethics Committee of the American Psychiatric Association was investigating whether Michael had violated the ethics of his profession by speaking to the press, the police, the prosecution and the Katz family about the supposedly confidential sessions with Gail and Bob in 1983. He feared both losing his license and a civil lawsuit from Bob after an acquittal." Crowley adds that the "night before the last day of the trial, Michael spoke to a journalist about his fears that his stance for justice—rather than law—might ruin his life, and those of his children" (347). Indeed, three days after the conviction, Bierenbaum's lawyer sent Stone a letter threatening to sue him for breach of confidentiality and unethical behavior. We never learn the outcome of the Ethics Committee's investigation or the threat of a civil lawsuit, but both events probably explain Stone's total silence following the first pretrial hearing of the case. He declined to speak with us about any aspect of the Bierenbaum story.

Michael Stone is now a professor of clinical psychiatry at Columbia College of Physicians and Surgeons in New York City. As he explains at the end of his 2009 book *The Anatomy of Evil,* an article about his research on the criminal mind, written by Benedict Carey and published in the *New York Times* on February 8, 2005, "gave me a presence in this field that granted me access to journalists and to professionals in mental health and law enforcement around the world." Immediately after the publication of the article he was invited to serve as the host of a television program dedicated to "educating the public about the often neglected 'why' question of evil" (361). In his acknowledgments, Stone expresses gratitude to writers of true-crime biography, including Kieran Crowley.

In his 1993 book *Abnormalities of Personality: Within and Beyond the Realm of Treatment,* Dr. Stone doesn't mention Bierenbaum, but he seems to refer to him in several places. The psychiatrist rejects the suggestion of a number of writers who believe that the word *psychopath* should be omitted as a diagnosis because of its pejorative connotations. "One catches the aroma, in these communications, of sanctimoniousness. Worse, it seems never to have occurred to these authors that there are persons who are 'too far gone'—in whom malice is too deeply entrenched—to be treatable by any method whatsoever, by any therapist no matter how skilled" (304). The therapist must be "comfortable with his own anger," Stone continues, "including his own sense of 'righteous indignation,' when some action or statement evokes these feelings" (305). Almost certainly the psychiatrist was thinking about his own anger and righteous indignation toward Robert Bierenbaum, who, at the time of the writing of the book, was a free man. Near the end of *Abnormalities of Personality* Dr. Stone expresses frustration about the extent to which psychopaths have manipulated psychiatrists. "In some respects psychiatry is more vulnerable than the law to the blandishments of psychopathic persons, since the mental set of psychiatrists, psychologists, etc., inclines us to believe what our patients tell us. Persons of a psychopathic bent have little difficulty manipulating mental health professionals" (477).

In his 2006 book *Personality Disordered Patients: Treatable and Untreatable*, Dr. Stone refers to his study of "92 uxoricides (wife murderers) in the research literature." Again, he makes no mention of Bierenbaum, but he does describe the characteristics of the psychopathic personality, all of which point in the direction of Robert Bierenbaum: "glibness, grandiosity, manipulativeness, deceitfulness, callousness, and lack of remorse" (232). And in *The Anatomy of Evil*, he names several wife murderers and then, as if he is speaking precisely about Bierenbaum, states, "Sometimes the plot is so bizarre that horror author Stephen King wouldn't touch it . . . only it isn't a novel, it's real. The murders are often staged with such inventiveness, originality, and care that the police (so the killer thinks) will never even figure out that there was a murder, let alone put their finger on the 'who' of the 'whodunit'" (187).

A Socially Conscious Attorney

Gail Katz Bierenbaum's death had a profound effect on her sister, heightening her determination to defend victims of domestic violence. Alayne Katz received her law degree from George Washington University in 1985, the year of her sister's disappearance. More than fifteen years later the Pace University School of Law's Women's Justice Center named their offices "Gail's House" and their advisory board the "Friends of Gail," and started a fund in Gail's name. Alayne Katz has been recognized for her outstanding commitment to victims of domestic violence; in 2011 she was the recipient of the Gail Katz Memorial Award from the Pace Women's Justice Center.

We knew that our interview with Alayne might be wrenching for her, but we could not have predicted that it might be helpful for her twenty-two-year-old son. She told us at the beginning of our conversation that although her son had known about the details of Gail's death, he knew hardly anything about Gail's life. "Annoyed" that we had thought Gail was a "druggie" who had dropped out of college, as she had been portrayed in some quarters, Alayne realized that she had never spoken in depth to her son about her sister's many wonderful qualities. Even Alayne was surprised that she had never before had this discussion with her son, especially since he resembles Gail in so many ways. Like her, he intends to become a psychologist, and he even has her initials tattooed on his back. And so for two hours before our interview, Alayne spoke to him about her sister. He wanted to hear, in his words, the "end of the story," including how she ended up with someone like Bob. "I told my son the truth, the whole truth, and nothing but the truth," the lawyer told us, wryly.

The Effect of the Bierenbaum Decision on Psychotherapy

Crowley asserts in the epilogue to *The Surgeon's Wife* that the "doctor-patient privilege has been bruised by the case because a precedent has been established that a New York court can call therapists into court and ask them if a particular person was their patient and if they warned that person's spouse about potential violence—and the psychiatrists must answer. It is a frightening prospect for most psychiatrists and many patients and will remain a hot topic for some time" (309). But Crowley misreads the psychiatric and legal implications of the story he has narrated. In retrospect, Judge Crocker Snyder's landmark ruling to bar the psychiatrists from testifying at the trial was far more momentous than the jury's conviction of Robert Bierenbaum. The court held that the "permissible breach of the privilege by a psychiatrist to provide Tarasoff warnings does not abrogate the physician-patient privilege" (Snyder, 4). Additionally, the court took note of the fact that the psychologist-patient privilege is "broader" than the physician-patient privilege (5).

After the Bierenbaum case, a limited release of psychiatric information in the face of imminent danger would not waive confidentiality in a trial in New York State. As the *New York Times* reported on September 13, 2000, psychiatric experts not connected with the Bierenbaum case applauded the judge's decision. "Dr. Paul S. Appelbaum, chairman of the department of psychiatry at the University of Massachusetts medical school, said, 'Were privilege statutes to be waived merely because of the disclosure of information,' they 'would all but become worthless.'"

As we pointed out in Chapter 1, this same issue has arisen several times in the federal courts, and while the decisions in the federal circuits conflict, the majority of those federal appellate courts, as of this writing, have taken a position similar to that taken by Judge Snyder. It is entirely possible that for the federal courts, in view of the "circuit split" regarding the so-called contours of the Jaffee psychotherapist privilege in the federal courts, this issue will eventually reach the Supreme Court.

There are important parallels between the Buried Bodies case (Chapter 2) and the Disappearing Wife case. Confidentiality lies at the center of both stories. The two cases implicate lawyers and psychotherapists in agonizing

dilemmas over whether their primary allegiance is to their client or patient, both of whom have been described as sadistic killers, on the one hand,
or to the bereaved families of victims of horrific crimes, on the other. Lawyers and psychotherapists are permitted to break confidentiality if they believe their clients or patients are about to commit a serious crime, but in
the case of Robert Garrow, the crimes to which he secretly confessed were
in the past, and for this reason his lawyers were under no obligation to disclose this information when he was tried for a different murder. Robert
Bierenbaum never confessed to the murder of his wife, but once Dr. Stone
issued a Tarasoff warning to Gail, professional ethics compelled him to
maintain a discreet silence, as did the other therapists who treated either
the husband or wife. Such a silence is understandably difficult, almost impossible, but it is essential for the lawyer-client and therapist-patient
relationship.

The real heroes and heroines of these two stories may well be those who
realized that their greatest loyalty was to the principle of confidentiality,
which serves as the foundation of the trusting relationship that is essential to both the practice of law and the practice of psychotherapy. Just as
the Buried Bodies case is often taught in law school, where an awareness
of the lawyer-client privilege is present in every course, so do we believe
that the Disappearing Wife case should be an essential part of psychotherapists' training.

Judge Snyder's decision on the privilege issue made clear that the confidentiality of psychotherapy is taken very seriously in the courts of New
York State, where the psychotherapist-patient privilege is not waived even
when the patient freely consents to the therapist's Tarasoff warning to an
endangered third party. Beyond that, and possibly of even greater significance, Judge Snyder's invocation of the U.S. Supreme Court's *Jaffee* decision, quoting directly from the decisions itself, illustrates that the Jaffee
privilege, technically applicable only in federal courts, has a wider reach,
now influencing the administration of state privilege law as well.

The Bierenbaum story represents the closing of a circle. The Supreme
Court had based the absolute Jaffee privilege on the existence of privilege
laws in all fifty states, some of which were considerably weaker than the
new federal privilege. By invoking the Jaffee privilege in her decision, Judge
Snyder strengthened New York's exception-ridden physician-patient privilege, at least in regard to the communication between psychiatrists and
their patients, a relationship that is protected by no other privilege law in

New York State. The Jaffee privilege thus comes back into a state court decision where it then strengthens, ironically, the state law.

Take-home lesson: The issuing of a warning letter, even with the patient's permission, does not in itself abrogate the patient's privilege to prevent his or her psychotherapist from testifying in court about the psychotherapy. The actual warning letter, however, can be used as evidence.

II. *United States v. Sol Wachtler*: "This Chief Judge Is Either Crazy or Criminal"

What to look for: If prominent people avoid psychiatric treatment because they fear public disclosure and stigma, can the result of their untreated mental disorder have disastrous ramifications for their future?

Sol Wachtler begins his memoir *After the Madness: A Judge's Own Prison Memoir* with two apt epigraphs, one by Montaigne: "There is no man so good, who, were he to submit all his thoughts and actions to the laws, would not deserve hanging ten times in his life"; the other by Hillel: "Do not judge a person until you have been in his position—you do not understand even yourself until the day of your death." He might have chosen a third, by the English writer Cyril Connolly: "Whom the gods wish to destroy they first call promising."

Wachtler's life was more than promising before his calamitous self-destruction. In 1972 he was elected to the New York State Court of Appeals, the youngest judge to sit on New York's highest court. In 1985 Governor Mario Cuomo appointed him chief judge of the Court of Appeals. For twenty years he served on that court—a court whose decisions are said to be second in importance only to those of the U.S. Supreme Court. One might disagree with Wachtler's judicial decisions, but his brilliance and ambition were undeniable. Wachtler's 450 opinions earned him the reputation of being not only a good judge but also a great one, as Alan Dershowitz observed in a *New York Times* op-ed piece, a judge demonstrating true Solomonic wisdom. One of Wachtler's landmark decisions was declaring the state's blue laws, the Sunday closing laws, unconstitutional; another was strengthening laws protecting freedom of expression; another was outlawing discrimination against women in the workplace. A moderate Republican who often held liberal social views, including being pro-choice and against the death penalty, he was a tireless advocate for children's and women's rights, authoring the opinion that overturned New York's marital exemption to the rape statute, dating back to seventeenth-century England, which held that women were the property of their husbands. "A marriage license should not be viewed as a license for a husband to forcibly rape his wife with impunity," Wachtler wrote with characteristic insight and eloquence in *People v. Liberta* in 1984. "A married woman has the same right to control her own body as does an unmarried woman. If a husband feels 'aggrieved' by his wife's refusal to engage in sexual intercourse, he should seek relief in the courts governing domestic relations, not in 'violent or forceful self-help.'"

Wachtler's lifetime of public service, judicial reform, and humanitarian work earned him fifteen honorary degrees from appreciative law schools, universities, and social service organizations. He wielded enormous power as the head of the nation's largest unified court system. When Governor Cuomo threatened to cut the court's budget during a fiscal crisis in 1991, Wachtler took the unusual step of filing a lawsuit, creating an unprecedented constitutional confrontation between the executive and judicial branches of the government. In this "clash of the Titans," as *Newsweek* described it, Wachtler prevailed. He was widely believed to be the likely Republican candidate for governor of New York in 1994, an election that many people expected him to win. His name was often mentioned as a future appointee to the U.S. Supreme Court or even as a candidate for president.

Kafkaesque

Wachtler's fall from grace was one of the most bizarre and dramatic reversals of fortune in the last decade of the twentieth century. While driving home on the Long Island Expressway on November 7, 1992, Wachtler, who the night before had met with the board of the New York State Bar Association, was forced off the road by a van and two other vehicles. Out jumped six men who surrounded his car. "My heart pounded uncontrollably," he recalls in his memoir. "An abduction! I was being taken by terrorists! I had been warned about this possibility by court security personnel: as Chief Judge I was a likely target. And here it was—it was happening!" (98).

No doubt Wachtler would have preferred being taken hostage by terrorists to being apprehended by FBI agents. Then began an ordeal that seemed more nightmarish than any Kafka story. Quickly handcuffed, searched, and read his rights, the chief judge was whisked off to the federal

courthouse in lower Manhattan, where he learned that a criminal complaint had been filed against him in New Jersey charging him with a myriad of crimes, including conspiracy to commit extortion, interstate
racketeering, blackmail, and threats to kidnap. In a series of diabolical telephone calls and letters written over a period of several months, Wachtler
had pretended he was a private investigator, "David Purdy," who demanded
twenty thousand dollars from Wachtler's former mistress, the wealthy socialite Joy Silverman. If she didn't immediately send him the money, "Purdy"
threatened to tell her husband that she was having an extramarital affair
with another man, David Samson, a New Jersey marital lawyer, who had
replaced Wachtler as her lover. Purdy also called Samson's wife and claimed
he had compromising photographs and audiotapes of her husband's affair
with Silverman.

It seemed inconceivable that the chief judge could write such vile and
despicable letters, but he made no attempt to deny his guilt. After being
booked, photographed, and fingerprinted, Wachtler was removed to the
psychiatric unit of Long Island Jewish Hospital, where he had served on
the board of trustees for the past quarter of a century. There he was chained
to his hospital bed despite the fact that he was guarded by two armed marshals. Later the marshals removed his handcuffs, but then, in one of the
many astonishing violations of usual protocol, he was manacled to his bed
for three days and three nights except when going to the bathroom or
shower. He found himself in a state of complete physical and psychological
helplessness. "It was worse than dying, which at least leaves accomplishments and reputation intact, whereas I, still living, was faced with the horror of leaving nothing save a squalid episode that would thrust into obscurity
any good I had ever done" (108). Wachtler readily accepted responsibility
for his aberrational behavior and was immediately contrite. It was apparent
to many people that the chief judge was not in his right mind, but Michael
Chertoff, the U.S. attorney from New Jersey who was zealously prosecuting the case, believed otherwise.

As Wachtler relates poignantly in his memoir, the decomposition of his
life began slowly and imperceptibly when he tried, half-heartedly, to end
his affair with Silverman, who was seventeen years his junior and the stepdaughter of his wife's uncle. Wachtler was tied to Silverman first financially and then romantically. When her stepfather died, leaving her an
inheritance of three million dollars, Wachtler was asked to serve as the
trustee of her estate. He began an affair with Silverman in 1987 that lasted
for four-and-a-half years.

Wachtler and Silverman often spoke with Eleanor Sloan, an unlicensed Philadelphia therapist who saw the two in her Manhattan office. She advised them on practical issues, such as how they might leave their spouses to marry each other. Seeing Sloan with Silverman seemed safe to the man who always worried about public scrutiny of his private life. "Sloan wasn't the sort of therapist who sits back and listens to a patient and only on occasion offers an interpretation or a suggestion," Linda Wolfe writes in *Double Life*, the story of Wachtler's affair with Silverman. "She was a talker, a bright, directive woman, with strong opinions about what would be best for her clients." Wachtler told Wolfe in a 1993 interview that he had never seen a therapist before. "Never thought I was in need of one. But Sloan was going to instruct me in how to leave my wife. How best to do it. She said, 'It will take you two years to leave your wife, that's how long the process takes'" (126). Sloan served as both a friend and advisor to Wachtler and Silverman, allowing them to sleep together in her Philadelphia home.

Despite his attraction to Silverman, Wachtler remained consumed with guilt over his infidelity to his wife of forty years, Joan, with whom he had four children. He reluctantly ended the affair with Silverman in October 1991, but when she told him that she had fallen in love with a younger and wealthier man, Wachtler soon became depressed. Immediately after the breakup he began making hang-up calls to Silverman, and then he started mailing her and her daughter letters from New Jersey in April 1992.

Wachtler continued to see Sloan after his relationship with Silverman ended. He never spoke with her about therapy issues, and she never advised him to seek psychiatric help. It's not clear what, if anything, he learned from his sessions with her. He never admitted his growing depression or his harassment of Silverman, which Silverman herself eventually conveyed to the therapist. Instead, he told Sloan that he was upset Silverman thought him capable of such behavior. "I've *got* to dispel this notion Joy has that the person who sent her those terrible letters was me. I've got to see her, talk to her about this" (Wolfe, 191). Eventually Sloan stopped seeing him and returning his telephone calls.

The Stigma of Mental Illness

Wachtler's growing depression included persistent headaches and weakness on the left side of his body. Convinced he was dying of brain cancer, he refused to have a diagnostic CAT scan or MRI because of a claustrophobic fear of entombment. Anxious to avoid appearing ill, he began visiting his internist privately on Sunday mornings, but he didn't reveal to her his adulterous affair, his increasing addiction to prescribed medication, or the extent to which his life was spinning out of control. The internist diagnosed him as suffering from depression and urged him to see a psychiatrist, but he refused, fearful of the stigma of mental illness. Haunted by the public specter of Thomas Eagleton, the U.S. senator from Missouri who was forced to withdraw as the Democratic candidate for vice president in 1972 because of the revelation that he had been treated for depression, Wachtler worried that his candidacy for governor of New York would be similarly ruined if the media discovered he was in psychiatric treatment. Evidently he felt that seeing a psychiatrist was more dangerous than a married high-profile public official being seen repeatedly in public with another woman. "I could not suffer the stigma that society imposes on someone who seeks to remedy a defect of the mind. A stigma which follows the taking of therapy, medication, and treatment. To seek such a remedy would be to publicly confess to such a defect, which my vanity and ambition would not permit" (Wachtler, 9).

In our June 12, 2013 telephone interview, Wachtler elaborated on his earlier skepticism and mistrust of psychiatry. "Legal training does not allow for a full understanding of psychiatry." The word *insanity*, he reminded us, is a legal term that arose long before the creation of psychiatry as a medical specialty. Before his experience with mental illness, Wachtler felt strongly that there was a tension between the two disciplines. "Psychiatry seemed very amorphous to me."

Wachtler cited his dissent in *Torsney v. Gold* as an example of his wariness of psychiatry. The case involved a police officer, Robert Torsney, who in 1976 was charged with second-degree murder for killing a fifteen-year-old male. During his trial the following year, Torsney admitted killing the youth but claimed he was not guilty by reason of insanity. The jury accepted his defense, and he was institutionalized for psychiatric evaluation. The commissioner of the Department of Mental Hygiene determined that Torsney was no longer mentally ill and sought his release. The court agreed, but an appellate court reversed that decision on the grounds that Torsney had failed to establish he was no longer mentally ill or a threat to himself or others. The case then went to the Court of Appeals, which, in a split ruling, affirmed the order of the committing court. Wachtler wrote a dissenting opinion in which he argued that the presumption of insanity continues until the contrary is proven. Joined by two other members of the court, Wachtler asserted that Torsney had failed to demonstrate he was no longer suffering from the symptoms associated with the wrongful act. As the law professor Carole F. Barrett wrote in 1980, "Judge Wachtler maintained that the determination of fitness for release should not be based on a psychiatric model, but rather should be the product of 'reasoned judgment based on careful scrutiny of the record.'"

Wachtler's dissent in *Torsney v. Gold* does not prove bias, but it was consistent with his mistrust of the insanity plea before his own experience with madness. The case only deepened his conviction that law and psychiatry were incompatible. He does not believe, however, that his stigmatization of mental illness or his mental instability influenced any of his legal decisions. After his imprisonment, several losing litigants sought to have their decisions in which Wachtler had participated set aside. The court reviewed each case and concluded there was no basis for reversal.

Wachtler was certainly not alone in allowing the stigma of mental illness to prevent him from seeing a psychotherapist. People with mental illness or substance abuse disorders have encountered discrimination even after laws were passed banning such discriminations, such as the Americans with Disabilities Act of 1990. Citing accounts reported to the National Coalition of Mental Health Professionals and Consumers, Inc., the psychiatrist Marcia Kraft Goin gives four examples of patients whose fear of the loss of privacy in psychotherapy drove them out of treatment, including "Todd," a married officer in a pharmaceutical company who became depressed when his three-year extramarital affair with another man ended. Fearful that his employer or wife would find out if this private information was revealed to his insurer, he insisted on absolute confidentiality from his therapist. "The psychiatrist said that Todd was in 'terrible shape' and that he should be hospitalized, but in order to get approval for the hospitalization, the psychiatrist would have to speak with the managed care company first, and Todd would not allow him to do so." A month later, Goin reports, Todd committed suicide (77–78).

The fear of the loss of privacy arising from psychotherapy may have also prevented Deputy White House Counsel Vincent Foster from seeking psychological help when he became severely depressed during the early months of the Clinton administration in 1993. The target of several hostile editorials in the *Wall Street Journal*, Foster left a suicide note that included the words "The WSJ editors lie without consequence." In an article entitled "An Interior Pain That Is All but Indescribable" published in *Newsweek*, William Styron wrote about the clear signs of clinical depression that preceded Foster's suicide along with his fear of public exposure. "Like many men, in particular certain highly successful and proudly independent men, Vincent Foster may have shunned psychiatry because, already demoralized, he felt it would be a final capitulation of his selfhood to lay bare his existential wounding in front of another fallible human being. When my own depression engulfed me, I had to overcome a lifelong skepticism and mistrust of the psychiatric profession in order to seek help." Styron adds that Foster had been given the names of two psychiatrists whom he apparently never called. "Among the most troubling details in his sad chronicle is the one concerning his consultation by telephone, only the day before his death, with his family physician back in Little Rock, who prescribed an antidepressant. This long-distance procedure would seem to be appallingly insufficient, and not only because of the absurd insufficiency of antidepressant medication at that critical moment. Foster was near the brink. He needed to see a skilled practitioner who most likely would have insisted that he go to a hospital, where he would be safe from himself."

The stigma of mental illness remains an ongoing problem, and it may deny lawyers admission to the bar. In an op-ed article appearing in the *New York Times* on August 5, 2013, Melody Moezzi writes that one of the questions in the Certificate of Fitness application from the Georgia Office of Bar Admissions asks lawyers whether in the past five years they have been diagnosed with or treated for a psychiatric disorder. Many applicants find themselves confronting a terrible dilemma when answering this question. "Recognizing that the admission of a mental illness on a fitness application could be harmful to their careers, some applicants with mental disorders either delay or forgo treatment, while others lie under oath, risking perjury charges." Moezzi admits that she herself narrowly escaped the quandary. She had been given a diagnosis of major depressive disorder before her application to the Georgia bar in 2006, a category that was not on the list of psychiatric disorders at the time, but after she was sworn in two years later, she received a "correct diagnosis" of bipolar disorder, which was on the list. She was doubly lucky: major depressive disorder was subsequently added to the list.

Moezzi didn't need to perjure herself, but others are less fortunate. Some people who acknowledge having a psychiatric disorder on the list are denied admission to the bar; others face lengthy delays, requests for their medical records, or the granting of provisional licenses. Moezzi, a human rights activist whose memoir *Haldol and Hyacinths* describes her life after being diagnosed with bipolar disorder, points out that there's no evidence that lawyers with a mental illness are less ethical or competent than other lawyers. That's why, she states indignantly, these stigmatizing questions are offensive and counterproductive. "No one should have to experience such humiliation," she concludes, adding "mental health inquiries are hypocritical and an embarrassment to the legal profession."

Drug-Induced Madness

Wachtler was thus understandably reluctant to acknowledge suffering from mental illness. He never told his physician he was taking powerful medications that he had convinced other doctors to prescribe for him, including Tenuate, an amphetamine-like drug he used to increase his energy level, which had decreased as a result of depression. Tenuate is a "weight loss" drug intended for the short-term treatment of obesity. In dosages greater than recommended, it is known to cause psychotic symptoms in some individuals. Within a four-month period he took 1,400 Tenuate pills. Nor did he tell his physician about a variety of over-the-counter medications he was taking. She ordered him to stop taking them, and he promised to do so. "I lied" (Wachtler, 10). She prescribed Halcion as a substitute for the Unisom, Percogesic, and codeine he was taking for sleep; she also prescribed the antidepressant Pamelor. According to the biographer John M. Caher, the author of *King of the Mountain*, an account of Wachtler as both a renowned jurist and criminal defendant, he was also taking, for headaches, the corticosteroid Celestone, associated with "psychic derangements" including personality changes (278). Wachtler estimates that between the spring of 1991 and the day of his arrest, he took over five thousand pills. All of these drugs, he later discovered from the various psychiatrists who treated him, induced and exacerbated a diagnosed manic-depressive (bipolar) disorder.

It is possible that these drugs contributed to if not directly caused Wachtler's increasingly aberrant behavior. One might have expected him to feel relieved that he had ended the affair with Silverman. He was, after all, the one who jilted her, but he reacted like a spurned lover to her disclosure that she had fallen in love with another man. Wachtler sought to win her back by conducting a bizarre plan of writing intimidating letters and making threatening phone calls. To authenticate the fictional David Purdy, Wachtler called a YMCA in Houston, where he imagined a man like Purdy might live. He disguised himself as Purdy by wearing a Stetson hat, string tie, and cowboy boots. Then, impersonating an obese, toothless man with a Texas drawl, Wachtler visited the Manhattan apartment where Silverman lived and talked with the doorman. The disguise was so convincing that the doorman never realized he was speaking with the chief judge, whom he otherwise would have recognized immediately. Wachtler created another character, "Theresa O'Connor," a devout Catholic living in Linden, New Jersey, who pretended she was protecting Silverman from Purdy's predatory behavior. To authenticate the new impersonation, Wachtler drove to "O'Connor's" community and spoke with a pastor to learn about the members of his church. Alternating between these demonic and angelic characters, Wachtler was not a "multiple personality" in a clinical sense, but he felt compelled to act out two characters who were both so far from his own life.

How did Wachtler, with his busy life as chief judge and future gubernatorial candidate, delivering hundreds of speeches a year throughout the state and country, have the time to create these imaginary characters and, in the process, live two lives—or perhaps three? "Easy. When you are in a manic state you have boundless energy—you have to in order to be capable of doing all the wondrous things you were capable of doing" (15–16). What did he delude himself into believing he could accomplish through his attempted extortion of Silverman? He hoped to win her back by encouraging her to seek his legal counsel and help, which he had been providing her for decades. The plan was impossible from the beginning, if only because he knew, on some level, that *he* had wanted to end the affair with her. But he seems to have forgotten this; the idea of being a spurned lover was intolerable. The plot to win back Silverman didn't succeed, he admits ruefully. "Instead it brought her to the FBI and, ultimately, me to prison and ruin" (5).

Jekyll-Hyde

Ironically, the man who did so much in his capacity as chief judge of New York's highest court to empower women and children, to protect the powerless from violence and illegal discrimination, created a cruel, misogynistic, and bizarre extortion scheme. How could he have sent an envelope containing a condom to Silverman's daughter even with the knowledge that the mother would probably intercept it, which is what happened? Why not a less terrifying plan to win back Silverman's love? Or, most reasonable of all, why not simply accept the end of a relationship that both parties no longer wanted? Wachtler never directly raises these questions, but the answer, we infer, lies in the Jekyll–Hyde aspect of human nature. Linda Wolfe certainly thought so, for she uses a passage from Robert Louis Stevenson's novel as an epigraph to her book describing Wachtler's relationship with Silverman: "Both sides of me were in dead earnest; I was no more myself when I laid aside restraint and plunged in shame, than when I laboured, in the eve of day, at the furtherance of knowledge or the relief of sorrow and suffering. . . . I thus drew steadily nearer to that truth, by whose partial discovery I have been doomed to such a dreadful shipwreck: that man is not truly one, but truly two." The theme of the dark double is ubiquitous in literature and no less pervasive in real life. Those who knew Wachtler the best were the most stunned when they discovered the existence of his evil alter ego, confirming Stevenson's and, four centuries before him, Montaigne's insights that within the best people lie the worst thoughts. "It's as if somebody else invaded his body," Judge Judith Kaye, Wachtler's judicial colleague who succeeded him as chief judge, remarked in disbelief (Wolfe, 225).

One of the striking aspects of Wachtler's double life was that while he was living in a delusional world, he was still able to function relatively well as chief judge. He quotes a statement by John S. McIntyre, then president of the American Psychiatric Association, who observed about Wachtler's case that it is not uncommon for those suffering from manic-depressive disorder to appear normal. "A patient may have severe mental illness which results in serious symptoms, including psychotic symptoms, in one area of his/her life and yet that person may function very effectively in a number of other spheres. This coexistence of excellent functioning in some areas and significantly disturbed thinking and behavior in one or more other areas is frequently true in bipolar disorder, especially the manic phase" (Wachtler, 12). After his arrest, Wachtler's relatives, colleagues, and friends realized that for some time he had been acting oddly, agitated, and easily distracted. Some of his speeches before his arrest had been rambling or inappropriate for the intended audience. On one occasion he told the members of his staff to assemble for an important announcement; he drove three hours from Long Island to Albany simply to tell them, tearfully, that he loved them very much.

Two Mad Judges

As Laura Kipnis points out in her book *How to Become a Scandal*, intriguing parallels exist between Wachtler and Daniel Paul Schreber, a prominent early nineteenth-century German chief judge who also displayed bizarre behavior. Freud never treated Schreber, but he was so fascinated by the jurist's account of his madness in *Memoirs of My Mental Illness*, published in 1903, that he wrote a major case study about it, "Psycho-Analytic Notes on an Autobiographical Account of a Case of Paranoia," published in 1911. "I don't imagine that Wachtler knew of Schreber," Kipnis observes, "but the similarities between the two are strangely numerous: both were Jewish, both suffered from bouts of lurid delusions, both had fantasies of persecution that, ironically, ensured their subsequent confinement, and both eventually wrote books describing their descents into madness." There are other similarities, Kipnis adds, including hypochondria, being tormented by a persecutor, and the "adoption of feminine personas with highly elaborate biographies" (96). Kipnis's account of Wachtler's story is often perceptive, but she revels in the spectacle of her subjects' downfalls, admitting she is a "scandal fan" (6). Do all people who write about scandal experience "malicious glee," as she claims? (22). She also largely ignores the question of Wachtler's mental illness.

"How Could I Have Written Those Detestable Letters?"

Joan Wachtler had suspected there was something "terribly wrong" with her husband, as she confided to her diary in October 1991.

> He's acting strangely. Very depressed, irritable, emotional. Moves his clothes in and out of our house. Spending little time at home. Came back from Florida after taking his mother down and told me it was the worst three days of his life—death, old people—he "feels he's dying." Doesn't understand what's happening to him. Feels "disassociated" with himself. Says a beautiful day is ugly. Hasn't slept in weeks even with pills—lost 15 pounds. Doesn't eat at all—drug related? (Wachtler, 7)

Wachtler never admits in his memoir that he had fallen out of love with his wife during the years of his affair with Joy Silverman, though, according to Wolfe, the Wachtler marriage had lost its passion and closeness. Extramarital infatuation became an all-consuming, irresistible temptation to him. He states that he had been faithful to his wife before Silverman came on the scene. Joan Wachtler's loyalty and devotion to him during the trial and throughout his imprisonment only heightened his guilt and shame toward her and their children. It would be impossible to exaggerate the intensity of Wachtler's shame over his actions, shame that would have been only slightly less intense even if he had not been treated during and following his arrest like a serial killer. He long believed that given the relatively modest salaries of state judges, he would not be able to leave a large inheritance to his children. What he thought he could give them was something far more important: a reputation for honesty and integrity through exemplary public service.

That legacy, Wachtler realized as soon as he was arrested, was forever destroyed. His overweening pride and ambition, he confesses in his memoir, had reached monstrous proportions, compelling the man of law to function outside the law. To a man like Wachtler, public adulation was everything; without that reputation, life would scarcely be worth living. One thinks of Othello's lament: "Reputation, reputation, reputation. O, I have lost my reputation! I have lost the immortal part of myself, and what remains is bestial." Thoughts of suicide assailed Wachtler, but he did not wish to burden his family with another tragedy. He knew from his maternal grandmother's suicide about the devastating legacy of self-inflicted death, and he vowed not to add to his family's suffering.

Wachtler describes his horror reading Linda Wolfe's account of his verbal violence against Silverman. "How could I have written those detestable letters?" (314). His remorse throughout his memoir is heartfelt and genuine, and he generally doesn't try to mitigate his guilt or responsibility. Yet he asks more than once why Silverman, who knew almost from the beginning that he was responsible for the threatening letters and telephone calls, didn't ask him to stop. Caher quotes part of an interview that Elaine Sheresky, the wife of Silverman's divorce lawyer, Norman Sheresky, gave to Jack Newfield, published in the *New York Post*, that undercuts Silverman's image as an innocent victim:

> In the first half of 1992, Joy told me she knew Sol was writing the harassing letters. . . . Joy told me this in a phone call she made to me. She knew it was Sol because the letters made reference to the contents of her apartment and Sol had been there and they had just broken up. Joy is being vindictive. She is used to getting her way. She did the chasing of him. She was the predator. She followed him to Albany.
>
> Joy is enjoying the scandal of it. She loves the excitement and glamour. She told me she might write a book about the case. The day Sol Wachtler was arrested she called me and was all excited. The next day she called me and asked if I thought the pictures of her in the newspaper were flattering enough. (Caher, 106)

Silverman prevailed on Norman Sheresky to tell Wachtler that she knew he was blackmailing her and was prepared to go to the FBI for help, but when the lawyer revealed this information, the chief judge replied, "She's crazy" (Wolfe, 185) and continued, albeit more slowly, his illegal behavior. Sometimes Wachtler's comments about Silverman are disingenuous. "One of the blurbs on the dustjacket of *Double Life*," he states in *After the Madness*, was "Lust, discovered late, destroyed Wachtler, but the woman who tempted and taunted him may be more guilty" (315). Though he immediately disagrees with this comment, declaring that he does not believe "Joy is guilty of anything," he has it both ways here. Caher reports that Wachtler "adamantly insists he was not a 'stalker,' arguing that he never physically harassed Silverman. However, under a broad definition he most certainly did stalk Joy Silverman, although from a distance" (291).

Would prolonged psychotherapy have helped Wachtler? There's no evidence that his sessions with Eleanor Sloan were useful, clinically or otherwise. He had no difficulty denying his harassment of Silverman. The psychoanalyst Ralph Fishkin has observed that "even if Sloan was an excellent therapist, Wachtler's failure to disclose the truth revealed that he was untreatable at that time" (personal communication, May 19, 2013). Wachtler had no difficulty lying to his physician that he had stopped taking his dangerous cocktail of over-the-counter and prescribed medications. Nevertheless, a psychiatrist might have recognized that his clinical symptoms were worsening and prescribed for him the proper medication to treat his mood disorder. And Wachtler might have been willing to analyze his anger toward Silverman with a therapist who was not also her close friend, as Sloan was.

Diminished Capacity

Wachtler's mental health at the time of his aberrant behavior was the key legal issue of the case. With the consent of the prosecution, he was taken for a psychiatric evaluation to the Payne Whitney Psychiatric Clinic of the New York Hospital–Cornell Medical Center, where he was examined by two distinguished psychiatrists, William A. Frosch, the chair of the Department of Psychiatry, and Frank T. Miller, who was later named chair of the Department of Affective Disorders. The two psychiatrists concluded in the "Miller Report" that "Judge Wachtler's severe mental illness is best categorized as a drug induced and exacerbated bipolar disorder" (Wachtler, 7). They believed that the "combination of prescription medications had initiated an increasingly intense pattern of manic behavior" (13). Dr. Donald Klein, one of the nation's top psychopharmacologists, also examined Wachtler, concluding: "It was during the period of high, chronic consumption of Halcion and Tenuate that Judge Wachtler's judgment became gravely impaired. . . . Similarly, the chronic use of high dose, high potency benzodiazepines [the class of drugs which includes Halcion] is associated with states of disinhibition [and] with impaired foresight and social judgment" (10). Another psychiatrist, Sanford Solomon, also believed that the combination of drugs produced "toxic mania" (Wolfe, 231). Robert Spitzer, professor of psychiatry at Columbia University and lead author of the third edition of the American Psychiatric Association's *Diagnostic and Statistical Manual of Mental Disorders* (1980), concluded after examining Wachtler that had he "not suffered from a severe major depression he never would have engaged in the behavior that ultimately led to his arrest" (Caher, 281).

But the prosecution's two expert witnesses disagreed. The psychiatrist Steven S. Simring conceded that Wachtler "probably had periods of depression," but he found "no objective history of psychotic symptoms or cognitive disturbance." Simring's conclusion was that Wachtler suffered from the malady of "lovesickness" (Caher, 281). The clinical psychologist Louis B. Schlesinger also ruled out mental illness, diagnosing Wachtler, in Caher's words, as a "basically sound person with a personality disorder compounded by compulsive and narcissistic traits" (284). Schlesinger later observed in an interview with Wolfe that "there's a clear psychopathic streak to Wachtler" (233).

There were three major conscious reasons for Wachtler's decision to reject the insanity defense and plead guilty. First, he knew at the time that what he was doing was wrong, despite his impaired judgment. Second, as a judge he was "leery of the psychiatric defense and far more inclined to err on the side of caution" (Caher, 276). Caher does not mention Wachtler's dissent in *Torsney v. Gold*, fifteen years earlier, but the judge's deep skepticism of psychiatry had not yet changed to sympathy. Wachtler acknowledges in his memoir his long-standing belief in the fundamental incompatibility between law and psychiatry. "The law says that a person should be punished for his unlawful acts. Psychiatry assumes that behavior is caused by forces within the person or by the environment acting on the person; if a person is unable to control his or her conduct it would be uncivilized to punish that person" (113). Finally, Wachtler pleaded guilty to avoid the shame of having his relatives and judicial colleagues testify at his trial. He told us that it was unbearable for him when his daughter Lauren, whom Chertoff suspected might be an accomplice, was subpoenaed by the grand jury.

Wachtler didn't believe his actions should go unpunished, but he knew that a provision of law allows a sentencing judge to take into consideration a defendant's "diminished capacity." The concept of diminished capacity arose from the mental health field, and though it is not embedded in the law, it appears in sentencing guidelines. (As we suggest in Chapter 8, diminished capacity was Prosenjit Poddar's defense when he pleaded not guilty by reason of insanity to the charge of murdering Tanya Tarasoff.) Wachtler never pleaded diminished capacity, as Kipnis claims (84), but he hoped the judge would use this provision to reduce his sentence. In a deal arranged with the prosecutor, Wachtler, who had been indicted on five felony counts, pleaded guilty on March 31, 1993, to a single count of "transmitting in interstate commerce threats to kidnap" (Caher, 107), that is, sending threatening letters through the U.S. mail. Judge Anne Thompson told Wachtler she would accept his guilty plea only if he stated unambiguously that he was mentally competent when he committed the crime. He had little choice but to agree, though not before invoking his psychiatrists' findings. "At the time I made the threats, I was conscious and aware of my actions and understood the making of such threats was wrong and illegal. I have been told by several psychiatrists that I suffered from a mental illness during that year, but I do not believe this would or should excuse what I did. I was able to appreciate the nature and quality of my acts" (Wolfe, 241).

As Ralph Fishkin points out, "having to claim that he was mentally competent when he had already realized he had diminished capacity might have been simultaneously an act of stoic courage and submission, but it also suggests the total sickness of our judicial system and the mental health system as they interrelate to each other. This promotes stigma and the fear of being stigmatized" (personal communication, May 20, 2013).

Despite the affidavits of six eminent psychiatrists who all agreed that he was in a manic-depressive state during the time of his criminal behavior, the prosecution rejected the claim of diminished capacity and insisted that there be no reduction in his sentencing. It was largely Chertoff who decided that the government would rule out the use of diminished capacity to reduce Wachtler's prison sentence. "There was no snap," Linda Wolfe quotes Chertoff as saying during the plea hearing. "Wachtler's acts weren't the product of a severe mental illness. They were the product of anger."

Chertoff could have stopped at that point but didn't, making a statement that appeared to heap scorn not only on Wachtler but on everyone who suffers from mental illness. "This was a man capable of going up on the bench and conducting lucid, erudite oral arguments—he wasn't a man who was staying home in a bathrobe, or going around like a screaming banshee" (242). Joan Wachtler, a certified social worker, was so enraged by Chertoff's stigmatization of mental illness that she wrote a letter to the *New York Times* that was published on April 8, 1993:

> Michael Chertoff's characterization of a manic-depressive as someone "staying home in a bathrobe or going around like a screaming banshee" destroys the progress made by the medical-psychiatric community and the entire mental health profession in educating the public about mental illness.
>
> Mr. Chertoff with this stereotyped negative bias has made a retrograde contribution to the mental health movement, setting it back many decades to a time before the advent of clinical assessment, diagnosis, and treatment with psychotherapy and medication. He has redrawn the archaic picture of any person with a mental illness as an unproductive citizen—an out-of-control raving maniac (Wolfe, 242–243).

Wolfe notes, in understated language, that Chertoff had little sympathy for psychiatric arguments. "By profession and philosophy, he believed there was good and evil in the world—good and evil and right and wrong—and that not every evil act needed to be explained as an illness, something to be treated" (235). His statement to a colleague, shortly before Wachtler's arrest, "This chief judge is either crazy or criminal" (200), left little doubt about the prosecutor's judgment of the case.

A Zealous Prosecutor

Michael Chertoff remains a key figure in the Wachtler story, and it's likely that the case would have had a different outcome with a less zealous prosecutor. The son of a rabbi, Chertoff graduated magna cum laude from Harvard College and then from Harvard Law School. After clerking for Supreme Court Justice William Brennan, he served as U.S. attorney for the District of New Jersey from 1990 through 1994. Considered to be "tough" on criminals, Chertoff earned his reputation by prosecuting organized crime and political corruption. "Chertoff hung on to Wachtler like a pit bull," Caher observes (104).

Others have also condemned Chertoff's prosecutorial tactics. In a review of *After the Madness* published in the *New York Times* on April 13, 1997, David J. Rothman agrees with Wachtler's criticisms of Chertoff. "Why else wait weeks after his culpability was established to arrest him, why else do it on an expressway rather than a few minutes later at home, why else make certain that the press was in full force to take photographs at his arraignment, why else force indignities on him while his case was going forward?" In a letter published in the *New York Times* on May 11, 1997, Chertoff rejects Rothman's allegations of misconduct and concludes that Wachtler "is entitled to get on with his life" but "not entitled to rewrite history, either at the expense of his victim or, for that matter, at the expense of those who caught him."

Chertoff later served on the U.S. Court of Appeals for the Third Circuit and, in 2005, was appointed by President George W. Bush to the post of Secretary of Homeland Security, succeeding Tom Ridge. Chertoff was also one of the coauthors of the USA Patriot Act, the controversial 2001 law that greatly expanded the power of law enforcement agencies to gather intelligence by limiting the public's right to remain private, an act that, in Elaine Scarry's view, "crafts a set of conditions" in which the Constitution is turned upside down: "our inner lives become transparent, and the workings of the government become opaque" (14).

No stranger to controversy, Michael Chertoff returned to the public spotlight during the "Bridgegate Conspiracy," the decision made by staff members and political appointees of New Jersey Governor Chris Christie to close two toll lanes of the George Washington Bridge, allegedly to punish a Democratic mayor who refused to endorse the Republican governor for reelection. The lane closings of the country's busiest bridge for four days in September 2013 created a massive traffic jam that inconvenienced thousands of motorists and called into question Governor Christie's involvement in the scandal and his viability as a Republican candidate for the 2016 U.S. presidential race. David Samson, who, we recall, was Joy Silverman's lover during the time of Chertoff's pursuit of Wachtler, had become in the intervening years the chair of the Port Authority of New York and New Jersey when the decision was made to close the toll lanes. Samson hired Chertoff to represent him as legal counsel in the scandal, which had attracted national attention. As a U.S. attorney, Chertoff had denied that he was engaged in a political vendetta against Wachtler, but as a lawyer, Chertoff was willing to defend a man, David Samson, who might be implicated in a different kind of political vendetta.

Resigning from the Court

Wachtler's actions shocked his colleagues on the Court of Appeals, but his decision to resign immediately and to plead guilty as soon as possible, no matter what the result of his guilty plea, did much to spare his colleagues from further embarrassment. They were generous in their forgiveness of him, as can be seen in a diary entry, later included in a memoir privately published, written by Judge Joseph W. Bellacosa, who was a member of the Court of Appeals until 1999, when he resigned to become dean of St. John's Law School. The court's forgiveness depended partly on the fact that Wachtler's swift resignation, without preconditions, limited the damage to its reputation:

> The Chief Judge meanwhile started exhibiting stresses of a different kind of high anxiety from his up-to-then undiagnosed bipolar swings. The effects on him led up to a singular aberrational and totally uncharacteristic personal breakdown of his sterling personal character, all of the off-the-bench variety. That then led to his arrest, guilty plea in Federal Court in New Jersey, his immediate incarceration in two of the toughest Federal prisons in North Carolina and Minnesota, an extra measure of disproportionate punishment inflicted on him because of who he was and for some indecipherable maliciously-motivated reasons.
>
> The Court had to weather a complex and sensitive series of transitional challenges in a scandal-media-driven milieu. It did so thanks to all the members of the Court pulling together as one, a credit to institutional strengths, and also as a result of the unrecognized action of Chief Judge Wachtler himself, resigning promptly, without holding on to his rank and bargaining chip of any potential plea bargain benefits. He instinctively took this drastic step, motivated, despite countervailing pressures, by his concerns for the Court and us as friends and colleagues to get us out of the eye of the storm swirling about him, media-wise, prosecution-wise, and otherwise. It also spared the Court and its members and staff from having to deal with the United States Attorney, and especially the Court from having to confront any official action against its own Chief Judge, during the pendency of the federal charges that were brought. His selfless action spared the Court from the maelstrom generated by a self-righteous and over-the-top zealous New Jersey U.S. Attorney, Michael Chertoff, driven by his own demons of self-righteousness and disproportionate justice.

The Judge Is Sentenced

Judge Thompson admitted that she was confused by the conflicting psychiatric testimony and the meaning of diminished capacity. "The Court cannot resolve the apparent conflict between the highly regarded doctors," she observed. "Nor can the Court rationally explain the defendant's behavior" (Wolfe, 254). Ruling out mitigating circumstances due to diminished capacity, she sentenced Wachtler on September 9, 1993, to fifteen months in prison, three more months than the defense wanted and three fewer months than the prosecution requested. He was also sentenced to two years of supervised release. He entered prison on September 28, 1993, at the age of sixty-three, and was released to a Brooklyn halfway house on August 29, 1994.

Fair and Equal Treatment?

Reading *After the Madness* as well as Caher's *King of the Mountain*, one cannot avoid the conclusion that Wachtler's bizarre behavior resulted in no less equally bizarre treatment by government prosecutors, law enforcement officers, and correctional administrators. Joy Silverman, who was politically connected as a result of Wachtler's influence and the large amounts of money she gave to Republican politicians, turned to her friend William Sessions, the head of the FBI, who then used all of its vast resources to monitor him. Weeks before his arrest law enforcement officials had enough evidence to prosecute him for less-serious criminal offenses. Wachtler remained convinced, with justification, that prosecutor Michael Chertoff "was not interested in stopping me, through arrest or otherwise, until my criminal conduct increased to more serious, dramatic, and publicity-worthy proportions" (Wachtler, 20).

There were many irregularities with the case, beginning with Wachtler's arrest on the Long Island Expressway, just minutes from his residence in Manhasset, a small town on the north shore of Long Island. Surely he could have been arrested when he reached home or when he was at work. He was not a suicidal terrorist ready to blow himself and others up, nor was he an alien who might flee the country. "Even John Gotti, charged with multiple murders, was allowed the dignity of self surrender" (Wachtler, 197). Another irregularity was that because the victim and perpetrator were both from New York, where the crime took place, the U.S. attorney from New York should have handled the case. Instead, he wanted to have nothing to do with it. After Wachtler was arrested and booked, he was denied bail, which would be customary in such a case. "At the time it struck me as being perverse that accused rapists and perpetrators of violent crimes were allowed out on bail, while I was deprived of my freedom" (36). During his psychiatric evaluation at Long Island Jewish Hospital, he was shackled to his bed, heightening his mortification. Following the evaluation, he was confined to his home for over six months and forced to wear an electronic device to his ankle. During most of this time, two armed guards, whom Wachtler was required to hire at his own expense, were stationed in his kitchen. Wachtler told us that he spent a thousand dollars a day to pay for these guards. "They don't treat murderers this way."

Judge Thompson recommended that Wachtler serve his time at a low-security federal camp in Pensacola, Florida, but three days before he was to begin his sentence, the Bureau of Prisons inexplicably ordered him to Butner, a prison designed for serious criminals, where the living conditions were much harsher. Federal prison camps like the one in Pensacola are for first offenders and white-collar offenders, "short-timers." By contrast, federal correctional institutions like the one at Butner are for hardened prisoners serving longer terms. (Bernie Madoff is serving a 150-year sentence at Butner.) Butner contained guard towers, razor-wire fences, and overcrowded cells, and solitary confinement is routinely used for prisoners deemed to be a threat to harm themselves or others.

Why was the change of prisons ordered? "I was told later that the order came from high sources in Washington who wanted me to endure this high-security prison in Butner, North Carolina, rather than the less onerous demands of a federal prison camp" (Wachtler, 27). Caher supports Wachtler's theory, presenting credible evidence that the "inarguably harsh and unusual treatment" may have been requested by his enemies, either Michael Chertoff or Senator Alfonse D'Amato, his bitter rival in New York State Republican politics (Caher, 299). "With Wachtler out of the game, D'Amato, after decades of scrounging for power, was to become the most important and powerful Republican in the state of New York. The prosecutor who nailed Wachtler later secured a plum assignment, compliments of Al D'Amato" (26).

Similar Crimes with Different Outcomes

Facing intense criticism about the sensationalistic way he handled the case, Chertoff insisted at the time that Wachtler's prosecution and imprisonment demonstrated that "no one is above the law, and all violators of the law must be treated equally" (Wachtler, 197). The opposite seems true, however. Without minimizing the despicable letters he wrote to Silverman, Wachtler points out that had they not been sent from New Jersey across a state line to New York, his crime would not have been a federal one but only, under New York State law, a misdemeanor.

Wachtler gives two examples in *After the Madness* of high-profile cases resembling his own but with different outcomes. Judy G. Russell, an assistant U.S. attorney from New Jersey, was involved in an extradition trial in which six anonymous threatening communications were received, five by her and one by the judge. It was believed at the time that terrorists sent the letters, and, as a result, bodyguards were assigned to Russell and the judge. Sharpshooters were placed on the courthouse roof, and the defendants were shackled. It turned out that Russell had sent the letters herself, though she later claimed to have no memory of doing so. Russell was allowed to commit herself voluntarily to a psychiatric hospital—without the indignity of being shackled to her bed!—after which she was able to return home. The judge spared her from prison because of the danger she might face from criminals she had prosecuted. The judge also ordered her court records sealed to spare her from publicity.

Wachtler cites another case, without providing specific details, of a powerful New York State judge who harassed his victim but did not land up in jail. The details are available on the Internet. As the *New York Times* reported on March 26, 1987, Bronx Surrogate Bertram R. Gelfand was removed from the bench as a result of what the State Commission on Judicial Conduct called his "repeated abuse of judicial authority" during a long affair with a female law assistant, Irene Gertel, who later became an assistant district attorney in Queens. The commission found that during his eight-year affair, Gelfand "dismissed and rehired her six times 'not because of the quality of her work but because he was trying to control her personal life and force her to meet his personal demands for fidelity.'" Reading the "Determination" of the State Commission on Judicial Conduct, one sees that Gelfand's case paralleled Wachtler's in several ways, including leaving "obscene and otherwise offensive" messages on the victim's telephone answering machine and using a false name to a doorman in an attempt to gain entrance to the apartment house. In many ways, Gelfand's behavior was more odious and vindictive than Wachtler's. Gelfand accompanied Gertel on a visit to her psychiatrist and told the psychiatrist that she had been lying about her relationships with other men. Even more outrageous was that prior to the visit, Gelfand drafted a letter, which he had Gertel sign, stating that she would be liable to him for one hundred thousand dollars if she revealed to anyone that he had accompanied her to the therapy session. Gelfand also attempted to prevent Gertel from obtaining other employment in the court system after firing her. Unlike Wachtler, who immediately resigned from the court and readily confessed his crimes, Gelfand showed a "repeated lack of candor" in his testimony to the commission and refused to resign. He lost his judgeship but never faced criminal charges.

In a bizarre case that attracted international attention, the NASA astronaut Lisa Marie Nowak was arrested at the Orlando International Airport on February 5, 2007, when she allegedly pepper-sprayed U.S. Air Force Captain Colleen Shipman, who was romantically involved with astronaut William Oefelein, with whom Nowak had a two-year affair. Nowak was subsequently charged with attempted kidnapping, burglary with assault, and battery. Stating her intention to use an insanity defense, she was allowed to enter a guilty plea to lesser charges and was sentenced to a year's probation. Nowak, like Russell and Gelfand, received lighter and more compassionate treatment than did Wachtler.

A Media Feeding Frenzy and "Circus" Prosecution

The media predictably sensationalized the Wachtler case. The gossip columnist for the *New York Post*, Cindy Adams, referred to Silverman as "Sol Wachtler's fatal attraction" (Wolfe, 243). The media also stigmatized Wachtler's mental illness and rushed to judgment. In an article published in *U.S. News & World Report* on December 7, 1992, John Leonard objected to the "psychologizing of crime" and claimed that the Wachtler case was smothered in "dubious psychiatry." Leonard cited the example of John Money, a well-known psychologist and sexologist who wrote in an op-ed piece published in *New York Newsday* that Wachtler was manifesting advanced symptoms of a "devastating illness" called Clerambault-Kandinsky Syndrome, that is, "erotomania," a psychotic delusion, named after a French psychiatrist, in which one believes that another person is secretly in love with him or her. Money's diagnosis, which was made without examining Wachtler and never confirmed by those therapists who did, allowed Leonard to conclude that "people at the top of society are far more likely to get away with psychologized and neurologized excuses than people in rough neighborhoods." Leonard predicted at the end of his article that the "public is being prepared for the fact that Judge Wachtler is extremely unlikely to serve a day in jail (or even to be brought to trial), though everyone knows that any nonjudge convicted on these charges would certainly earn some time in the slammer."

Massive news leaks surrounded the case, as may be seen in a long article written by Lucinda Franks, "To Catch a Judge," published in the December 21, 1992, issue of the *New Yorker*, only six weeks after Wachtler's arrest. Appearing in a column in the magazine entitled, ironically, "Annals of Surveillance," the article never reveals how the journalist herself managed to eavesdrop on Chertoff and his associates to report their private conversations and thoughts about the case as it unfolded. Chertoff emerges as the hero in Franks's article, a fair but tough-minded prosecutor who publicly operates on the assumption that Wachtler is being set up for the crime but privately thinks that "the worst was probably true: the most powerful judge in the state was tormenting and shaking down a woman and her child" (61). Franks not only reproduces Chertoff's comments to his colleagues but also endorses his interpretation of Wachtler as a "very ambitious" and "conspiratorial" person who had all the personality characteristics of the Silverman extortionist. "He has calibrated this series of threats very carefully over a period of months, like a Grand Master moving pieces around a chessboard. He's turned the screws very slowly, each call and letter a different, a little nastier, cranking things up and enjoying it, but delaying the final act." Franks reinforces this interpretation when, immediately after this damning characterization, an FBI agent labels Wachtler "clearly a sociopath" (61).

Near the end of her article Franks tries to rule out extenuating circumstances to explain Wachtler's behavior, including the possibility of mental illness. "A battery of medical and psychiatric tests have been given to Wachtler, and, reportedly, they have not turned up any significant abnormalities" (66). The word *reportedly* is one of the few times in the article when Franks concedes she relies on hearsay, hearsay that we know, in this case, could not be further from the truth. She ends the article by reporting, this time accurately, Chertoff's determination not to allow this case "to slip away on the basis of murky claims of Wachtler's emotional troubles. This was not a crime of impulse. This was not a crime of passion. This was not just a terrible mistake" (66).

Lucinda Franks never acknowledges in her clearly biased *New Yorker* article that her husband is Robert Morgenthau, the Manhattan district attorney. Nor does she acknowledge, as Caher was able to learn from a confidential source, that Wachtler and Morgenthau "despised" each other (136). Wachtler briefly cites the Franks article for contributing to what the

New Jersey Law Journal called in an editorial the "circus" prosecution of the case: "The leaks were so detailed that *The New Yorker* was able to publish a fascinating blow-by-blow description" of the events leading to Wachtler's arrest (Wachtler, 295). But more damaging than the leaks was Franks's refusal, reflecting the prosecution's refusal, to consider the possibility that a person may be legally sane in the judgment of the law but also acting under the influence of diminished capacity, a condition that, as a result of a serious mood disorder and powerful medication, may result in severely distorted judgment.

Club Fed

Both the prosecution and the judge ruled that Wachtler's behavior was not caused by diminished capacity, but the government later decided he *was* mentally ill and could be treated only at a federal correctional institution like the one at Butner, where, as we noted earlier, mentally ill patients were often placed in solitary confinement for long periods of time. Butner prison authorities ordered Wachtler into "protective custody" after he was stabbed twice in the back while lying on his bed. They believed, along with the FBI, at least initially, that the wounds were self-inflicted, even though the position of the stab wounds made this unlikely. "'Protective custody' is another euphemism for 'seclusion,' which is a euphemism for the hole, which is an aphorism for hell" (Wachtler, 119). Wachtler was thrice victimized at Butner, first by an inmate who stabbed him in the back, then by the authorities who sent him into solitary confinement for almost a month for his own "protection," and then by the media, such as the *New York Daily News*, which reported from an unidentified Butner source that Wachtler's wounds were self-inflicted.

Wachtler was later transferred to the Federal Medical Center at Rochester, Minnesota, which had a mental health unit. There he was told by one of the prison doctors that a blood test indicated the possibility of a cancerous lesion or tumor and that a bone scan at the neighboring Mayo Clinic would be necessary. To travel to the Mayo Clinic required, however, that he be chained and shackled. Wachtler was so horrified by the thought of being treated like a wild animal that he refused to go. "I would rather allow a cancer to go undetected and kill me than to again subject myself to leg shackles and chains" (184).

Far from being a "Club Fed," the term some people used to describe what they thought was the preferential treatment Wachtler had received in the criminal justice system, the two prisons in which he was incarcerated were hardly appealing. He uses the word "Dante-esque" to characterize the moment when raw sewage gushed into his cell at Butner (38)—an experience also described by Joseph Lifschutz during his brief incarceration, as we reported in Chapter 3. Wachtler describes how the inmates line up for their medication like the patients in Ken Kesey's *One Flew Over the Cuckoo's Nest*, benumbed, zombie-like. There's a catch-22 absurdity in his situation. Had he been declared mentally ill during his trial, he would not have been found guilty; had he been declared mentally competent, he would not have been sent to prisons like Butner and Rochester, where he had to endure "protective custody" that makes one crazy. There were other absurdities. Despite spending the last forty years of his life in and around law libraries, Wachtler was told that he could not work in the small prison law library because his presence would "create too much of an attraction and disturbance" (68). He recalls reading somewhere that the "law is like a cobweb that catches small flies but lets the big bugs fly through" (252). He may be considered a big bug, but most of his fellow inmates are small ones, especially those serving long terms for nonviolent drug offenses.

In this grim, dehumanizing world, where prisoners who are treated like garbage soon become garbage and where violence begets violence, Wachtler recalls Viktor Frankl's observation about prison life: "everything can be taken from a man but one thing: the last of the human freedoms—to choose one's attitude in any given set of circumstances, to choose one's own way" (91). He survived by biding his time as patiently as possible and by observing, whenever possible, the dark humor of his situation, as when he recites for comic relief the different ways in which he's addressed: Washler, Washtler, Ashtler.

Keeping a journal during and following his imprisonment also helped Wachtler preserve his sanity. The journal entries form the basis of his memoir. Wachtler is a thoughtful and graceful writer, and his memoir abounds in references to other literary writers whose fictional and nonfictional stories helped him understand and cope with the surreal events of his life. Reading *Darkness Visible*, William Styron's moving chronicle of the events leading to his major depression and near suicide in 1985 when he was sixty, Wachtler, who was also sixty when he was struck by depression, identified with many aspects of Styron's story, including a link between depression and a preoccupation with an imaginary illness—in his own case, brain cancer. Another similarity between Styron and Wachtler, though the latter doesn't comment on it, is that both were addicted to Halcion, which is capable of precipitating not only severe depression but also suicidal ideation. Halcion produced a suicidal breakdown in another writer, Philip Roth, as he admits in his 1993 novel *Operation Shylock*. Roth's former wife Claire Bloom provides additional details of the Halcion-induced breakdown in her 1996 memoir *Leaving a Doll's House* (see Chapter 5).

The book that turned out to be most life transforming for Wachtler was *Touched with Fire*, written by Kay Redfield Jamison, a clinical psychologist and tenured professor of psychiatry at Johns Hopkins University who herself has suffered from and written about severe bipolar disorder. *Touched with Fire* explores the link between mood disorders and creativity. Reading her discussion of manic-depressive disorder was a revelation to Wachtler, even more therapeutic, he writes, than the lithium and antidepressants he was taking to control his own bipolar disorder. "Had my most recent manic focus been on positive goals rather than destructive aberration," he comments wryly, "I might be writing entries from the governor's mansion rather than a prison cell" (19).

Wachtler would have identified even more closely with Jamison's memoir *An Unquiet Mind* (1995), where she talks about her own experience with manic-depressive disorder. In mania, Jamison writes, the "ideas and feelings are fast and frequent like shooting stars, and you follow them until you find better and brighter ones." But the exhilaration and euphoria soon give way to overwhelming confusion. "Memory goes. Humor and absorption on friends' faces are replaced by fear and concern. Everything previously moving with the grain is now against—you are irritable, angry, frightened, uncontrollable, and enmeshed totally in the blackest caves of the mind" (67). Wachtler would have been especially interested in Jamison's discussion of the stigma of mental illness, including her fear that by writing about her history of manic-depressive illness, she would be compromising her professional career. "I dread the fact that my suicide attempt and depressions will be seen by some as acts of weakness or as 'neurotic.' Somehow, I don't mind the thought of being seen as intermittently psychotic nearly as much as I mind being pigeonholed as weak and neurotic" (203).

Wachtler learned much while in prison, and though he would have preferred that this new phase of his education had arisen in a different way, he made the best of the situation. He began his political career by taking advantage of—indeed, exploiting—the conservative law-and-order movement of the 1970s, which sought to deal with crime through strict law enforcement, tough laws, and mandatory sentence guidelines. Agreeing then with the "rage to punish" movement, he now believes that such an approach is counterproductive and wasteful, particularly with respect to the nonviolent drug offenders who constitute half the prison population. He argues persuasively if at times repetitiously that mandatory sentencing guidelines deny judges the benefit of their wisdom and discretion by ruling out an understanding of the context and circumstances of a crime as factors in determining a just sentence. He's especially critical of the draconian penalties for drugs that were instituted by his benefactor, Nelson Rockefeller. "Whether you are for or against the legalization of controlled substances— and I, for one, am opposed to legalization—we must realize the costs to society when we use our prisons to warehouse low-level drug dealers and users for long sentences. In most instances, one-, two-, or three-year maximum sentences would provide a sufficient deterrent" (173).

Protective Custody

Some of the decisions Wachtler made about the criminal justice system as chief judge now strike him as naïve or wrong. He still believes in the constitutionality of strip-searches in prison, but whereas in the beginning he felt that these searches were not necessarily another form of punishment, now he feels otherwise. As chief judge he wrote the opinion upholding the constitutionality of solitary confinement, adding the qualification, "Merely confining an inmate in a segregated cell does not constitute cruel and unusual punishment. There are, of course, some cells that are so subhuman as to constitute such punishment even for a very brief confinement. One day in some cells might be constitutionally intolerable" (30). This qualification, he adds dryly, which was based on intuition, is now confirmed for him by his own experience. Rejecting the widespread belief held by corrections officers that solitary confinement encourages reflection, privacy, and safety, Wachtler argues that it is instead another form of torture.

Wachtler's judgment is supported by an intensive year-long study conducted by the New York State Civil Liberties Union, which concludes, as Taylor Pendergrass reports in an *Albany Times-Union* op-ed article, that such extreme forms of isolation and deprivation decrease safety for prison staff and prisoners. "These conditions cause grave emotional and psychological harm, including severe depression and uncontrollable rage. For the vulnerable, particularly those suffering from mental illness, extreme isolation can be life-threatening. The deprivation of necessities, such as food, exercise and basic hygiene, intensifies the suffering." The report also notes, as

Wachtler's own experience confirms, that contrary to popular belief, extreme isolation is not limited to violent offenders. "We spoke with a prisoner who spent six months in extreme isolation for refusing to return a food tray, and another who got four months for getting a homemade tattoo." Wachtler reports that five inmates were put "in the hole" simply because they led a food boycott.

Privacy for Others

As chief judge, Wachtler made a number of landmark rulings in favor of privacy rights for others that must have seemed galling when he was in prison. He had defended a wife's right not to have sex with her husband, but he invaded his former mistress's privacy and threatened her (and her young daughter) with violence when she chose another man. Wachtler's sensitivity to women's issues was evident in a unanimous opinion he wrote in 1990 that allowed the prosecution to use the "rape trauma syndrome" to explain a sexual assault victim's hesitation to testify about the attack and the assailant. "Because cultural myths still affect common understanding of rape and rape victims and because experts have been studying the effects of rape upon its victims only since the 1970s, we believe that patterns of response among rape victims are not within the ordinary understanding of the lay juror" (Caher, 330). Wachtler's demonic "Purdy" embodies, verbally, the sexual violence that often results in rape trauma syndrome.

Wachtler was a proponent of shield laws that protected reporters from having to disclose the identity of their sources, which allowed journalists like Lucinda Franks to write slanted stories like "To Catch a Judge" without disclosing the identities of the prosecutors and law enforcement officers who pursued the chief judge. In an historic decision, the Court of Appeals struck down a Long Island school district's requirement that all probationary teachers undergo mandatory drug testing. The urinalysis requirement violated the teachers' privacy rights and employment contract, Wachtler's court unanimously concluded. In prison, however, he repeatedly was subjected to such drug tests, one of the many indignities inmates experience. In *People v. DeBour* Wachtler upheld Louis DeBour's conviction for weapon possession, stating that the police were justified in frisking him when it appeared he had a gun, but in a later case, when police frisked a suspect based only on an anonymous tip, Wachtler voted to overturn the conviction and found that the rights of citizens would be dangerously compromised if authorities were allowed to conduct searches without legitimate evidence. In prison Wachtler was subjected to constant searches at the whim of authorities. Inmates, he discovered, had no privacy rights.

"Judge Not"

Wachtler admitted in an interview in *Psychology Today* in July 1997 that there is something about being a judge that limits genuine self-examination, at least in his own case. "We're taught that judges sit at the right hand of God. After a while, some judges start believing this. People call you 'your honor,' and when you sit down, they sit down. No one interrupts a judge, but a judge interrupts anyone. People think you're the font of all wisdom. You can't live like that without being affected." He confessed in the same interview that vanity was one of the reasons that prevented him from seeking psychological help. "Here I was, a manic depressive. I would check into a hotel room under an assumed name and stay there crying for two days without ever pulling the shade up." Asked how he would have judged his own crime if he were the sentencing judge, he replied, "I would have probably said, 'He's 63 years old, he's never committed a wrong act before, he's given up his judgeship, visited psychiatric clinics, he's stable on medication, so why imprison him?'" Agreeing with the interviewer that judges need "psychological literacy," Wachtler conceded that it is a quality even great jurists have lacked. "To show you how psychologically illiterate the judiciary can be, the famous Supreme Court Justice Oliver Wendell Holmes said that a mentally retarded woman should be subject to a hysterectomy because her mother was an imbecile, and she was an imbecile, and three generations of imbeciles are too much. That's Oliver Wendell Holmes!"

Reducing the Stigma of Mental Illness

An eloquent advocate of psychological literacy for the judiciary, Wachtler now devotes himself to reducing the stigma of mental illness and incarceration. "Twenty years after my self-destruction," he told us, "the stigma of being a convicted felon still plagues me." He would be encouraged to know about the ongoing efforts to educate judges, lawyers, and probation officers about the role of mental illness in the criminal justice system. Richard Warner reports that the chief judge in Boulder, Colorado, requested four training sessions for all district court judges; probation officers and private lawyers were also invited to attend. "Outcomes were positive. For example, a pre-/post-test revealed that the judges' knowledge of schizophrenia improved from 47% accuracy to 74% accuracy, and some reported imme-diate changes in their sentencing practice for adults and juveniles. Subse-quently, the judge requested two more training sessions on juvenile disorders, which were provided by a child psychiatrist" (168).

It's difficult to think of anyone who has worked harder than Wachtler to promote the integration of law and psychiatry. He created the Law and Psychiatry Institute to educate judges on the effects of mental illness on behavior. "Most judges don't know the difference between schizophrenia and a hangnail." The judges are enthusiastic about the program, but it re-mains underfunded. He helped create the first Mental Health Court in New York State; now there are forty. He also helped create the Veterans Mental Health Court for returning veterans suffering from mental illness, partic-ularly post–traumatic stress disorder, who have committed nonviolent crimes.

Wachtler is a strong advocate of confidentiality in psychotherapy. He knows that, just as the fear of public exposure prevented him from seeking professional help, others may avoid psychotherapy for the same reason. He believes that psychotherapists must do everything possible to maintain con-fidentiality while at the same time protecting the public from patients who threaten to harm themselves or others. "The lay public," he wrote to us in an e-mail,

> which is not enmeshed in the nuances and need for confidentiality, would expect a doctor who knew of his patient's dangerousness to make every effort to protect an innocent prospective victim. Those members of the lay public make up the juries and, if the truth be known, are the trial judges as well. A great irony because, as your book demonstrates, although the public, the legislators and the judges may approve and indeed encourage this breach of confidentiality, it is the fear of such a thing which keeps people away from psychotherapy and a possible cure.

Seeking Redemption

"Perhaps there could be a life for me after all," Wachtler writes near the end of *After the Madness*; "the redemption I hoped for might be at hand" (349). He knows the danger of self-serving testimony, and he quotes Mon-taigne's droll observation: "A man never speaks of himself without losing something. What he says in his disfavor is always believed, but when he commends himself, he arouses mistrust" (195). The man whose life and career had seemed so promising before his cataclysmic fall has been staging a comeback. Wachtler has sought redemption in many ways, including through his writings. His memoir *After the Madness* resembles Oscar Wilde's *The Ballad of Reading Gaol* (1898), a heartfelt poetic account of crime and punishment. Kipnis describes *After the Madness* as a "nonstop carnival of humiliation and disgrace" (95), but it is more than a mea culpa. It abounds in wise and compassionate proposals to reform the criminal justice system.

Blood Brothers

Wachtler put his talent for storytelling to good use when he teamed up with his former law clerk David Gould to craft a legal thriller, *Blood Brothers*, published in 2003. The novel focuses on two youths who live in a small rural town in race-torn Georgia in the 1950s. Luke Lipton, the brilliant son of Brooklyn-born parents transplanted to the South, forms an unlikely alliance with T. C. Simmons, a rough, narrow-minded youth who is a member of the Ku Klux Klan. The two outsiders, who could not be more culturally different, find themselves mysteriously drawn to each other, but their friendship ends when Luke discloses that he's Jewish. Forty years later, Luke, who has become a partner in a prestigious New York City law firm, returns to Georgia to defend his estranged blood brother, who has been indicted for the unsolved murder of a black man.

   Blood Brothers is not exactly a roman à clef, a "novel with a key," in which actual people are disguised as fictional characters, but there is little doubt that Wachtler used many of his experiences both as chief judge and convicted criminal in the novel. Most of *Blood Brothers* is narrated from Luke's point of view, but the novel switches to third person when describing T. C.'s strip searches in prison, an "exercise that is administered often and routinely to prisoners. Strip searches are designed more to humiliate the prisoner than to insure the security of the prison facility" (113). The worst part of prison is solitary confinement, the "hole," where the inmate undergoes further humiliation.

   Parts of *Blood Brothers* recall Wachtler's traumatic experience in prison, but other parts evoke sharp political satire, as when the novel describes the U.S. Senator from New York, Allan Kirk, who bears an uncanny resemblance to former Senator Alfonse M. D'Amato. The parallels between the fictional and real senators are striking. Both have a genius for raising large sums of money to destroy those whom they perceive as a threat to their vaulting ambition; both are underestimated by those who mistake sleaze for incompetence; both are on the verge of being indicted for graft and corruption; both embody malignity and cruelty. The three-term D'Amato was defeated for reelection in 1998 by Charles Schumer; the fictional senator is denied a fourth term when he is involved in a freakish accident with a trained circus elephant wearing a large GOP FOR KIRK cloth saddle. Spooked by laughter from the senator's sycophants, the elephant begins defecating on the hapless politician, who falls and is trampled to death.

There's nothing mirthful about Eddie Rabison, the Atlanta district attorney in *Blood Brothers* who appears to embody Wachtler's bitter anger toward Michael Chertoff. "Chertoff" means "devil" in Russian, and Rabison indeed comes across as evil incarnate. "I had done my research on this prosecutor," Luke tells us, "who represented all I detested in this species" (173). Rabison is a bully, a man given to frequent tirades, a favorite of the press to whom he leaks confidential grand jury information, and a person who confuses the power of his office with his own egotistical power. Rabison gains notoriety when he disparages the insanity defense of a mentally ill defendant. "Nobody hears voices I don't hear" (173). Rabison is also the darling of publications like the *New Yorker*. "A national magazine referred to him 'as a courageous warrior who will not rest until justice is done'" (177).

Blood Brothers reveals Wachtler's devotion to justice as well as his understanding of the forces that subvert justice. Sometimes he slyly calls attention to his own experience, as when Luke refers to the "former Chief Judge of New York [who] once made the often quoted statement that if a prosecutor wanted to, he could indict a ham sandwich" (306), a statement that also appears in *After the Madness* (292). Other times he offers insights into human nature: "A noted psychiatrist once said that a person can tolerate tragedy far more than he can tolerate uncertainty" (315). *Blood Brothers* is intriguing not mainly for its overt autobiographical implications but for its shrewd analysis of the ways in which one struggles, successfully or not, to overcome one's destructive past.

Judicial Enlightenment

Apart from writing fiction, Wachtler has sought redemption through the creation of a nonprofit business, Comprehensive Alternative Dispute Resolution Enterprise (CADRE), which aims to resolve civil disputes through mediation or binding arbitration. He continues to speak publicly about reforming the criminal justice system and helping the mentally ill. In 2000 he received an award from the Mental Health Association of New York State. "When you talk to the judge, he will tell you his credibility is diminished," the association's chief executive, Joe Glazer, told a *New York Times* reporter on October 20, 2000. "But in the world of mental illness, his credibility is enhanced every time he stands in front of the microphone." Wachtler is currently on the adjunct faculty at Touro Law School in Long Island, and in 2007 he received his law license back fourteen years after his conviction.

Wachtler served thirteen months in prison, but as one might infer from his statements, he has given himself a life sentence, without the possibility of release, for his criminal behavior. He struggles everyday to keep his demons at bay. "I always feel that depression breathing down my neck." He visits his psychiatrist two or three times a month for psychotherapy and monitoring of his medication. He remains haunted by the dark events that occurred more than two decades ago, a prisoner of the past. He told us about a recurrent nightmare in which he finds himself in the robing room of the Court of Appeals, on which he had long served with distinction. There are seven chairs in the room, one for each of the seven judges. As they walk to their chairs, led by Chief Judge Judith Kaye, Wachtler must sit in a solitary chair in a corner of the room, isolated from his former colleagues. Recently he went back to his old court and discovered, to his astonishment, that there was indeed a chair in the corner—the same one that regularly invades his sleep.

Wachtler's dream suggests that whereas he was once an insider, chief judge of one of the country's most influential courts, he is now an outsider, able to see only what he has lost. The former judge has himself been judged, found guilty, and banished from the world he once loved so dearly. Yet this is only part of the story, for as an outsider he may have done more to promote the union of law and psychiatry than if he were still on the Court of Appeals. The man who stigmatized mental illness has become a passionate proponent of judicial enlightenment, with the destigmatization of mental illness as his major goal, the theme around which he has defined his new identity. Many, perhaps most, people in his situation would have spent the remaining years of their lives in shame and disgrace, but he has chosen to call attention to his fall and, in doing so, has made possible the rise of a new career.

Wachtler's postprison life illustrates the ways in which stigmatized people can accept and respect themselves and thus reach a turning point in their moral development. No longer does he need to hide his failures. Writing in 1963, the great American sociologist Erving Goffman observed in his book *Stigma*, "in the published autobiographies of stigmatized individuals, this phase in the moral career is typically described as the final, mature, well-adjusted one—a state of grace" (102). Drawing attention to his situation, Wachtler is, again in Goffman's words, "consolidating a public image of his differentness as a real thing and of his fellow stigmatized as constituting a real group" (114).

Sol Wachtler's need for redemption is biblical, recalling an observation the Canadian writer Margaret Atwood makes in her book *Payback* about the Old Testament hero Samson, who loses his power by divulging the secret of his vulnerability to a treacherous woman. Samson redeems himself from his enemies and tormentors, regaining the freedom of his soul though at the expense of his life. "How fascinating that we say a person 'redeems himself,'" writes Atwood, "when he's been found guilty of a disgraceful action and then balances it out with a good or noble one. There's a pawnshop of the soul, it appears, where souls can be held captive but then, possibly, redeemed" (59). Everything the former chief judge has done since he confessed to disgracing himself has been to earn redemption, balancing his evil actions with many good and noble ones.

It is too soon to evaluate Sol Wachtler's importance in New York State jurisprudence, but perhaps the best estimate comes from his former friend and bitter foe, Mario Cuomo. "He should always be remembered as a great judge," Cuomo told Caher in an interview on November 10, 1997. "Everybody's life is a book, his a long and full one. This [scandal] is a page. It is one he wishes he could rewrite, but he can't. But it is only one page. To judge the whole book by one page is unfair" (27).

Take-home lesson: The fear of disclosure and stigma can have a profound and disastrous effect on a person's life by deterring him or her from seeking out appropriate help for a mental disorder because of the worry that psychotherapy will not remain strictly confidential.

Conclusion

In his 2009 book *In Confidence*, a study of when to protect secrecy and when to require disclosure, Ronald Goldfarb offers an example of a confidential disclosure that may be a cry for help:

> A Greek Orthodox priest asked a psychiatrist friend in his parish what he should do about a dilemma. The priest had found a job for one of his parishioners, who was subsequently fired for misconduct. In anger, the parishioner told the priest that he was going to burn down his employer's home. Not sure whether this was bluster and hyperbole or a real threat, and not wanting to divulge a pastoral confidence, the priest sought the psychiatrist's advice. Call the police, the psychiatrist advised, if there is any chance that the parishioner might be capable of doing what he threatened. So the priest did. The police gave the man a choice—to be incarcerated for his threats or to commit himself for mental health care. He chose the latter option. When he was released after treatment, he thanked the priest for intervening—it was the help he needed. (6–7)

Goldfarb doesn't footnote this vignette, as he does the hundreds of other confidentiality dilemmas he documents in his scholarly study, and one wonders whether the illustration is apocryphal in its dubious authority, easy resolution, and happy ending. The intervention benefits all three people—parishioner, priest, and psychiatrist—and an implied fourth person as well, the parishioner's employer, whose house is *not* burned down. The penitent-clergy confessional privilege is broken, with apparently no untoward consequences. Moreover, the parishioner receives much-needed psychotherapy and never worries that what he discloses to his psychiatrist will be violated, as occurred with the confidential confession to the priest.

As the privacy stories in our book demonstrate, however, it is often difficult if not impossible to know in advance when a confidential disclosure represents a cry for help or something else. What might that something else be? Sometimes a confidential communication is a test to see whether the recipient will maintain a discreet silence. Sometimes a confidential disclosure to a mental health professional, even when it may sound like a threat of violence, betokens the turbulent emotions experienced by nearly every patient in psychotherapy. Neither priests nor therapists can accurately predict acts of violence. Some, though not all, confidential communications are intended to be broken. And some disclosures are taken to the grave—prematurely. Goldfarb points out wryly that there are few if any inviolable secrets. "Benjamin Franklin, a founding father of our country in the eighteenth century, and Carlos Marcello, a founding father of the American Mafia in the twentieth century, have both been quoted as stating, 'Three may keep a secret if two of them are dead'" (1).

The Conundrums of Confidentiality

Goldfarb is aware of the many conundrums of confidentiality. He argues that confidentiality should be encouraged and expanded wherever possible, but he recognizes that the law allows for, even requires, disclosure in certain exceptional circumstances. How can we tell when a situation is exceptional? Kierkegaard's observation comes to mind here: we can understand our life backward, but we must live it forward. If anything, Kierkegaard may be too optimistic, for sometimes we cannot understand our lives even in retrospect.

Goldfarb's anecdote raises many of the privacy questions that we have explored throughout our book, all of which were far more problematic than his vignette of a confidential disclosure representing a cry for help. Psychiatry and religion do not share the same assumptions or practices, as Jack Richard Ewalt and Dana L. Farnsworth point out in *Textbook of Psychiatry*, but they share many of the same concerns. "The relationship of psychiatrists and clergymen to those they help is based on concern, sensible involvement, sympathy, and a respect for the dignity of the individual." This relationship, Ewalt and Farnsworth continue, derives from a long tradition of privileged communication based on custom if not law. "The patient or parishioner is encouraged to place full confidence in the psychiatrist or clergyman—in fact, this confidence is almost absolutely necessary to effective treatment—and he may rest assured that what he discloses in the course of consultation will remain completely confidential" (299). Ewalt and Farnsworth made this observation in 1963, thirty-three years before the *Jaffee v. Redmond* decision, yet they knew that the penitent-priest privilege and the patient-therapist privilege are not identical. The privilege of confidentiality applies to both parishioner and clergy, whereas it extends only to the patient, not the therapist.

There's also another difference. "In Confession the sinner tells what he knows," Freud declares in *The Question of Lay Analysis*; "in analysis the neurotic has to tell more" (189). Joseph Lifschutz, we recall, claimed that the closest parallel to the psychotherapist's need for absolute confidentiality is the Catholic confessional, which is sealed with a vow of silence. In a secular society, Lifschutz maintained, the psychotherapist requires the same independent privilege as a priest. Lifschutz was willing to go to jail to defend the principle. The California Court of Appeals disagreed with his legal position, however, concluding that an absolute principle created by the legislature for clergy did not imply that a weaker privilege for psychotherapists was a denial of equal protection under the law. Later court rulings and statutory laws have decreed that the patient, not the therapist, owns the privilege, once again disagreeing with Lifschutz's—and Anne Hayman's—position that the therapist should not break confidentiality even if the patient requests it. Having defended the principle of confidentiality in psychotherapy, Lifschutz decided that discretion was the better part of valor and found a way to reveal only the bare minimum of his patient's treatment a decade earlier.

Goldfarb's clinical vignette lacks the emotions of a good story, particularly the darker emotions of anger, sorrow, resentment, and guilt that almost always appear in privacy stories. The clinical vignette also lacks conflict and therefore human interest. The task of a short-story writer is not to solve a problem, Chekhov famously remarked, but to state it correctly. Goldfarb gives us a solution without dramatizing a problem.

One can imagine, however, making Goldfarb's vignette into a story. The Greek Orthodox priest seemed to have no reluctance in asking the psychiatrist friend in his parish for help, but suppose the priest worried about the consequences of violating confidentiality. If he were a Roman Catholic priest living in France four hundred years earlier, when canon law had an *absolute* seal of confession, the outcome would have been dramatically different, as one can see from the chilling example of medieval canon law that Daniel W. Shuman and Myron F. Weiner cite in their book *The Psychotherapist Privilege: A Critical Examination*:

At Toulouse, in 1579, an innkeeper murdered a guest and buried the body
in the cellar: he confessed the crime to a priest who, seduced by a reward
offered for the detection of the murderer, denounced the criminal to the
magistrates; under torture the culprit confessed the crime, adding that
none but the confessor could have betrayed him; an investigation ensued,
which resulted in the Parliament of Toulouse releasing the criminal and
hanging the priest, after he had been degraded by the bishop. (50)

Torment

One can only imagine the medieval priest's dilemma, his human story, as
he struggled to figure out whether to honor or betray the innkeeper's
confession. The dilemma also tormented Frank Armani. He defended a
different kind of murderer, Robert Garrow, who buried bodies not in a
cellar but in an abandoned mineshaft. The pressure to break confidentiality to disclose a crime may at times seem unbearable for priests, psychotherapists, and lawyers. A former classmate of Armani, Goldfarb quotes a
statement the embattled defense lawyer made in a television interview.
"'It's a terrible thing to play God.' Armani [was] anguished over the agony
of the victims' parents: 'Your mind screaming one way. Relieve these
parents! . . . One sense of morality wants you to relieve the grief.' His torment, as well as the families', must have been unbearable; any attorney
would feel that way" (72).

Karen Beyer also confronted a tormenting dilemma. After the police
officer Mary Lu Redmond shot and killed Ricky Allen Sr. to prevent a
homicide, his mother, Carrie Jaffee, brought a civil rights lawsuit against
the officer and the village in which she was employed. Traumatized by the
shooting, Redmond had been receiving counseling from Karen Beyer. The
two worked together for nearly two years, until the social worker received
a subpoena for her testimony and psychotherapy notes. Unlike Joseph Lifschutz and Anne Hayman, who both believed they were entitled to the psychotherapy privilege, Beyer did not share this belief. Moreover, Redmond
never allowed Beyer to disclose information about the police officer's treatment. Under extreme pressure to reveal the contents of her psychotherapy
notes, Beyer resisted because she lacked Redmond's consent and because
of the devastating effect such a release would have on Redmond. The stage
was thus set for a momentous legal case that found its way to the U.S. Supreme Court, a case that frightened the social worker because of the possibility that the court would decide against confidentiality and privilege.

Karen Beyer's story is noteworthy for several reasons. First, the subpoena was the beginning of the end of her therapeutic relationship with Redmond. Second, the therapist had the strong support of the American Psychoanalytic Association and other professional groups, which submitted an unprecedented fourteen amici curiae briefs in support of psychotherapy confidentiality. And finally the Supreme Court ruled in 1996, in the historic *Jaffee v. Redmond* case, that there is indeed an absolute patient-therapist privilege in the federal courts.

Torment is a unifying theme in several of the privacy stories we have examined. The torment arose regardless of whether one decided to maintain privileged information, allowing therapy to continue, or break confidentiality, thus ending the therapeutic relationship. Privacy stories implicate the therapist, along with the reader, in moral, ethical, legal, personal, and professional conundrums that require Solomonic decision making. The dilemma arises when a patient's right to privacy threatens to interfere with another person's privacy. The decision to breach confidentiality often has life-or-death consequences. The psychologist Lawrence Moore agonized over whether to disclose to the UC Berkeley police that his patient, Prosenjit Poddar, threatened to harm another student, Tanya Tarasoff. Moore notified the police, who questioned and then released the Indian engineering student. Moore's colleagues ostracized him for breaching confidentiality, but subsequent events confirmed his worst fears. Would Prosenjit have been able to defuse his violence if he had remained in therapy with Moore? Would society be better served if those who make threats remain in therapy? These are only two of the many unresolved questions in the Tarasoff story. The 1974 California Supreme Court decision was vacated in 1976, when the court declared there is a "duty to protect." The "duty" varies or is even nonexistent from state to state, and there have been few recent cases in which therapists have been found liable for not issuing a Tarasoff warning. Nevertheless, *Tarasoff II* continues to have a significant effect on American psychotherapists.

Gail Katz Bierenbaum's psychologists and psychiatrists also experienced agonizing dilemmas. They knew the young wife was in grave danger from her volatile husband and urged her to be careful. Her mysterious disappearance in 1985 seemed to justify their suspicions. One of her psychiatrists, Michael Stone, gave interviews to newspapers and magazines about his conviction that Robert Bierenbaum was the murderer. When the case was reopened more than a decade later, Stone was eager to testify, then inexplicably changed his mind and has never commented on the case in his many subsequent books. The victim's sister, Alayne Katz, regarded Dr. Stone as a hero in speaking publicly about the case, supporting her own conviction that Bierenbaum had murdered her sister. As a lawyer, however, Alayne Katz knew that the psychiatrist was breaching confidentiality. The judge refused to agree to the prosecution's request to compel the therapists to testify. The Bierenbaum case has become a landmark in upholding the patient-therapist privilege even when a Tarasoff warning has been issued.

The *Tarasoff* and *Jaffee* rulings are the two poles around which many privacy stories revolve. Mental health professionals must ask themselves whether a patient's potential for violence is strong enough to justify an exception to the federal patient-therapist testimonial privilege. A therapist is not required to warn a person of a possible threat, but the therapist may be liable in a civil suit if a patient carries out such a threat.

Martin Orne's decision to give Anne Sexton's therapy tapes to her daughter, Linda Gray Sexton, and to cooperate with the poet's biographer, Diane Wood Middlebrook, created a fierce controversy that shows no signs of abating. If anything, the controversy may intensify with the publication of Dawn Skorczewski's *An Accident of Hope*, which demonstrates Orne's inability to articulate forthrightly his reasons for keeping the tapes. Whether or not one agrees with the psychiatrist's decision, he suffered in a variety of ways as a result of his harrowing experience. So, too, did Linda Gray Sexton suffer, vilified for her decision to expose her mother's dark secrets after her death.

Tormenting ambiguities also surround the swift and stunning fall of Chief Justice Sol Wachtler, who, fearing the public stigma of going to a psychiatrist, chose instead to act out his increasingly delusional and grandiose fantasies. The *fear* of the loss of privacy prevented him from seeing a psychotherapist. He, too, was tormented by the decision whether to conceal or reveal his madness to those who might have been in a position to help him. Would he have sought a therapist's help if mental illness were not stigmatized? Would a therapist have helped him understand and control his dark demon? Wachtler's story is a cautionary tale of the ways in which undiagnosed and untreated mental illness may result in "diminished capacity" affecting nearly every aspect of a person's life.

What Ifs

Privacy stories raise endless what ifs. The priest and the psychiatrist in Gold-farb's vignette agree to notify the police about the parishioner's threats, but such agreement can never be taken for granted. What would have happened if the priest and psychiatrist disagreed over what to do? What would have happened if one of them, the psychiatrist, for example, breached confidentiality in another way—by writing an article about the parishioner and publishing it in a professional journal?

Why might a psychiatrist write about the parishioner? We can imagine such a story. Perhaps the police decided, in their infinite wisdom, to commit the parishioner to the psychiatrist's care. The two men were, after all, members of the same church; both knew and presumably respected the priest. The psychiatrist might be willing to treat the parishioner as a way to help both church and community. Let's suppose that the psychiatrist discovered in therapy how the parishioner's confidential disclosure to the priest revealed a deep narcissistic injury arising from a traumatic childhood injury. Suppose the psychiatrist's specialty is the link between aggression and creativity, and it turns out that the parishioner is a gifted novelist who has already established a reputation as an angry young man. The priest's and psychiatrist's timely intervention, along with appropriate mental health treatment, appears to have helped heal the parishioner's wound. The title of the psychiatrist's article? Because the parishioner vowed to burn down his employer's house, the psychiatrist decided to call his article "The Angry Act."

Now let us suppose that the parishioner, recognizing the opportunity to transmute a humiliating childhood event into an engaging fiction, published the story and then discovers, to his mortification, that the psychiatrist has appropriated his own words. Anyone who reads the psychiatrist's case study and the parishioner's novel will be able to uncover the writer's identity. The psychiatrist denies responsibility and insists that he doesn't need permission to write about his patients. A far-fetched breach of confidentiality? Yes, but no more improbable than Dr. Kleinschmidt's breach of confidentiality when writing about Philip Roth's life.

What if Dr. Kleinschmidt had respected Roth's privacy and sought his permission to publish a case study? What if he had asked the novelist to make the necessary disguises to preserve his anonymity? What if Dr. Kleinschmidt had taken a relational approach to therapy, emphasizing the extent to which both analyst and analysand co-created the story of treatment? Since we are in the realm of pure speculation, what if Dr. Kleinschmidt didn't write "The Angry Act"? If the novelist did not have his own privacy violated in analysis, would he have been less willing to violate the privacy of real-life people in his thinly disguised stories? Would he be less defensive about the link between his life and fiction?

Loss of Confidentiality—and Confidence—in Psychotherapy

"The Latin roots of 'confidence' and 'confidentiality,'" Tim Bond points out, "are *confidere*, meaning 'to have full trust'" (123). Significantly, in the privacy stories we have described, the loss of confidentiality in the patient-therapist relationship resulted in the loss of confidence in psychotherapy itself, along with the termination of treatment. A similar outcome appears in the two cases in which psychiatrists published their patients' case histories without permission. Like Roth, Harriet Werner felt unmasked and betrayed by her psychiatrist's breach of confidentiality. Both Roth and Werner felt that their therapists had stolen their words—and, in effect, had stolen their identities and exposed them to the public. Unlike Roth, Harriet Werner sought redress not through fiction but through the court system, where she ultimately prevailed after a bruising and costly fight.

A Sophie's Choice

Goldfarb notes that privileges are not favored in the United States and are justified "only if there is a transcendent good that outweighs the search for truth" (32). He agrees with the need to preserve a balance between the confidentiality of the patient-therapist relationship and the laws mandating the reporting of problems such as domestic violence, child sexual abuse, suicide, and homicide. Sometimes therapists find themselves confronting a "Sophie's choice," as Goldfarb reveals in the most personal passage in his book:

> My daughter, a social worker, counseled a young girl whose father was abusive. New York law required that she (my daughter) report the abuse to a government agency. If he was reported, the father would lose his temporary citizenship and work status, which would be a serious blow to the whole family, her young ward pleaded. My daughter feared that disclosure would be more harmful than consulting with the family and attempting to cure the problem. But she had no choice in the matter; the law was clear and she followed it—against her professional judgment. (116)

Goldfarb never tells us the outcome of his daughter's difficult decision. Should she in an act of principled civil disobedience have disregarded an unjust law in order to remain true to her professional training and experience? Should she have been prepared to go to jail, as Joseph Lifschutz did? Some of the therapists in our book have confronted such choices, struggling with their dual allegiance to patient and society. Recall Anne Hayman's observation: "In such a situation, the analyst could be faced with a choice between two dreadfully 'wrong' decisions—either of doing nothing to prevent an anticipated harm somewhere or of mistreating and damaging an analysand by some relaxation of the rule of confidentiality" (304). A dual allegiance, as we have seen, is often a *divided* allegiance. A therapist caught in the dilemma of a dual or divided allegiance often feels, as Robert L. Pyles suggests, like a "double agent" (263). How do therapists choose between following the mandate of the law, on the one hand, and trusting what they have learned from their professional education and experience, on the other? Goldfarb is himself in a difficult situation here, intent on describing the vexed nature of confidentiality dilemmas while at the same time respecting his daughter's privacy. One wonders whether he had his daughter's permission in writing about her.

Protecting and Breaking Secrets

Privacy stories oscillate between protecting and betraying secrets. Sissela Bok, whose 1982 book *Secrecy: On the Ethics of Concealment and Revelation* remains the best study of this elusive subject, captures the paradoxical nature of hidden information: "secrecy both protects and thwarts moral perception, reasoning, and choice. Secret practices protect the liberty of some while impairing that of others. They guard intimacy and creativity, yet tend to spread and to invite abuse. Secrecy can enhance a sense of brotherhood, loyalty, and equality among insiders while kindling discrimination against outsiders" (xvi). Any inquiry into the ethics of secrecy must consider, as Bok explains, the conflicts arising from difficult choices: "between keeping secrets and revealing them; between wanting to penetrate the secrets of others and to leave them undisturbed; and between responding to what they reveal to us and ignoring or even denying it. These conflicts are rooted in the most basic experience of what it means to live as one human being among others, needing both to hide and to share, both to seek out and to beware of the unknown" (xvi). Recall Freud's vivid statement in *Fragment of an Analysis of a Case of Hysteria*: "He that has eyes to see and ears to hear may convince himself that no mortal can keep a secret. If his lips are silent, he chatters with his finger-tips; betrayal oozes out of him at every pore" (77–78). A patient's willingness to share secrets with a psychoanalyst contributes to the feeling of specialness, closeness, or even intimacy that fosters the development of the positive transference essential to the power of the analytic treatment to bring about therapeutic change.

Bok reveals, on the penultimate page of *Secrets*, that researching the book convinced her of the urgency of the subject. "The conflicts over secrecy may be perennial, but the accelerating pace of technological innovation and the present worldwide political tensions are now unsettling the already precarious standards for keeping, probing, and revealing secrets. New techniques, from ever more sophisticated devices for eavesdropping to computerized data banks, have vastly enlarged the amount of information at the disposal of those with the know-how and the resources to acquire it" (284). Relevant as this statement is to the early 1980s, it becomes even more relevant—and disquieting—three decades later.

The Crisis of Confidentiality

Privacy stories involve wrenching legal and psychological decisions whose implications and consequences may not become apparent for years. Readers may not always agree with the decisions made by the therapists in these privacy stories. Therapists may spend the rest of their lives second-guessing their decisions. Probably all mental health professionals agree, however, on the importance of confidentiality in the patient-therapist relationship. Therapists also agree that confidentiality is at the center of psychotherapy, "*constitutive of the process itself,*" as Jonathan Lear suggests (5). Moreover, there is agreement over the urgency of the situation by an ever-increasing number of mandatory reporting laws. "The subject of confidentiality is currently of burning interest," Otto F. Kernberg warns: "the mental health professions have experienced an encroachment by the courts and by third-party payers on the confidential nature of the patient-therapist interaction. The ethical commitments, autonomy, responsibility, and safety of the psychoanalyst are all gravely endangered by the challenge to confidentiality" (80). Kernberg's urgent statement in 2003 seems even more true a decade later.

As we discussed in Chapter 1, support for strict confidentiality of psychotherapy is eroding in parts of the mental health professions. This trend is the result of the attempt by the mainstream of the psychiatric profession to abandon its psychotherapy roots and become more "medical," that is, more like "regular doctors." This is evidenced by such initiatives as "The Decade of the Brain"; close ties between the pharmaceutical companies and the leadership of the American Psychiatric Association, which for many years was dependent to a significant degree on funding from "Big Pharma" in the form of journal advertising; the permeation of continuing education sessions by Big Pharma–sponsored "education" sessions that often resembled (albeit lightly disguised) infomercials for particular products; and the new definition of mental disorders in the fifth edition of the *Diagnostic and Statistical Manual*, which implies that every mental disorder is based on an underlying physical dysfunction in the brain. In taking this route, contemporary psychiatrists are much more likely to view the information they gather from patients as being merely "medical data" and therefore not entitled to the more stringent protection that the legal tradition, culminating in the *Jaffee* decision, sees as necessary for the practice of psychotherapy.

Another symptom of the loss of confidentiality in psychotherapy is the American Psychiatric Association's recent decision to "sunset" (eliminate) its Committee on Confidentiality. There is now no body within the American Psychiatric Association devoted solely to that major concern. Even within the American Psychoanalytic Association, the Committee on Confidentiality was "sunsetted" in 2010, partly out of member lack of interest and partly because of the failure to identify a member willing to chair the committee. As a result of protests from a small number of members of the American Psychoanalytic Association, the committee was reconstituted later the same year but with no members from the prior committee. Sadly, much institutional memory has been lost.

As we have pointed out, the erosion of strict confidentiality may also be seen in the new tendency to view clinical psychotherapy data simply as "medical information." This trend has been fostered by the nationwide transition to networked electronic medical records in which the concept of confidentiality has been redefined to permit dissemination of clinical information from one practice to all the other practices involved in the care of a particular patient. An even more striking example of the erosion of strict confidentiality is the proliferation of "addictionologists," medical practitioners who use a "medical model" to treat addictions on an outpatient basis. Addictionologists advocate for the weakening of the special and stringent federal protection for addiction treatment information (CFR 42 part 2) so that their clinical practices can be integrated into the new networked information systems.

In the well-known landmark 1977 Supreme Court decision in *Whalen v. Roe* (429 U.S. 589 [1977]), the court decided unanimously that a New York State law requiring prescriptions of controlled substances to be reported to the state did not violate patients' constitutional right to privacy. The court stated that extreme precautions had been taken to protect the data: the single computer on which the data were recorded was in a locked room with limited access, and the computer was not networked to any other computer. But in a concurring opinion published separately, Justice Brennan was troubled by the central computer storage of the collected data. "The central storage and easy accessibility of computerized data vastly increase the potential for abuse of that information, and I am not prepared to say that future developments will not demonstrate the necessity of some curb on such technology." The technology of central data storage is increasing at dizzying speed, with ominous implications for loss of privacy.

New mandating regulations are also precipitously increasing. I-Stop, a New York State law signed in 2012 and taking effect in 2013 and 2015, requires doctors to report within twenty-four hours via the web all prescriptions for controlled substances and to check the now widely available patient prescription database before writing each prescription. Touted as a national model for improved communication between health care providers and law enforcement agencies, I-Stop is designed to catch patients engaging in "doctor shopping" to obtain multiple prescriptions. Among the substances covered by this provision are the benzodiazepine tranquilizers often prescribed for psychotherapy patients. As of 2015 *all prescriptions* written in New York State are required to be electronic.

To our knowledge, the reexamination of these new legal requirements as possible constitutional rights violations anticipated by Justice Brennan in 1977 has never taken place. The lack of awareness of the steady erosion of privacy has blunted the development of public outrage that could lead to legal challenges to the I-Stop law. Another recent New York State law, more controversial than I-Stop, requires psychotherapists to report to the authorities, again via the web, any patient who owns a gun and is deemed "likely" to commit a violent crime. This law, which has been challenged, lowers the traditional standard protecting patients' confidentiality, which required reporting only if such a violent act was deemed both "credible" and "imminent." These new laws have been introduced and passed despite the growing awareness of the insecurity of computer systems networked via the Internet.

The erosion of the support for strict confidentiality of psychotherapy in New York State also appears to reflect a general decline in the concern with privacy in our general culture. Frank Rich, in an article in *New York* magazine about the revelations of the National Security Agency's gathering of phone records of all American citizens without probable cause, points out that the reaction among the populace to this news was "ho hum." The country's growing indifference to privacy, Rich contends, is a recent phenomenon. "If one wanted to identify the turning point when privacy stopped being a prized commodity in America, a good place to start would be with television just before the turn of the century. The cultural revolution in programming that was cemented by the year 2000 presaged the devaluation of privacy that would explode with the arrival of Facebook and its peers a few years later." (24) According to Rich,

> Many of us not only don't care about having our privacy invaded but surrender more and more of our personal data, family secrets, and intimate yearnings with open eyes and full hearts to anyone who asks and many who don't, from the servers of Fortune 500 corporations to the casting directors of reality-television shows to our 1.1 billion potential friends on Facebook. Indeed, there's a considerable constituency in this country—always present and now arguably larger than ever—that's begging for its privacy to be invaded and, God willing, to be exposed in every gory detail before the largest audience possible. We don't like the government to be watching as well—many Americans don't like government, period—but most of us are willing to give such surveillance a pass rather than forsake the pleasures and rewards of self-exposure, convenience, and consumerism. (24)

There are other reasons to resist the loss of privacy, including, as the George Washington University law professor Daniel J. Solove points out, the freedom to change one's mind.

> Many people are not static; they change and grow throughout their lives. There is a great value in the ability to have a second chance, to be able to move beyond a mistake, to be able to reinvent oneself. Privacy nurtures this ability. It allows people to grow and mature without being shackled with all the foolish things they might have done in the past. Certainly, not all misdeeds should be shielded, but some should be, because we want to encourage and facilitate growth and improvement.

One of the causes and effects of the crisis of confidentiality is that there is little if any consensus over when, if ever, confidentiality should be breached. Should confidentiality be absolute or qualified? If the latter, should therapists warn patients at the beginning of therapy, as Otto Kernberg recommends?

> In the treatment of patients with borderline personality organization carried out at the Personality Disorders Institute of the Weill Medical College of Cornell University, Department of Psychiatry, we inform our patients that complete confidentiality is guaranteed, with the exception of circumstances in which the therapist is convinced that the life of the patient or of someone else is at risk, at which point the therapist would take whatever measures he or she deems necessary to protect the patient and others from danger. (81)

Other questions persist. How can therapists publish case studies, the kind of research that is essential for the advancement of knowledge, without violating their patients' privacy? Freud never solved this problem, and there is still no agreement over how this problem can be overcome, if at all. There is no agreement over how much information, if any, therapists should share with third-party payers. Nor is there agreement over the extent to which confidentiality decisions should be made by the legislative branch of government, through the enactment of statutes, or by the judicial branch, through legal rulings.

Another Threat to Confidentiality

In recent decades a schism has appeared, and then widened, in assumptions about the "causes" of mental disorders. As the cultural anthropologist T. M. Luhrmann shows in her 2000 book *Of Two Minds*, the world of psychiatry is split by two competing therapeutic models, with vexing implications for patients and therapists alike. Several professional and advocacy groups, such as the American Psychiatric Association and the National Alliance on Mental Illness, have promoted the idea that mental disorders are for the most part "brain disorders." These organizations have emphasized treatments based on a physical/medical rather than psychological approach. In their view, the treatment of mental disorders is a medical procedure not unlike that provided for other physical "illnesses." These professional and advocacy groups maintain that a "brain disease" model of mental illness will diminish stigma. While records of medical treatments are protected to some degree by law and custom, they are shared with other health professionals, insurers, and certain others without patient consent. By contrast, the current protection of psychotherapy records is much more stringent. Indeed, the Supreme Court's *Jaffee* decision, and the half-century of legal scholarship and court decisions preceding the new federal psychotherapy privilege, were based on the sharp distinction between the privacy needs of medical and psychotherapy treatment. As the court wrote in its majority opinion:

> Treatment by a physician for physical ailments can often proceed successfully on the basis of a physical examination, objective information supplied by the patient, and the results of diagnostic tests. Effective psychotherapy, by contrast, depends upon an atmosphere of confidence and trust in which the patient is willing to make a frank and complete disclosure of facts, emotions, memories, and fears. Because of the sensitive nature of the problems for which individuals consult psychotherapists, disclosure of confidential communications made during counseling sessions may cause embarrassment or disgrace.

If mental disorders were conflated with physical illnesses, would the preservation of special confidentiality in psychotherapy inadvertently contribute to stigma? Does promulgating the view that mental disorders are "brain disease" diminish stigma? Patrick Kennedy, the former U.S. congressman from Rhode Island from 1995 until 2011 and the primary sponsor of the Mental Health Parity and Addiction Equity Act of 2008, stumbled into these questions in an interview with Wallace McKelvey published in the *Press of Atlantic City* on September 14, 2013. "My legislation was simple," Kennedy asserted. "We ought to treat the brain like every other organ in the body. So if insurance covers in-network, out-of-network . . . pharmacy benefits and emergency room services (for physical ailments), then you need to cover all of those services for schizophrenia, for bipolar disorder, for addiction." Kennedy, the son of the late U.S. Senator Ted Kennedy of Massachusetts, believes that mental illness is a physical illness caused mainly if not entirely by a malfunctioning brain. He concedes that the mentally ill feel stigmatized and demand higher privacy standards to avoid being discriminated against if they go into treatment. But he makes the paradoxical and dubious argument that the mentally ill end up harming themselves by demanding confidentiality.

Patrick Kennedy points to himself and his family as examples of those who have been harmed by stigma. While championing in Congress mental health parity, he was active in his own addiction. He admits that he chose not to seek treatment because of the fear of what others might think. In his view, the biggest challenge in the mental health community lies in clients' "own alienating stigma." Ted Kennedy was traumatized by his two brothers' assassinations, Patrick Kennedy reveals, but he adds that his father never sought psychological help. Why? "Because the stigma was so great, he thought it would mean people would think less of him if he sought treatment. It was shocking, but we lived in the cone of silence." Ted Kennedy's fear of stigma prevented him from seeking psychological treatment, and for years he engaged in self-destructive behavior after suffering from severe trauma.

Patrick Kennedy offers a scenario in which strict confidentiality in psychotherapy might actually prove fatal to people like himself. "Let's say I'm a recovering addict, which I am. I have a high tolerance to opioids because I was really addicted to Oxycontin. Let's say I get wheeled in to AtlantiCare in another couple months and no one knows that I am someone who's a recovering addict. If they put me under and the anesthesiologist doesn't know the way my brain works, I'm a dead man walking. That could totally cost me my life."

An impassioned advocate for mental health reform, Patrick Kennedy received in 2003 awards from both the American Psychoanalytic Association and the American Psychiatric Association. But there are several problems with his argument for opposing confidentiality in psychotherapy, beginning with the belief that eliminating higher privacy standards in information technology will eradicate stigma. As we saw with New York Chief Judge Sol Wachtler, stigma remains a powerful obstacle to receiving mental health treatment. The passage of a single law, even one that places mental illness under the umbrella of health insurance, will not cause stigma to disappear overnight.

But even if all the stigma associated with mental disorders magically disappeared, another compelling reason for privacy remains. Without confidentiality, patients would not feel free to make the intimate revelations that are essential for psychotherapy. As the U.S. Supreme Court decided in *Jaffee*, a patient speaking with a psychotherapist, no less than a client speaking with a lawyer, must be able to rely on the promise of confidentiality to fulfill the intended nature of the relationship. Or as Jerome Beigler observed pithily, "As asepsis is to surgery, so is confidentiality to psychiatry" (273).

Another problem with Kennedy's argument is the belief that mental illness is a brain disease and should be treated like the other organs in the body. A study by Bernice A. Pescosolido and her colleagues published in the *American Journal of Psychiatry* in 2010 found that holding a "neurobiological conception" of mental illness and alcohol dependence increases public support for treatment but "either was unrelated to stigma or tended to increase the odds of a stigmatizing reaction" (1325). The study also found that "focusing on genetics or brain dysfunction in order to decrease feelings of blame in the clinical encounter may have the unintended effect of increasing client and family feelings of hopelessness and permanence" (1329).

As Elliot C. Valenstein has shown in his book *Blaming the Brain*, the evidence to support the various biological and chemical theories of mental illness is shaky at best and flawed and unconvincing at worst. The pharmaceutical industry has promoted quick-fix pharmacology, though the safety and effectiveness of many psychoactive drugs remain questionable. Valenstein ends his book with an unsettling metaphor. For reasons that have little to do with science, the biochemical preoccupation with mental illness is pursuing a path fraught with many dangers. "It is like a ship without any navigational guidance being driven forward by a powerful motor through a sea with many uncharted reefs" (241).

This is an old debate. Most reasonable people understand that the brain is "the organ of the mind." They realize that mental disorders can, completely in some instances ("general paresis of the insane" is a classic example), and to greater or lesser degree in other instances, be attributable to physical, chemical, or other physiological disorders in the brain. Most mental disorders appear to be "caused" by a combination of a person's physical disorder, or at least a genetic predisposition to develop the disorder, and life experiences.

Freud knew this. Aware of the criticism that he had improperly ignored the physical aspect of mental disorders, he replied that psychoanalysis had chosen its path because the existing knowledge of the physical contribution was too inadequate in the early twentieth century to lead to rational physical treatments. Freud never denied the existence of a physical component as a causal factor in most cases of mental illness. He was, after all, trained as a neurologist. He understood that genetic endowment and life experience work together to bring about development. He also acknowledged the extent to which constitutional factors and life experience (or chance) varies from person to person so that patients with the same disorder can be thought to be placed at different points on a spectrum. Thus, the cause of a particular patient's disorder may be virtually entirely physical at one extreme or almost entirely psychological at the other. And Freud was willing to revise his theory in light of new scientific advances. In "The Dynamics of Transference" he writes, "We shall estimate the share taken by constitution or experience differently in individual cases according to the stage reached by our knowledge; and we shall retain the right to modify our judgement along with changes in our understanding" (99, note 2).

Because many psychiatric patients are now being treated by a combination of medication and some form of "talk therapy," their records contain data of supposedly varying sensitivity. Trying to thread this needle, in creating the "psychotherapy notes" provision for special protection of such notes that we described in Chapter 1, the Department of Health and Human Services attempted to address the problem of combination treatments by splitting the "medical record" into two parts, one of which contains a list of certain items in the record that cannot be specially protected by the stringent rules for psychotherapy information, such as "medication prescription and monitoring" and "results of clinical tests." These are presumably placed in a part of the record that is shared with other "providers" and insurers. In Patrick Kennedy's hypothetical example, information about drugs he was presently taking could be shared among his physicians, but information about the content of his psychotherapy sessions could not without his explicit agreement.

Miranda Warnings?

We stated in the Introduction that some psychotherapists are now issuing "Miranda" warnings, advising their patients not to make certain statements that may be used against them. To determine how analysts feel about Tarasoff or Miranda warnings, Paul posted the following questions on the American Psychoanalytic Association Members List:

> My question is, do you, at the onset of a psychoanalytic treatment, go through a list of all the things that a patient could say during sessions which you may be required by law to report to one government agency or another? Do you also list for the patient all the situations which could arise in which you are allowed, but not required, to report certain of the

patient's utterances to others? Or do you tend to handle such issues as analytic "grist for the mill" on an ad hoc basis when they arise? How?

If you do give such warnings at the beginning of treatment (which some legal writers have analogized somewhat sarcastically to "Miranda Warnings"), do you feel that giving such a warning contradicts Freud's advice about never making a collusive "deal" with a patient that certain information which comes to the patient's mind may go unspoken?

I (and perhaps other readers of this list) would be more interested in *what you actually do* rather than theoretical speculations.

We divided the replies into "ad hoc" and "Miranda warning," and while the sample, given its size and the way it was collected (not randomized), cannot be used for any statistical inference, it is sufficient for our purpose in that it demonstrates that opinion among the members of the American Psychoanalytic Association is divided. "I most certainly don't give a Miranda warning at the beginning of treatment," one person responded, "and I'd be very surprised if you hear from any analysts who do." Surprised, indeed! Several analysts *did*, in fact, give Miranda warnings, ranging from simply informing patients of the limits of confidentiality to giving them a list of "office procedures and policies" describing the circumstances in which the analyst is required to "not protect their confidential information to me."

We suspect that most psychotherapists would probably agree with the statement made by a person who *does* issue a Miranda warning that patients generally don't pay much attention to such admonitions and that the only issue about which they appear concerned involves information to insurance companies. Nevertheless, that some analysts feel compelled to issue a Miranda warning is disturbing. The results of Paul's query suggest not only that analysts are split over the need for Miranda warnings but also that such warnings may well become routine in the future, a development about which few if any psychotherapists feel sanguine. The need to give a Miranda warning is, ironically, a cautionary tale about the present state of psychotherapy and the direction toward which it is headed.

Tellingly, a Miranda warning appears in *The Sopranos*, the spellbinding HBO series created by David Chase that captured the imagination of millions of viewers from 1999 through 2007. As Glen Gabbard notes in his book *The Psychology of* The Sopranos, the principle of confidentiality is compromised in the first therapy session when the psychiatrist Jennifer Melfi (played by Lorraine Bracco) informs the mobster Tony Soprano (James Gandolfini) that she will be forced to contact the police if he reveals anything he knows about a future act of violence. "Is Dr. Melfi right to tell Tony about the limits of confidentiality?" Gabbard asks. "Probably so. At the same time, she is also squelching his capacity to reveal who he is to her. What ensues after her warning is a watered-down version of his life that is used to great comic effect by the writers" (49).

Gabbard points out two ironies arising from the threat of breached confidentiality in *The Sopranos*. "From a mobster's perspective, there is no distinction between a therapist who is duty-bound to contact the authorities and a rat who squeals. Both deserve rat poison" (50). From a therapist's point of view, limiting a patient's freedom of expression violates the fundamental rule of psychoanalysis. "If an area of the patient's life is off limits to the analyst, all his shameful secrets and guilty pleasures may take shelter in that concealed sector of the psyche." The result, Gabbard adds, is "compartmentalization," which dooms Tony's therapy (51). Regarded as one of the greatest television series of all time, *The Sopranos* received an award in 2001 from the American Psychoanalytic Association for its artistic depiction of psychoanalysis and psychoanalytic psychotherapy; at the same event Lorraine Bracco was honored for creating the "most credible psychoanalyst ever to appear in the cinema or on television."

Tony Soprano pays for his therapy sessions with cash, and thus he doesn't have to worry about the loss of privacy arising from the intrusion of third-party payers. But not all psychotherapists are in Dr. Melfi's or Tony Soprano's situation. While at the present time many, perhaps most, psychoanalysts are not "covered entities" as defined by the HIPAA "Privacy Rule" (a set of privacy regulations governing how clinical information is handled), it is possible that before long most "health care providers," including psychoanalysts, will become "covered" (by the federal rules) and then fall under the requirements of the HIPAA privacy regulations. The privacy regulations mandate that a covered provider must give, or at least offer, each patient at the onset of treatment a so-called Notice of Privacy Practices, a written document that spells out the ways in which patient information is required to be, or may be, disclosed without the patient's consent. In addition, the same notice must be posted in the provider's waiting room, and if the provider maintains a website, the notice must be posted there as well. Analysts seem to have no stated policy as to how they intend to deal with this eventuality.

Disciplinary Differences

Disagreement over whether Miranda warnings are necessary is only one of the many differences emerging from the privacy stories in our book. We ourselves have not always agreed with each other about the judgments made by the therapists in these privacy stories. To give but one example, psychoanalysts and literary scholars reach strikingly different conclusions about the Anne Sexton–Martin Orne controversy. Most of the mental health professionals who have commented on the case have been dismayed by Orne's willingness to cooperate with Sexton's biographer. By contrast, most of the literary scholars who have commented on the case have welcomed the opportunity to see the ways in which her therapy and poetry were inextricably intertwined. Another disciplinary difference is that psychoanalysts are generally reluctant to "psychoanalyze" living authors even if using only publicly available information, regarding the practice as unethical. Indeed, the code of the American Psychiatric Association states that it is "unethical for a psychiatrist to offer a professional opinion unless he or she has conducted an examination and has been granted proper authorization for such a statement." By contrast, the practice of offering psychological interpretations of living authors and their writings has long been acceptable to literary critics and historians, who try to achieve a delicate balance between the right to privacy and the right to freedom of expression.

Despite these disciplinary differences, we believe that readers will appreciate the complexity of the decisions made by the characters in these privacy stories. In an age of unprecedented technological change, confidentiality is being challenged in ways that could not have been imagined a few years earlier. We hope that by telling the human stories behind noteworthy breaches in the patient-psychotherapist relationship, we have demonstrated that confidentiality matters.

Works Cited

"Affidavit of Harriet G. Werner." 33 NY 2d 902.

Akhtar, Salman. *Good Stuff: Courage, Resilience, Gratitude, Generosity, Forgiveness and Sacrifice.* New York: Jason Aronson, 2013.

Alibrandi, Tom, with Frank H. Armani. *Privileged Information.* New York: Harper, 1984.

American Medical Association. *Code of Medical Ethics: Current Opinions and Annotations.* Chicago: American Medical Association, 1997.

American Psychiatric Association. *Manual of Psychiatric Peer Review.* Washington: American Psychiatric Association, 1976.

American Psychoanalytic Association. "Reporting Information for Claims Review of Psychoanalysis." In *Manual of Psychiatric Quality Assurance,* ed. M. Matteson, 237–238. Washington: American Psychiatric Press, 1992.

———. "Practice Bulletin 3: External Review of Psychoanalysis." *The American Psychoanalyst* 34 (1999).

———. "Principles and Standards of Ethics for Psychoanalysts." New York: American Psychoanalytic Association, 2001.

Amin v. Rose. New York Law Journal (December 7, 2000): 31, col. 1.

Anonymous [Anne Hayman]. "Psychoanalyst Subpoenaed." *Lancet* (October 16, 1965): 785–786.

"APA to Oppose Analyst's Publishing Case History." *Psychiatric News: Official Newspaper of the American Psychiatric Association* 9 (1974).

Aronowitz, S. "Following the Psychotherapist-Patient Privilege Down the Bumpy Road Paved by *Jaffee v. Redmond.*" *Annual Survey of American Law* (1998): 307–348.

Atwood, Margaret. *Payback: Debt and the Shadow Side of Wealth.* Toronto: Anansi, 2008.

Avishai, Bernard. *Promiscuous:* Portnoy's Complaint *and Our Doomed Pursuit of Happiness.* New Haven, Conn.: Yale University Press, 2013.

Ayala, Elaine. "Duhl Helped Pioneer the Family Therapy Movement." *San Antonio Express-News* (January 4, 2011).

Barnes, Julian. *The Sense of an Ending*. New York: Vintage, 2011.

Barrett, Carole F. "CPL 330.20: Persons Involuntarily Committed Pursuant to CPL Entitled to Procedural and Substantive Safeguards Guaranteed Involuntary Civil Detainees." *St. John's Law Review* 54, no. 3 (1980). http://scholarship.law.stjohns .edu/lawreview/vol54/iss3/8.

Beck, James C. "The Current Status of the Duty to Protect." In *Confidentiality Versus the Duty to Protect: Foreseeable Harm in the Practice of Psychiatry*, ed. James C. Beck. Washington: American Psychiatric Press, 1990.

Bellacosa, Joseph W. *Wondrous Journeys and Leaps of Faith: Joseph's Memoir*. Privately published.

Berg, Leida, and Harold Steinberg. *In Search of a Response*. New York: Tiresias, 1973.

Berman, Jeffrey. "Review of *Tales from the Couch: Writers on Therapy*, ed. Jason Shinder." *Psychoanalytic Psychology* 18 (2001): 743–755.

———. "Revisiting Philip Roth's Psychoanalysts." In *The Cambridge Companion to Philip Roth*, ed. Timothy Parrish, 94–110. New York: Cambridge University Press, 2007.

———. *The Talking Cure: Literary Representations of Psychoanalysis*. New York: New York University Press, 1985.

Beyer, Karen. "First Person: *Jaffee v. Redmond* Therapist Speaks." *The American Psychoanalyst* 34 (2000). http://www.jaffee-redmond.org/articles/beyer.htm.

Beigler, Jerome S. "*Tarasoff* v. Confidentiality." *Behavioral Science and the Law* 2 (1984): 273–289.

Binder v. Ruvell, Civil Docket 52 C 25 35, Circuit Court of Cook County, IL (1952). Reported in full in the *Journal of American Medical Association* 150 (1952): 1241–1242.

Biskupic, Joan. *American Original: The Life and Constitution of Supreme Court Justice Antonin Scalia*. New York: Farrar, Straus and Giroux, 2009.

Bloom, Claire. *Leaving a Doll's House*. Boston: Little, Brown and Company, 1996.

Blum, Deborah. *Bad Karma: A True Story of Obsession and Murder*. New York: Atheneum, 1986.

Bok, Sissela. *Secrets: On the Ethics of Concealment and Revelation*. New York: Pantheon, 1982.

Bollas, Christopher, and David Sundelson. *The New Informants: The Betrayal of Confidentiality in Psychoanalysis and Psychotherapy*. Northvale, N.J.: Jason Aronson, 1995.

Breger, Louis. *Freud: Darkness in the Midst of Vision*. New York: Wiley, 2000.

Brenner, Ira. "Making Extraordinary Monetary Arrangements." In *Unusual Interventions: Alterations of the Frame, Method, and Relationship in Psychotherapy and Psychoanalysis*, ed. Salman Akhtar, 3–29. London: Karnac, 2011.

Breuer, Josef, and Sigmund Freud. *Studies on Hysteria.* In *The Standard Edition of the Complete Psychological Works of Sigmund Freud,* vol. 2. London: Hogarth, 1955.

Brown, Kenneth S. "The Medical Privilege in the Federal Courts—Should It Matter Whether Your Ego or Your Elbow Hurts?" *Loyola of Los Angeles Law Review* 38 (2004): 657–706. http://digitalcommons.lmu.edu/llr/vol38/issue2/4.

Bruccoli, Matthew J. *Some Sort of Epic Grandeur: The Life of F. Scott Fitzgerald.* New York: Harcourt Brace Jovanovich, 1981.

Buckner, Fillmore, and Marvin Firestone. "'Where the Public Peril Begins': Twenty-Five Years After Tarasoff." *Journal of Legal Medicine* 21 (2000): 187–222.

Bullough, Vern L. "Sexton's Daughter, Biographer, and Psychiatrist." *Society* 29 (1992): 12–13.

Burka, Jane B. "Psychic Fallout from Breach of Confidentiality: A Patient/Analyst's Perspective." *Contemporary Psychoanalysis* 44 (2008): 177–198.

Caher, John M. *King of the Mountain: The Rise, Fall, and Redemption of Chief Judge Sol Wachtler.* New York: Prometheus, 1998.

Carton, Sharon. "The Poet, the Biographer, and the Shrink: Psychiatrist-Patient Confidentiality and the Anne Sexton Biography." *University of Miami Entertainment and Sports Law Review* (1993): 117–165.

Catlin, Roger. "Legendary Writer Opens Up." *Hartford Courant* (March 17, 2013).

Chan, Kevin W. "*Jaffee v. Redmond*: Making the Courts a Tool of Injustice?" *Journal of American Academy of Psychiatry Law* 25 (1997): 383–389.

Clemens, Norman A. "Privacy, Consent, and the Electronic Mental Health Record: The Person vs. the System." *Journal of Psychiatric Practice* 18 (2012): 46–50.

Cline, Sally. *Zelda Fitzgerald: The Tragic, Meticulously Researched Biography of the Jazz Age's High Priestess.* New York: Arcade, 2012.

Colburn, Steven E., ed. *Anne Sexton: Telling the Tale.* Ann Arbor: University of Michigan Press, 1988.

Couser, G. Thomas. *Vulnerable Subjects: Ethics and Life Writing.* Ithaca, N.Y.: Cornell University Press, 2004.

Crowley, Kieran. *The Surgeon's Wife.* New York: St. Martin's, 2001.

Davies, Robertson. *The Manticore.* New York: Penguin, 1976.

DePaulo, Lisa. "Intimations of Murder." *Vanity Fair* (September 2000).

District of Columbia. "District of Columbia Mental Health Information Act of 1978." DC Code, 1978, sec. 6–2001 et seq.

Doe v. Roe, 42 A.D.2d 559 (1973).

Doe v. United States, 493 U. S. 906 (1989).

Dubbleday, C. "The Psychotherapist-Client Testimonial Privilege: Defining the Professional Involved." *Emory Law Journal* 34 (1985): 777–826.

Dubovsky, Steven. "Are Psychiatric Patients Really Dangerous?" *NEJM Journal Watch* (March 16, 2009). http://www.jwatch.org/jp200903160000001/2009/03/16/are-psychiatric-patients-really-dangerous?.

Edwards, Griffin Sims. "Doing Their Duty: An Empirical Analysis of the Unintended Effect of *Tarasoff v. Regents* on Homicidal Activity." http://papers.ssrn.com/so13/papers.cfn?abstract_id=1544574, 2010.

Ewalt, Jack Richard, and Dana L. Farnsworth. *Textbook of Psychiatry.* New York: McGraw-Hill, 1963.

Fish, Stanley. *How to Write a Sentence and How to Read One.* New York: Harper, 2011.

Fleming, John G., and Bruce Maximov. "The Patient or His Victim: The Therapist's Dilemma." *California Law Review* 62 (1974): 1025–1068.

Forrester, John. "Trust, Confidentiality, and the Possibility of Psychoanalysis." In *Confidentiality: Ethical Perspectives and Clinical Dilemmas,* ed. Charles Levin, Allannah Furlong, and Mary Kay O'Neill, 20–27. Hillsdale, N.J.: Analytic Press, 2003.

Franks, Lucinda. "To Catch a Judge." *New Yorker* (December 21, 1992).

Freud, Anna. *The Ego and the Mechanisms of Defense.* Rev. ed. New York: International Universities Press, 1977.

Freud, Sigmund. "The Dynamics of Transference" (1912). In *The Standard Edition of the Complete Psychological Works of Sigmund Freud,* vol. 12. London: Hogarth, 1958.

———. *Fragment of an Analysis of a Case of Hysteria* (1905). In *The Standard Edition of the Complete Psychological Works of Sigmund Freud,* vol. 7. London: Hogarth, 1953.

———. *The Interpretation of Dreams* (1900). In *The Standard Edition of the Complete Psychological Work of Sigmund Freud,* vols. 4–5. London: Hogarth, 1953.

———. "On Beginning the Treatment: Further Recommendations on the Technique of Psycho-Analysis" (1913). In *The Standard Edition of the Complete Psychological Works of Sigmund Freud,* vol. 12. London: Hogarth, 1958.

———. *On the History of the Psycho-Analytic Movement* (1914). In *The Standard Edition of the Complete Psychological Works of Sigmund Freud,* vol. 14. London: Hogarth, 1957.

———. *An Outline of Psycho-Analysis* (1940). In *The Standard Edition of the Complete Psychological Works of Sigmund Freud,* vol. 23. London: Hogarth, 1964.

———. *The Question of Lay Analysis* (1926). In *The Standard Edition of the Complete Psychological Works of Sigmund Freud,* vol. 20. London: Hogarth, 1959.

———. "Screen Memories" (1899). In *The Standard Edition of the Complete Psychological Works of Sigmund Freud,* vol. 3. London: Hogarth, 1962.

Fridhandler, B. "The Tale of Joseph Lifschutz." http://www.fridhand.net/articles/Lifschutz.pdf.

Friedman, Neil. *Remarkable Psychotherapeutic Experiences: A Client's Report.* New York: Xlibris, 2005.

Friend, Tad. "The Harriet-the-Spy Club." *New Yorker* (July 31, 2000).

Furlong, Allannah. "The Why of Sharing and Not the What: Confidentiality and Psychoanalytic Purpose." In *Confidentiality: Ethical Perspectives and Clinical Di-*

lemmas, ed. Charles Levin, Allannah Furlong, and Mary Kay O'Neil, 39–49. Hillsdale, N.J.: Analytic Press, 2003.

Gabbard, Glen O. "Disguise or Consent: Problems and Recommendations Concerning the Publication and Presentation of Clinical Material." *International Journal of Psycho-Analysis* 81 (2000): 1071–1086.

———. *The Psychology of* The Sopranos*: Love, Death, Desire, and Betrayal in America's Favorite Gangster Family.* New York: Basic Books, 2002.

Gay, Peter. *Freud: A Life for Our Time.* New York: Norton, 1988.

George, Diana Hume. "Introduction." In *Sexton: Selected Criticism*, ed. Diana Hume George. Urbana: University of Illinois Press, 1988.

Gergen, Mary. "The (H)orne of a Dilemma." *Society* 29 (1992): 21–23.

Gillard, Gary. *Supertext and the Mind-Culture System: Freud, Levi-Strauss, Bateson.* Ph.D. dissertation, Murdoch University, 1994. http://www.garygillard.net/writing/supertext/ch2./html.

Ginsberg, Brian D. "Therapists Behaving Badly: Why the Tarasoff Duty Is Not Always Economically Efficient." *Willamette Law Review* 43 (2006): 31–68.

Goffman, Erving. *Stigma: Notes on the Management of Spoiled Identity.* Englewood Cliffs, N.J.: Prentice-Hall, 1963.

Goin, Marcia Kraft. "Clinical Issues in Mental Health Care." In *Privacy and Confidentiality in Mental Health Care*, ed. John J. Gates and Bernard S. Arons, 71–90. Baltimore, Md.: Paul H. Brookes, 2000.

Goldman, Ronald. *In Confidence: When to Protect Secrecy and When to Require Disclosure.* New Haven, Conn.: Yale University Press, 2009.

Goldner, Jesse. "Legal Briefs: Claims of Psychotherapist-Patient Privilege in Child Custody and Child Abuse Cases after *Jaffee v. Redmond*." *Register Reports* (Spring 2011). http://www.nationalregister.org/trr_spring11_goldner.html.

Goodman, S. "Report of the Secretary of the Executive Council." *Bulletin of the American Psychoanalytic Association* 29 (1973): 435–453.

Gopnik, Adam. "Man Goes to See a Doctor." In *Tales from the Couch: Writers on Therapy*, ed. Jason Shinder, 18–37. New York: Morrow, 2000.

Gostin, L. O. "National Health Information Privacy: Regulations Under the Health Insurance Accountability and Portability Act." *Journal of American Medical Association* 285 (2001): 3015–3021.

Gourevitch, Philip. "A Cold Case." *New Yorker* (February 14, 2000): 42–60.

Graber, Glenn C., Alfred D. Beasley, and John A. Eaddy. *Ethical Analysis of Clinical Medicine.* Baltimore, Md.: Urban, 1985.

Gurevitz, Howard. "Tarasoff: Protective Privilege Versus Public Peril." *American Journal of Psychiatry* 143 (1977): 289–291.

Gutheil, Thomas G. "Moral Justification for Tarasoff-Type Warnings and Breach of Confidentiality: A Clinician's Perspective." *Behavioral Sciences and the Law* 19 (2001): 345–353.

Hale, N. G. *The Rise and Crisis of Psychoanalysis in the United States: 1917–1985*. Vol. 2. Oxford: Oxford University Press, 1995.

Harris, George C. "The Dangerous Patient Exception to the Psychotherapist-Patient Privilege: The Tarasoff Duty and the Jaffee Footnote." *Washington Law Review* 74 (1999): 33–68.

Haven, Cynthia. "Diane Middlebrook, Professor Emeritus and Legendary Biographer, Dies at 68." *Stanford Report* (December 15, 2007).

Hayman, Anne. "A Psychoanalyst Looks at the Witness Stand." In *Confidentiality: Ethical Perspectives and Clinical Dilemmas*, ed. Charles Levin, Allannah Furlong, and Mary Kay O'Neill, 293–308. Hillsdale, N.J.: Analytic Press, 2003.

Herbert, Paul B. "The Duty to Warn: A Reconsideration and a Critique." *Journal of American Academy Psychiatry Law* 30 (2002): 417–424.

Herbert, Paul B., and Kathryn A. Young. "Tarasoff at Twenty-Five." *Journal of American Academy Psychiatry Law* 30 (2002): 275–281.

Hobhouse, Janet. *The Furies*. New York: New York Review of Books, 1993.

Hoover, J. Edgar. "Let's Keep America Healthy." *Journal of the American Medical Association* 144 (1950): 1094–1095.

Horowitz, Frances Degen. "Confidentiality and Privacy." *Society* 29 (1992): 27–29.

Hughes, Samuel M. "The Sexton Tapes." *Pennsylvania Gazette* 90 (December 1991): 20–28, 39.

Imwinkelried, Edward L. "An Hegelian Approach to Privileges Under Federal Rule of Evidence 501: The Restrictive Thesis, the Expansive Antithesis, and the Contextual Synthesis."*Nebraska Law Review* 73 (1994): 511–523.

In re Doe, 964 F.2d 1325 (2d Cir. 1992).

In re Zuniga, 714 F. 2d 632 (CA6).

Insel, Thomas R. and Remi Quirion. "Psychiatry as a Clinical Neuroscience Discipline." *Journal of the American Medical Association* 294 (2005): 2221–2224.

Jaffee v. Redmond, 51 F.3d 1346 (1995).

Jaffee v. Redmond, 116 S.Ct. 1923 (1996).

Jaffee v. Redmond, 518 U.S. 1 (1996).

Jamison, Kay Redfield. *An Unquiet Mind*. New York: Knopf, 1995.

Jenkins, Peter. "Introduction." In *Legal Issues in Counselling and Psychotherapy*, ed. Peter Jenkins, 1–11. London: Sage, 2002.

Jones, Ernest. *The Life and Work of Sigmund Freud*. 3 vols. New York: Basic Books, 1953–1957.

Jong, Erica. "Anne Sexton's River of Words." *New York Times* (August 17, 1991).

"Judge Clears Psychiatrist of Refusing to Testify." *New York Times* (June 2, 1970, 49).

"Judge Not." *Psychology Today* 30, no. 4 (July 1997). http://web.ebscohost.com.db-gate.nysed.gov/ehost/delivery?sid=59b7.

Kakutani, Michiko. "Books of the Times; A Poet's Life Through a Sexual Lens." *New York Times* (August 13, 1991).

Kantrowitz, Judy Leopold. *Writing About Patients: Responsibilities, Risks, and Ramifications.* New York: Other Press, 2006.

Kernberg, Otto. "Some Reflections on Confidentiality in Clinical Practice." In *Confidentiality: Ethical Perspectives and Clinical Dilemmas,* ed. Charles Levin, Allannah Furlong, and Mary Kay O'Neill, 80–83. Hillsdale, N.J.: Analytic Press, 2003.

Kipnis, Kenneth. "In Defense of Absolute Confidentiality." *Virtual Mentor* 5 (October 2003). http://virtualmentor.ama-assn.org/2003/10/hlaw2-0310.html.

Kipnis, Laura. *How to Become a Scandal: Adventures in Bad Behavior.* New York: Picador, 2011.

Kleinschmidt, Hans J. "The Angry Act: The Role of Aggression in Creativity." *American Imago* 24 (1967): 98–128.

Kroll, Jerome. "The Silence of the Tapes." *Society* 29 (1992): 18–20.

Lear, Jonathan. "Confidentiality as a Virtue." In *Confidentiality: Ethical Perspectives and Clinical Dilemmas,* ed. Charles Levin, Allannah Furlong, and Mary Kay O'Neill, 3–17. Hillsdale, N.J.: Analytic Press, 2003.

"Legal Right Goal for Psychiatrist: Coast Doctor Once Jailed, Seeks to Protect Records." *New York Times* (December 14, 1969).

Leibovici, Solange. "'A Man, His Destiny, and His Work': Nathan Zuckerman Caught Between Biography and Autobiography—A Talk Presented at the Thirtieth International Conference on Psychology and the Arts." University of Porto, Portugal, June 28, 2013.

Leonard, John. "The Psychologizing of Crime." *U.S. News & World Report* 113, no. 22 (December 7, 1992). http://web.ebscohost.com.dbgateway.nysed.gov/ehost/delivery?sid=5549.

Lerman, Lisa G., Frank H. Armani, Thomas D. Morgan, and Monroe H. Freedman. "The Buried Bodies Case: Alive and Well After Thirty Years." http://www.americanbar.org/content/dam/aba/migrated/cpr/pubs/buried.auth checkdam.pdf.

Levin, Peter, dir. *Sworn to Silence.* 1987.

Lewin, Barbara R. "The Anne Sexton Controversy." *Society* 29 (1992): 9–11.

Lifschutz, Joseph. "The Anne Sexton Tapes: Confidentiality in Psychoanalysis and Psychotherapy." *The American Psychoanalyst* 26 (1992): 8.

Louisell, David W., and Kent Sinclair Jr. "Reflections on the Law of Privileged Communications—The Psychotherapist-Patient Privilege in Perspective." *California Law Review* 30 (1971).

Luhrmann, T. M. *Of Two Minds: The Growing Disorder in American Psychiatry.* New York: Knopf, 2000.

Lynn, David J., and George E. Vaillant. "Anonymity, Neutrality, and Confidentiality in the Actual Methods of Sigmund Freud: A Review of Forty-Three Cases, 1907–1939." *American Journal of Psychiatry* 155 (1998): 163–171.

Malcolm, Janet. *The Silent Woman: Sylvia Plath and Ted Hughes.* New York: Knopf, 1995.

McKelvey, Wallace. "A Conversation with Patrick Kennedy: Ex-Congressman Still Fights for Mental Health Causes." *Press of Atlantic City* (September 14, 2013). http://www.pressofatlanticcity.com/news/breaking/a-conversation-with -patrick-kennedy.

Moezzi, Melody. "Lawyers of Sound Mind?" *New York Times* (August 5, 2013).

Merton, Vanessa. "Confidentiality and the 'Dangerous' Patient: Implications of Tarasoff for Psychiatrists and Lawyers." *Emory Law Journal* 31 (1982): 263–343.

Middlebrook, Diane Wood. *Anne Sexton: A Biography.* Boston: Houghton Mifflin, 1991.

Moore, Mark. "Conducting the Treatment Outside the Office." In *Unusual Interventions: Alterations of the Frame, Method, and Relationship in Psychotherapy and Psychoanalysis,* ed. Salman Akhtar, 31–63. London: Karnac, 2011.

Mosher, Paul W. "Psychotherapist-Patient Privilege: The History and Significance of the United States Supreme Court's Decision in the Case of *Jaffee v. Redmond.*" In *Confidential Relationships: Psychoanalytic, Ethical, and Legal Concerns,* ed. Christine M. Koggel, Allannah Furlong, and Charles Levin, 177–206. Amsterdam: Rodopi, 2003.

———. "We Have Met the Enemy and He (Is) Was Us." In *Confidentiality: Ethical Perspectives and Clinical Dilemmas,* ed. Charles Levin, Allannah Furlong, and Mary Kay O'Neill, 229–249. Hillsdale, N.J.: Analytic Press, 2003.

Mosher Paul W., and Peter P. Swire. "The Ethical and Legal Implications of *Jaffee v. Redmond* and the HIPAA Medical Privacy Rule for Psychotherapy and General Psychiatry." *Psychiatry Clinics of North America* 25 (2002): 575–584.

Mossman, Douglas. "Critique of Pure Risk Assessment or Kant Meets Tarasoff." *University of Cincinnati Law Review* 75 (2006): 523–610.

National Council of State Legislatures. "Mental Health Professionals' Duty to Protect/Warn." http://www.ncsl.org/research/health/mental-health-professionals -duty-to-warn.aspx.

Nelken, M. "The Limits of Privilege: The Developing Scope of Federal Psychotherapist-Patient Privilege Law. *Rev. Litigation* 20 (2000): 1–43.

New Jersey. *Practicing Psychology Licensing Act.* New Jersey Stat. 45 (1985), 14B-2-45: 14B-46.

Nietzsche, Friedrich. *Beyond Good and Evil.* Trans. Helen Zimmern. In *The Philosophy of Nietzsche.* New York: Modern Library, 1954.

North Carolina Bar Association. *Rules of Professional Conduct,* Opinion #175, 1995.

Olinick, Stanley L. "The Galloping Psychoanalyst." *International Review of Psycho-Analysis* 7 (1980): 439–445.

Onek, Joseph N. "Legal Issues in the Orne/Sexton Case." *Journal of the American Academy of Psychoanalysis and Dynamic Psychiatry* 20 (1992): 655–658.

Orne, Martin. "Foreword." In *Anne Sexton: A Biography*, by Diane Wood Middlebrook, xiii–xviii. Boston: Houghton Mifflin, 1991.

———. "The Sexton Tapes." *New York Times* (July 23, 1991).

Oyez Project. *Jaffee v. Redmond*. http://www.oyez.org/cases/1990-1999/1995/1995_95_266

———. *Roe v. Doe*. http://www.oyez.org/cases/1970-1979/1974/1974_73_1446.

Pearson, Jesse. "The Follies of Documentary Filmmaking." *Vice*. http://www.vice.com/read/doc-v14n9.

Pendergrass, Taylor. "No Dignity in 'The Box.'" *Albany Times-Union* (October 2, 2012).

People v. Liberta, 64 N.Y.2d 152, 474 N.E. 2d 567, 485 N.Y.S. 2d 207.

People v. Poddar, 10 Cal. 3d 750, 1974.

People v. Stritzinger, 34 Cal. 3d 521, 1983. Kaus, J., Concurring and Dissenting.

Pescosolido, Bernice A., et al. "'A Disease Like Any Other'? A Decade of Change in Public Reactions to Schizophrenia, Depression, and Alcohol Dependence." *American Journal of Psychiatry* 167 (2010): 1321–1330.

Pierpont, Claudia Roth. *Roth Unbound: A Writer and His Books*. New York: Farrar, Straus & Giroux, 2013.

Prosser, William. "Privacy." *California Law Review* 48 (1960): 383–423.

Pyles, Robert L. "The American Psychoanalytic Association's Fight for Privacy." In *Confidentiality: Ethical Perspectives and Clinical Dilemmas*, ed. Charles Levin, Allannah Furlong, and Mary Kay O'Neill, 251–264. Hillsdale, N.J.: Analytic Press, 2003.

Rich, Frank. "When Privacy Jumped the Shark." *New York* (July 8, 2013): 22–28.

Roazen, Paul. "Privacy and Therapy." *Society* 29 (1992): 14–19.

"*Roe v. Doe*: A Remedy for Disclosure of Psychiatric Confidences." *Rutgers Law Review* 29 (1975): 190–209.

Rosenbaum, Max. "The Travails of Martin Orne: On Privacy, Public Disclosure, and Confidentiality in Psychotherapy." *Journal of Contemporary Psychotherapy* 24 (1994): 159–167.

Rothstein, Mervyn. "Philip Roth and the World of 'What If.'" *New York Times* (December 17, 1986).

Rossum, Ralph A. *Antonin Scalia's Jurisprudence: Text and Tradition*. Lawrence: University Press of Kansas, 2006.

Roth, Philip. *The Counterlife*. New York: Farrar, Straus and Giroux, 1986.

———. *Deception*. New York: Simon and Schuster, 1990.

———. *The Facts: A Novelist's Autobiography*. New York: Farrar, Straus & Giroux, 1988.

———. *I Married a Communist.* Boston: Houghton Mifflin, 1998.

———. *My Life as a Man.* New York: Holt, Rinehart and Winston, 1974.

———. *Operation Shylock.* New York: Vintage, 1994.

———. *Patrimony: A True Story.* New York: Simon and Schuster, 1991.

———. *Portnoy's Complaint.* New York: Random House, 1969.

———. *Zuckerman Unbound.* New York: Farrar, Straus & Giroux, 1981.

Rudnytsky, Peter L. "Book Review of *Writing About Patients: Responsibilities, Risks, and Ramifications* by Judy Leopold Kantrowitz." *Journal of the American Psychoanalytic Association* 55 (2007): 1406–1411.

———. "Interview with Peter Rudnytsky." In *Philip Roth—The Continuing Presence: New Essays on Psychological Themes,* ed. Jane Statlander-Slope, 149–167. Newark, N.J.: Northeast, 2013.

Saks, Elyn R., and Shahrokh Golshan. *Informed Consent to Psychoanalysis: The Law, the Theory, and the Data.* New York: Fordham University Press, 2013.

Scalia, Antonin. "God's Justice and Ours." *First Things* (May 2002). ww.first things.com/article/2007/01/gods-justice-and-ours.

———. *A Matter of Interpretation: Federal Courts and the Law.* With commentary by Amy Gutman, editor; Gordon S. Wood; Laurence H. Tribe; Mary Ann Glendon; and Ronald Dworkin. Princeton, N.J.: Princeton University Press, 1977.

Scarry, Elaine. *Rule of Law, Misrule of Men.* Cambridge, Mass.: MIT Press, 2010.

Searles, George J. *Conversations with Philip Roth.* Jackson: University Press of Mississippi, 1992.

Seidel v. Werner, 81 Misc.2d 220 (1975).

Seidel v. Werner, 50 A.D.2d 743 (1975).

Seidel v. Werner, 81 Misc.2d 1064 (1975).

Sexton, Anne. *Anne Sexton: A Self-Portrait in Letters,* ed. Linda Gray Sexton and Lois Ames. Boston: Houghton Mifflin, 1977.

———. *The Complete Poems.* Boston: Houghton Mifflin, 1981.

———. *No Evil Star: Selected Essays, Interviews, and Prose,* ed. Steven E. Colburn. Ann Arbor: University of Michigan Press, 1985.

Sexton, Linda Gray. *Half in Love (Surviving the Legacy of Suicide).* Berkeley, Calif.: Counterpoint, 2011.

———. *Searching for Mercy Street: My Journey Back to My Mother, Anne Sexton.* Boston: Little, Brown, 1994.

Shechner, Mark. "Roth's American Trilogy." In *The Cambridge Companion to Philip Roth,* ed. Timothy Parrish, 142–157. New York: Cambridge University Press, 2007.

Shopper, Moisy. "Breaching Confidentiality." *Society* 29 (1992): 24–26.

Shuman, Daniel W., and Myron F. Weiner. *The Psychotherapist-Patient Privilege: A Critical Examination.* Foreword by Ralph Slovenko. Springfield, Ill.: Charles C. Thomas, 1987.

Siegler, Mark. "Confidentiality in Medicine: A Decrepit Concept." *New England Journal of Medicine* 307 (1982): 1518–1521.

Simon, Robert J. "The Duty to Protect in Private Practice." In *Confidentiality Versus the Duty to Protect: Foreseeable Harm in the Practice of Psychiatry*, ed. James C. Beck. Washington: American Psychiatric Press, 1990.

Skinner, Curtis. "SAFE Act Registry of Mentally Ill Nets Few Gun Permit Holders." *New York World* (June 3, 2013). http://www.thenewyorkworld.com/2013/06/03/safe-act-registry.

Skorczewski, Dawn. *An Accident of Hope: The Therapy Tapes of Anne Sexton*. New York: Routledge, 2012.

Slovenko, Ralph. "Psychotherapist-Patient Privilege: A Picture of Misguided Hope." *Catholic University Law Review* 23 (1974): 649–673.

———. "The Psychotherapist-Patient Testimonial Privilege." *American Journal of Psychoanalysis* 57 (1997): 63–68.

———. *Psychotherapy and Confidentiality: Testimonial Privileged Communication, Breach of Confidentiality, and Reporting Duties*. Springfield, Ill.: Charles C. Thomas, 1998.

Slovenko, Ralph, and Gene L. Usdin. *Psychotherapy, Confidentiality, and Privileged Communication*. Springfield, Ill.: Charles C. Thomas, 1966.

Snyder. Criminal Term, Part 88.

"Social Work Month 2012 Toolkit." National Association of Social Workers. http://www.socialworkers.org/pressroom/swMonth/2012/toolkit/mentalhealth/talkingpoints.

Solove, Daniel. "10 Reasons Why Privacy Matters." http://www.linkedin.com/today/post/article/20140113044954-2259773-10-reasons-why-privacy-matters.

Soulier, Matthew F., Andrea Maislen, and James C. Beck. "Status of the Psychiatric Duty to Protect, Circa 2006." *Journal of the American Academy of Psychiatry and the Law* 38 (2010): 457–473.

Spence, Donald J. *Narrative Truth and Historical Truth: Meaning and Interpretation in Psychoanalysis*. New York: Norton, 1984.

Stanley, Alessandra. "Poet Told All; Therapist Provides the Record." *New York Times* (July 15, 1991).

Statlander-Slote, Jane, ed. *Philip Roth—The Continuing Presence: New Essays on Psychological Themes*. Newark, N.J.: Northeast, 2013.

Stein, Seth P., Robert L. Schoenfeld, and Nancy A. Hampton. "Brief of Amici, New York State Psychiatric Association, Inc. and American Psychoanalytic Association," September 6, 2000.

Stern, M. M. "Therapeutic Playback, Self Objectification, and the Analytic Process." *Journal of the American Psychoanalytic Association* 18 (1970): 562–598.

Stone, Alan. A. "The Tarasoff Decisions: Suing Psychotherapists to Safeguard Society." *Harvard Law Review* 90 (1976): 358–378.

———. *Law, Psychiatry, and Morality: Essays and Analysis*. Washington: American Psychiatric Association, 1984.

Stone, Michael H. *Abnormalities of Personality: Within and Beyond the Realm of Treatment*. New York: Norton, 1993.

———. *The Anatomy of Evil*. New York: Prometheus, 2009.

———. *Personality-Disordered Patients: Treatable and Untreatable*. Washington: American Psychiatric Association, 2006.

Styron, William. *Darkness Visible: A Memoir of Madness*. New York: Random House, 1990.

———. "An Interior Pain That Is All But Indescribable." *Newsweek* (April 18, 1994).

"Supreme Court Refuses to Decide Roe/Doe Case." *Psychiatric News* (March 19, 1975).

Svetanics, M. Leigh. "Beyond 'Reason and Experience': The Supreme Court Adopts a Broad Psychotherapist-Patient Privilege in *Jaffee v. Redmond*." *Saint Louis University Law Journal* 41 (1997): 719–759.

Swidler & Berlin v. United States, 524 U.S. 399 (1998).

Szasz, Thomas. *Liberation by Oppression: A Comparative Study of Slavery and Psychiatry*. Piscataway, N.J.: Transaction, 2003.

Talbot, Margaret. "Supreme Confidence: The Jurisprudence of Justice Antonin Scalia." *New Yorker* (March 28, 2005): 40–55.

Tarasoff v. Regents of University of California, 17 Cal.3d 425 (S.F. No. 23042 Supreme Court of California, July 1, 1976).

Thomas, Dorothy. "Edith Fisch." *Jewish Women's Archive: Jewish Women Encyclopedia*. http://jwa.org/encyclopedia/article/fisch-edith.

Turner, Christopher. "Marilyn Monroe on the Couch." *Telegraph* (June 23, 2010). http://www.telegraph.co.uk/culture/film/7843140/Marilyn-Monroe-on-the-couch.html.

United States Department of Health and Human Services, Office for Civil Rights. "Health Information of Deceased Individuals." September 19, 2013. http://www.hhs.gov/ocr/privacy/hipaa/understanding/coveredentities/decedents.html.

———. *Mental Health: A Report of the Surgeon General*. Rockville, Md.: U.S. Department of Health and Human Services, Substance Abuse and Mental Health Services Administration, Center for Mental Health Services, National Institutes of Health, National Institute of Mental Health, 1999.

———. "Standards for Privacy of Individually Identifiable Health Information; Final Rule." *Federal Register* 64 (1999): 59917–59966.

———. "Standards for Privacy of Individually Identifiable Health Information: Final Rule." *Federal Register* 65 (2000): 82461–82829.

———. "Standards for Privacy of Individually Identifiable Health Information: Final Rule." *Federal Register* 67 (2002): 53181–53273.

United States v. Auster, 517 F.3d 312, 315–16 (5th Cir. 2008).

United States v. Burtrum, 17 F. 3d 1299 (CA10).

United States v. Corona, 849 F. 2d 562 (CA11 1988).

United States v. Chase, 340 F.3d 978, 992 (9th Cir. 2003) (en banc).

United States v. Ghane, 673 F.3d 771, 787–88 (8th Cir. *2012*).

United States v. Glass, 133 F.3d 1356, 1360 (10th Cir. 1998).

United States v. Hayes, 227 F.3d 578, 583–87 (6th Cir. 2000).

United States v. Meagher, 531 F.2d 752 (5th Cir. 1976).

Valenstein, Elliot S. *Blaming the Brain: The Truth About Drugs and Mental Health.* New York: Free Press, 1998.

Wachtler, Sol. *After the Madness: A Judge's Own Prison Memoir.* New York: Random House, 1997.

Wachtler, Sol, and David Gould. *Blood Brothers.* Beverly Hills, Calif.: New Millennium, 2003.

Walker, Jesse. "Let the Viewer Decide." *Reason* 39 (2007): 50–56.

Warner, Richard. "Implementing Anti-Stigma Programmes in Boulder, Colorado, and Calgary, Alberta." In *Understanding the Stigma of Mental Illness: Theory and Interventions*, ed. Julio Arboleda-Florez and Norman Sartorius, 161–174. West Sussex: Wiley, 2008.

Warren, Robert Penn. *All the King's Men.* New York: Bantam, 1973.

Watson, Andrew S. "Levels of Confidentiality in the Psychoanalytic Situation." *Journal of the American Psychoanalytic Association* 20 (1972): 156–176.

Weissberg, Josef F. "Trust Is at Stake." *New York Times* (July 26, 1991).

Werner v. Werner, 70 Misc.2d 1051 (1972).

Wise, Erica. "Child Abuse/Neglect Reporting: Issues and Controversies." *Register Report* (Fall 2010), http://www.nationalregister.org/trr_fall10_wise.html.

Wolfe, Linda. *Double Life: The Shattering Affair Between Chief Judge Sol Wachtler and Socialite Joy Silverman.* New York: Pocket, 1994.

Wortis, Joseph. *Fragments of an Analysis with Freud.* New York: McGraw-Hill, 1975.

Young, Philip. *Ernest Hemingway: A Reconsideration.* University Park: Pennsylvania State University Press, 1966.

Index

335

PSYCHOANALYTIC INTERVENTIONS ❦

Esther Rashkin and Peter L. Rudnytsky, series editors